# Microarray Technology in Practice

# Microarray Technology
# in Practice

Steven Russell, Lisa A. Meadows and Roslin R. Russell
Department of Genetics and Cambridge Systems Biology Centre
University of Cambridge
Cambridge, UK

AMSTERDAM • BOSTON • HEIDELBERG • LONDON
NEW YORK • OXFORD • PARIS • SAN DIEGO
SAN FRANCISCO • SINGAPORE • SYDNEY • TOKYO

Academic Press is an Imprint of Elsevier

Academic Press is an imprint of Elsevier
525 B Street, Suite 1900, San Diego, CA 92101-4495, USA
30 Corporate Drive, Suite 400, Burlington, MA 01803, USA
32, Jamestown Road, London NW1 7BY, UK

First edition 2009

**British Library Cataloguing in Publication Data**
A catalogue record for this book is available from the British Library

**Library of Congress Cataloging-in-Publication Data**
A catalog record for this book is available from the Library of Congress
ISBN–13: 978-0-12-372516-5

For information on all Academic Press publications
visit our website at elsevierdirect.com

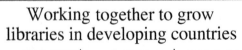

Working together to grow
libraries in developing countries

www.elsevier.com | www.bookaid.org | www.sabre.org

ELSEVIER      BOOK AID International      Sabre Foundation

# Contents

## 12   Other Array Technologies

## 13   Future Prospects

## Index                                                              441

# Foreword and acknowledgments

We have written this book because we believe that a single volume over viewing the current state-of-the-art in microarray technology is currently lacking. While there are many excellent books and reviews covering microarrays, most of the books were written a few years ago and do not cover the more recent developments in very high-density arrays for gene expression and genomic DNA analysis. As we hope to demonstrate, there has been tremendous progress in the widely available platform technologies so that whole-genome analysis is now available for virtually any organism with a sequenced genome at a cost affordable by many research laboratories. Hand-in-hand with technology developments, the analytical tools for data processing and analysis have also progressed and we hope that this book provides a comprehensive overview of the current approaches that is accessible to the beginner. To accompany the book we have generated a website that contains supplementary information, additional details on some of the analysis methods and a series of useful links. The site can be accessed at http://www.flychip.org.uk/mtip/

The preparation of this book has relied on advice and suggestions from several of our colleagues but we are particularly grateful to Bettina Fischer, Richard Auburn and Natasha Karp for their insightful comments on the manuscripts. Our other colleagues in the Cambridge FlyChip facility and more widely, provided an excellent environment for developing the technology and we thank Boris Adryan, Nuno Barbosa-Morais, David Kreil, Gos Micklem, Santiago Sevillano Matilla, Peter Sykacek, Natalie Thorne, Sarah Vowler and Rob White for intellectual and technical input. A special mention has to go to François Guillier whose unflappable systems administration has kept us honest. Steve couldn't have done any of this without the unfailing support of Michael Ashburner and John Roote: drinks are on me boys! Of course, we could not have considered approaching the compilation of this work without the support of our families, who have had to put up with our absence on many weekends and evenings so our heartfelt thanks to Fiona, Shuggie, Jonathan, James, Andrew, Struan and Isla. Lisa would particularly like to thank her mum, Ann, for all her help with childcare.

It is difficult to get to grips with the complexities of microarray analysis without getting your sleeves rolled up and doing experiments yourself. In the early days of the technology this was a relatively expensive business, requiring considerable infrastructure and Steve would like to particularly acknowledge the prescience of the UK Biotechnology and Biological Sciences Research Council, who invested considerable funding to establish core genomics infrastructure in the UK, with a special mention to Alf Game for his energy and enthusiasm.

Steve Russell is a Reader in Genome Biology in the Department of Genetics, University of Cambridge and founder of the Cambridge Systems Biology Centre. After a PhD at the University of Glasgow with Kim Kaiser (a true technological visionary) where he was trying to do fly functional genomics before functional genomics was really invented, he came to Michael Ashburner's lab in Cambridge and has never left! He established the FlyChip Drosophila microarray facility in 1999 and has been involved in fly functional genomics and microarray analysis ever since. He helped found the International Drosophila Array Consortium to make core array resources widely available to the worldwide research community.

Lisa Meadows is currently a Research Associate in the FlyChip microarray group, primarily focusing on the technical and 'wet lab' aspects of microarray technology. She established and led implementation of the FlyChip laboratory protocols including array printing, RNA extraction, amplification, labeling, hybridization, scanning and image analysis. Prior to this she obtained a strong grounding in molecular biology techniques and Drosophila research with a PhD from the University of Cambridge and postdoctoral research at the University of Mainz, Germany.

Roslin Russell is currently a Senior Computational Biologist at the Cambridge Research Institute, Cancer Research, UK, specializing in the experimental design and analysis of expression, ChIP-chip, CGH and SNP array data from various commercial platforms. She completed her PhD at the University of Cambridge, where she developed her own spotted microarrays for the study of the human major histocompatibilty complex (MHC) region in health and disease and gained experience in comparative microarray platform analysis. Following this, she participated in the establishment of a microarray facility for the Leukaemia Research Fund at the Genome Campus in Hinxton, Cambridge and then joined FlyChip where she led the development of a microarray bioinformatics and analysis pipeline using R and various packages in Bioconductor.

## Chapter 1

# Introduction

| | |
|---|---|
| 1.1. **Technology** | 1.3. **A Brief Outline** |
| 1.2. **A Brief History** | **References** |

Although there may have been a considerable amount of hype surrounding the emergence of DNA microarray technology at the end of the 20th century, the dawn of the 21st century has seen the realization of much of the initial promise envisaged for this technique. We now have a set of stable platform technologies that allow virtually any nucleic acid assay at a genome-wide scale for any organism with a sequenced genome, with the past 5 or so years seeing both the stabilization of the available platforms and the convergence in the data derived from different array technologies. In this book we gently introduce the reader to all aspects of modern microarray technology from designing the probe sequences on the array through to submitting the data to a public database. On route we hope to provide a comprehensive overview of the current state-of-the-art in the field and help guide the novice through the myriad of choices available for conducting experiments and analyzing the subsequent data. Although the primary focus is on the use of microarray technology for gene expression profiling, we describe platform technologies for other nucleic acid analysis, principally genome exploration and high-throughput genetics, as well as introducing more recent technologies for arraying proteins. We hope there is material in this volume that is useful for both those new to the field and those with more experience.

## 1.1 TECHNOLOGY

In his book '*The Sun, The Genome, and The Internet: Tools of Scientific Revolution*', Physicist Freeman Dyson argues that it is principally the development of new technologies that drive scientific progress rather than revolutionary new concepts (Dyson, 2001). Nowhere is this truer than in modern biology, where advances in molecular biology have opened up incredible vistas on the living world by providing a set of tools that have allowed us to begin deciphering the genetic code underpinning the molecular basis of life. Technologies such as

Microarray Technology in Practice

DNA cloning, nucleic acid hybridization, DNA sequencing and polymerase chain reaction, facilitate the isolation and analysis of virtually any nucleic acid molecule and have culminated in the determination of the complete human genome sequence along with the sequences of over 4000 other species. The emergence of DNA microarray technology promises much more in this vein of discovery with the ability to interrogate entire genomes or gene complements in a comprehensive high-throughput way.

While we can now determine the DNA sequence of any organism with relative ease, interpreting that sequence is an entirely different matter. From the sequence we may be able to assemble a reasonable description of the repertoire of protein coding genes encoded in a genome, however, understanding how those genes are deployed to generate the incredible diversity of cell types and organisms that characterize the living world remains a daunting challenge. Trying to decipher this complexity is one of the goals of functional genomics, a branch of molecular biology that may be defined as the study of gene functions and inter-relationships. If we take as an example the relatively simple eukaryote, the bakers yeast *Saccharomyces cerevisiae*, over 6000 genes, including the 100 or so that code for transcriptional regulators, need to be coordinated to allow the cell to survive and replicate in a variety of different environments. It is fair to say that despite its relative simplicity, we are only just beginning to scratch the surface in terms of understanding the gene regulatory systems that govern the behavior of this organism. When trying to decipher the complexity of living systems it must be recognized that defining gene functions and their interactions is only part of the story. As Denis Noble cogently argues in '*The Music of Life: Biology Beyond the Genome*' (Noble, 2006), it is the property of a biological system as a whole, the multilayered interactions between genes, their products and the cellular environment that generates biological function. Thus the rapidly developing science of integrative systems biology aims to understand biological processes in terms of such systems and in doing so provides robust mathematical descriptions of the networks and interactions underpinning the systems. Of course in order to completely describe a system one must have a reasonably complete inventory of the components of the system and it is therefore desirable to be able to accurately define the set of genes expressed in a particular system and how the expression of these genes changes over time or in response to a particular stimulus. Prior to the advent of microarray technology, comprehensive analysis of gene expression was not really possible since classical methods can only really accommodate the analysis of a few tens of genes rather than the tens of thousands encoded by a metazoan genome. In their current state, microarrays facilitate just such an analysis. We would argue that the combination of bottom up functional genomics and top down systems approaches will be necessary if we are to truly understand how the genetic information locked in the genome sequence is translated into the incredible complexity that is a cell and how collections of cells cooperate to elaborate multicellular organisms.

As Eric Davidson beautifully illustrates in *The Regulatory Genome: Gene Regulatory Networks In Development And Evolution* (Davidson, 2006): 'All morphological features of adult bilaterian body plans are created by means of pattern formation processes. The *cis*-regulatory systems that mediate regional specification are thereby the keys to understanding how the genome encodes development of the body plan. These systems also provide the most fundamental and powerful approach to understanding the evolution of bilaterian forms.' If we accept this line of reasoning, we need to understand how the information in the genome is dynamically deployed during development. In the first place, we must discern where and when genes are turned on and off: defining the spatio-temporal profile of the gene complement. Perhaps more importantly, we need to understand how the *cis*-regulatory systems regulating gene expression are encoded in the genome sequence and how regulatory molecules interpret this information to accurately control gene expression. The comparatively recent development of microarray-based methods for genome-wide mapping of in vivo transcription factor binding sites will undoubtedly contribute to such efforts and significantly advance our exploration of this complexity.

Along with our fascination with basic biology, the biosciences seek to uncover fundamental aspects of human biology and nowhere is this more evident than in our desire to understand the molecular basis underlying human disease. While some diseases are inherited in a straightforward Mendelian fashion, the majority of human conditions are much more complex traits influenced by many different genes and external environmental factors. Getting to grips with such complexity is very much akin to the challenges we face in understanding developmental processes: what genes are involved and how does the genome of particular cell types respond to changes in internal and external stimuli. Unfortunately, unlike the model organisms used for studying developmental problems, humans are not good experimental systems and we must therefore rely on genomic differences between individuals to provide us with clues about the genes involved in particular diseases. In this area we see an explosion in data accumulation over the past two or three years with the advent of microarray-based high-density genotyping platforms facilitating the identification of genetic differences in large-scale case-control studies. While this technology is in its infancy, the studies already performed promise considerable insights into complex multigenic traits. Although there is certainly a possibility that the new generation of ultra high throughput sequencing may overtake microarray based assays by allowing relatively cheap whole human genome resequencing, the possibility for using microarray-based assays as diagnostic or prognostic tools in the clinic is very attractive in terms of speed and cost. Of course not all human diseases have a genetic basis and infections by pathogenic organisms remain a major contributor to human mortality, especially in the developing world. In this case, as well as understanding the biology of the infectious agent and the affected human tissues we need to get to grips with the interactions between the host and the pathogen. This is a fascinating area of biology that is beginning to yield to functional genomics approaches with the prospect that weaknesses in the parasite can be harnessed for the development of effective therapeutic agents. This is a pressing need in the developing world: for example, it is estimated by the World Health Organization that there is a new person infected with tuberculosis every second with approximately one third of the world's population currently infected with the TB bacillus. Similarly, more than one million people die each year of malaria with the majority of disease occurring in sub-Saharan Africa: two children every minute are killed by this preventable infection. We desperately need to develop cheap and effective therapies that can combat such diseases and understanding the complex lifecycle of the responsible pathogens is certainly one step along the way.

It is, in our view, obvious that microarray technologies have much to offer in the exploration of biological complexity. While 'the secrets of life' are unlikely to be completely uncovered in our lifetime, the application of modern genome exploration methods are sure to shed some light on the complexity of life. We should bear in mind that it is only a little over 10 years since the publication of the first widely accessible microarray technology and in this time we have witnessed an incredible acceleration in our ability to explore gene expression and other areas of genome biology. A demonstration of this can be seen from some yearly PubMed searches using simple terms such as *microarray* or *genomics* (Figure 1.1), with well over 5000 papers a year now published containing these terms. It is clear that the technology is making a major impact on biology and we suspect that soon the use of DNA microarrays for expression profiling will be a universally deployed integral part of any molecular biology-based investigation.

## 1.2 A BRIEF HISTORY

There are many reviews and books that cover the brief history of microarray technology and it is not our intention here to extensively go over this ground. However, a few words about the basics are worth revisiting to set the scene before we overview the contents of the book. Nucleic acid microarrays are founded on the exquisite specificity inherent in the structure of the DNA duplex molecule (Watson and Crick, 1953) since complimentary single strands are able to recognize each other and hybridize to form a very stable duplex. It was Ed Southern (Southern, 1975) who first realized that this specificity could be used to detect specific sequences in a complex mixture by labeling a known DNA fragment (the probe) and using this to interrogate a fractionated sample of, for example, genomic DNA. The Southern blot technique was soon adapted so that specific RNA molecules in a fractionated cellular extract could be similarly detected using Northern blots (Alwine et al., 1977) and the routine analysis of mRNA transcripts was established. It was realized that the concept of using a

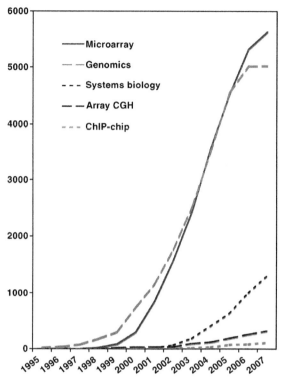

**FIGURE 1.1** Impact of microarray technology in PubMed citations. Number of publications (y-axis) per year retrieved from PubMed using search terms to capture the indicated areas.

labeled probe fragment to identify complimentary sequences was adaptable to parallel processing of DNA clones with Grunstein and Hogness (1975) providing the first demonstration of using hybridization to isolate specific clones from plasmid libraries. Together, these methods and their subsequent developments provide the foundation for virtually all aspects of current molecular genetics and the conceptual basis for DNA microarray technology.

In its current forms, microarray technology derives from two complimentary approaches developed in the 1990s. The first 'cDNA microarrays' were produced in Patrick Brown's laboratory in Stanford (Schena et al., 1995), utilizing gridding robots to 'print' DNA from purified cDNA clones on glass microscope slides. The slides were interrogated with fluorescently labeled RNA samples and the specific hybridization between a cDNA clone on the slide and the labelled RNA in the sample used to infer the expression level of the gene corresponding to each cDNA clone. Thus was born the spotted array, a technology that can be implemented in individual research labs and has provided an accessible route for high throughput gene expression profiling in many areas of biology. Since the

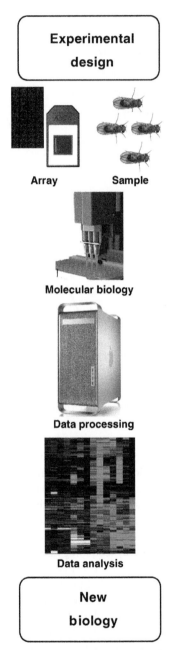

**FIGURE 1.2**   A microarray study. An overview of the stages involved in a typical microarray gene expression study. Experimental design is of paramount importance and is derived from a consideration of the biological question, the available samples and the type of microarray platform used. Molecular biology techniques are used to isolate and label samples, which are hybridized to the array

DNA fragments printed on the array correspond to the specific labelled molecule used to interrogate Southern or Northern blots, the clones on the array are referred to as probes and the labelled complex RNA mixture as the target. In parallel work at Affymetrix, in situ synthesis of defined oligonucleotides probes at very high density on glass substrates (Fodor et al., 1991) was shown to provide a reliable route for measuring gene expression (Lockhart et al., 1996) and set the scene for the development of the current generation of ultra high density micro-arrays now employed for gene expression, genome tiling and genotyping.

In essence, a microarray study consists of a series of defined stages as out-lined in Figure 1.2. An experimental question is formulated and a design phase combines the desired biological objectives with the type of array platform selected and the available biological samples to generate a robust experimental design. Some relatively straightforward molecular biology is used to extract and label the RNA samples, hybridize to the array and acquire the primary data. These data are processed and a variety of statistical and analytical tools used to identify changes in gene expression thus leading to new biological insights. In terms of the mechanics of a microarray study, there are two fundamental approaches that initially arose from the type of microarray platform used (Figure 1.3). In the case of the cDNA or spotted array, the analysis generally involves a comparative hybridization, where two RNA samples are separately labelled with different (generally fluorescent) reporters: both labelled samples are then combined and hybridized together on the same array. In this way these so-called dual channel arrays allow for a comparative analysis between samples (wild type and mutant for example). While the data for each sample are collected separately during the acquisition stage, they are generally combined during the analysis stage to yield a ratio of gene expression between one sample and another. Such a comparative approach allows for very sensitive detection of differences in gene expression between two samples. The alternative, or single channel approach, was developed in concert with the Affymetrix GeneChip platform though it is not unique to this technology. In this case, RNA from each biological sample is individually hybridized to an array and comparisons are made at the subsequent data analysis stage. Obviously single channel experiments generally require twice as many arrays as dual channel experiments though this is dependent upon the experimental design. A particular benefit of single channel experiments is that they can provide far more flexibility when comparing a variety of different samples hybridized to the same platform: it should be obvious that each channel of a dual channel experiment can be independently treated as a single channel, though caution must be exercised when doing this.

---

and the data acquired via a dedicated scanner. A variety of software tools are used to process and normalize the array data prior to statistical and meta-analysis to identify and classify differentially expressed genes. The end result is hopefully new insights into the biology under investigation.

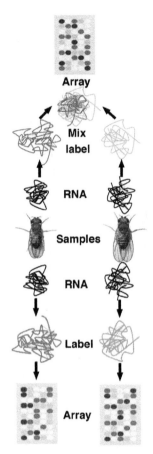

**FIGURE 1.3**   Types of microarray study. Starting with biological samples to be compared (in this case male and female fruit flies) broadly speaking two approaches are possible. The top half of the figure exemplifies the dual-channel or two color approach: RNA extracted from each sample is independently labelled with two different fluorescent dyes (represented by the different shades of gray). The labelled samples are combined and hybridized together on a single array, each fluorescence channel (representing each of the samples) is acquired separately and the data combined for a ratiometric analysis of expression differences between samples. In the bottom half of the figure, the single channel approach is shown. In this case, each sample is extracted and labelled independently with the same fluorescent reporter and singly hybridized to an array. Comparisons are carried out after data extraction.

As we note above, there are several excellent reviews and textbooks covering the development of microarray technology as well as providing detailed descriptions of particular experimental or analytical methods. The collections of reviews published by *Nature Genetics* under the 'Chipping Forecast' banner are particularly useful starting points for reviewing the development of the field.

The first was published at the beginning of 1999, the second at the end of 2002 and the third in 2005. Each of these supplements provides a series of authoritative reviews on the state-of-the art in various aspects of microarray technology or analysis: they are highly recommended. Other recommended books include *Microarray Bioinformatics* (Stekel, 2003), *Microarray gene expression data analysis* (Causton et al., 2003), *Microarray Data Analysis* (Kornberg, 2007) and the two volumes edited by Schena (2000, 2007). Focusing specifically on data analysis we particularly recommend *Guide to Analysis of DNA Microarray Data* (Knudsen, 2004) and *Statistics for Microarrays: Design, Analysis and Inference* (Wit and McClure, 2004). Other useful books and essential reviews are noted in the following outline section or in the main text of each chapter.

## 1.3 A BRIEF OUTLINE

We have organized this book into 12 chapters, each encompassing an important area of microarray technology, in the hope that the reader can either take the volume as complete introductory text or consult individual chapters depending on their particular experimental or analytical needs. Some of the general books on microarray analysis tend to leave the area of experimental design to the end. In our view this area is of such fundamental importance that we provide an introduction to experimental design up front. It is critical that those considering a series of microarray studies consider the experimental design at the very outset. Failure to do so can lead to data that is unsuitable for addressing the real question of interest. We cannot emphasize enough the need to get this crucial area right and direct even experienced molecular biologists to some introductory literature: Yang and Speed (2002), Simon et al. (2002) and Kerr (2003) provide some general guidance in this area with more focused examples on, for example, time course analysis provided by Glonek and Solomon (2004). In terms of replication and sample sizes, both Lee and Whitmore (2002) and Jørstad et al. (2007) overview the basics and provide a good starting guide.

The section introducing experimental design should really be read in conjunction with Chapter 7, which covers some of the basic and more advanced statistical methods currently applied to microarray data analysis. In general, a gene expression microarray study aims to identify differences in transcript levels between two or more experimental samples. Although this is trivial to write, the statistical issues underpinning the identification of differentially expressed transcripts are complex for a variety of reasons. Indeed there is no perfect 'one size fits all' solution and the selection of appropriate statistical tools must be done with reference to the experimental design. In Chapter 7 we cover a variety of classical and more modern univariate or multivariate statistical tools for identifying differentially expressed genes including the recently developed gene set and category analysis techniques that aid in the interpretation of large lists of differentially expressed genes. Since we are not statisticians, we are unencumbered by the frequentist versus Bayesian debate and provide examples of both types of approach where appropriate. Allison et al. (2006) provide an introductory review to microarray data analysis techniques with the reviews in Kimmel and Oliver (2006b) or the book by Speed (2003) giving more practical advice. Kanji (2006) has produced a good introductory statistical textbook and more advanced reviews and books are mentioned in Chapter 7.

Of course data must be generated before it can be analyzed and in this area we provide several chapters that break down the various stages of the experimental process, from designing and preparing or sourcing arrays, through sample extraction and labeling to data collection and preprocessing. Obviously any experimental data is only as good as the assay used to generate it and in this respect microarray data is no different. Up until comparatively recently, debates raged about the relative merits of different microarray platforms, high-density short oligonucleotide, long oligonucleotide, cDNA and so on. Recent detailed comparative studies by the Microarray Quality Control Project (MAQC: see Casciano and Woodcock (2006) and the accompanying articles in the same issue of *Nature Biotechnology*, especially Shi et al. (2006), for a thorough discussion) demonstrate a very high degree of inter-platform reproducibility. The key conclusion of these comprehensive studies is that the key factor underpinning cross-platform comparability is the probe sequence. It is now becoming clear that principled probe design is of paramount importance if semi-quantitative measures of gene expression are to be made (Kreil et al., 2006). When sensitive and specific probes are employed, microarray technology is capable of yielding results just as reliable as other transcript measurement technologies such as quantitative real-time polymerase chain reaction. We confidently predict that in a year or two, gene expression measures generated via validated array platforms will be accepted as robust indicators of transcript levels assuming the experiments that generate such data are appropriately controlled.

Chapter 3 provides an overview of the current state of the art with respect to probe design and describes the microarray platforms available for gene expression analysis. Although gene expression analysis is by far the most popular area where microarray technology is applied, it was realized from the very outset that microarrays could be deployed for other assays (Fodor et al., 1993). In Chapter 12 we explore the use of microarray technologies in other experimental arenas, focusing on tiling arrays for whole genome expression, comparative genomic hybridization or ChIP-array analysis and also describe platforms for high-throughput genotyping. We also introduce the increasingly popular adaptation of microarray platforms for screening living cells or for carrying out high throughput functional interaction assays, including protein–DNA and protein–protein interactions. These emerging technologies promise a substantial increase in our ability to assay protein or gene function in a high throughput way and offer the prospect of much more cost effective pipelines in drug screening or development pipelines. We emphasize that we do not provide laboratory manuals detailing particular experimental protocols, rather we hope to inform the reader of the methodology underlying each of the microarray platforms so that informed choices can be made. References to particular experimental techniques are provided at appropriate places in each chapter but for those who wish to immerse themselves in the subject, up to date methods are available in Schena (2007) and in Kimmel and Oliver (2006a).

Along with the microarray platforms themselves, there have been substantial improvements in the way that biological samples are extracted and labelled prior to hybridization to microarrays. It is now feasible to expression profile individual eukaryotic cells although one needs to bear in mind that any of the amplification procedures currently employed will inevitably alter the representation of the labelled pool with respect to the transcript abundance in the original sample (see Nygaard and Hovig, 2006 or Subkhankulova and Livesey, 2006 for some recent discussion of amplification methods). As with any molecular biology procedure, collection and labeling of samples for microarray analysis can be problematic and in Chapter 4 we overview the most widely used techniques for collecting and processing RNA, outlining some of the pitfalls commonly encountered. We provide suggestions regarding the types of controls that can be employed and direct readers to the relevant literature.

Whether arrays are prepared in house or obtained from a commercial source, the critical common steps are hybridization and scanning. Careful attention to hybridization conditions can substantially improve data quality, especially when attention is paid to calibrating a microarray to ensure optimal hybridization stringency is used, and we provide advice on some of the best methods for ensuring consistency in this step in Chapter 5. With the notable exception of Affymetrix GeneChip technology, which requires a specific scanning instrument, the majority of other microarrays can be scanned using a number of dedicated scanners. These are expensive optical instruments and ensuring they are properly calibrated and used with optimal settings can reap considerable dividends in terms of data quality and consistency. In the end, the images obtained from the scanner are the raw data of any microarray-based assay and reliably converting the image into a numerical representation of RNA abundance via the intensity of the hybridization signal is far from trivial. Not only are the scanner parameters critical, the software used to process the image in order to extract the intensity data from each array element and to determine background measurements can have profound effects on the final data. Unfortunately image processing is far from a solved problem and different techniques generate different primary data. In Chapter 5 we cover the different segmentation algorithms underpinning microarray image analysis and indicate where certain techniques may be more appropriate than others. The area of image analysis has recently been reviewed (Qin et al., 2005; Timlin, 2006) and these articles provide useful entry points into the field.

After data has been extracted from microarray images it needs to be processed prior to any statistical analysis and this area, perhaps more than any other in the microarray field, has seen considerable activity and controversy. Data must be 'cleaned', evaluated to ensure there are no obvious artefacts, and then normalized to allow reliable estimates of gene expression within and between arrays. There are many normalization methods employed in microarray analysis that, to a greater or lesser extent, are able to deal with the inevitable biases that occur in the primary data. Chapter 6 provides an extensive discussion of between and within array normalization methods, including those able to account for spatial and intensity variability. We provide a separate discussion of the various different pre-processing techniques employed in the analysis of the popular Affymetrix platform, including various alternatives to the standard Affymetrix software. Throughout Chapter 6, and indeed in the subsequent statistical analysis chapter, we concentrate on public domain tools that are freely available either through the open source Bioconductor suite (Reimers and Carey, 2006) or the open source R software environment (Deshmukh and Purohit, 2007). While there are commercial software packages available for microarray data analysis, some of these may have disadvantages: they can be expensive, it may not be entirely clear how particular algorithms are implemented and, in some cases, the approaches may be out of date since software update cycles can be lengthy. All of these issues are avoided by the use of free open source software that is constantly being improved by the scientific community. While there may be a learning curve associated with mastering Bioconductor or R, it is clear that these environments offer much more flexibility and control over the data analysis than some commercial tools. In addition, there are open source efforts to provide a user-friendly graphical user interface to some Bioconductor tools (Wettenhall and Smyth, 2004).

While statistical analysis of microarray data sets allows the reliable identification of genes of interest in a particular experiment, many hundreds of genes may pass the relevant tests and exploring such complex data can be a daunting prospect for any biologist more used to dealing with a handful of genes. The area of meta-analysis, which includes approaches such as clustering and dimensionality reduction techniques, facilitates the organization and presentation of complex data in an intuitive and informative manner. Such tools come to the fore when data from tens or even hundreds of array experiments need to be integrated and compared so that any underlying biology emerges. The type of analytical approach employed for meta-analysis is dependent upon the amount of additional information about the biological samples available and on the objectives of the analysis. Chapter 8 presents a wide range of techniques and the appropriate software tools available for their implementation. Classification and comparison approaches may wish to draw upon the results of genomic experiments published by others and in contrast to more traditional biological approaches this can present some problems. Microarray data is complex and the final data presented in a publication is strongly influenced by the pre-processing and statistical approaches used to analyze the primary data. For this reason it is imperative that those who generate and publish microarray data do so in such a way that it can be utilized by others. This requires transparency in terms of the analytical approaches used and in some cases should also involve making primary unpublished data available. In order to accomplish this, standardized methods for data curation and storage are required. Chapter 9 provides an overview of the public data repositories available for storing microarray data in a broadly accessible way and highlights the efforts of international collaborations aimed at standardizing the collection and annotation of microarray data. Important initiatives such as minimal information about a microarray experiment (MIAME, Brazma et al., 2001) and the Microarray Gene Expression Data Society (MGED; Ball et al., 2004) are crucial in this regard and it is to be hoped that all published microarray experiments result in data stored according to common standards. We provide a brief discussion of how to submit and retrieve microarray data using the public databases as an aid to researchers looking to publish their gene expression studies or obtain relevant data from the public domain. As with Chapters 6 and 7, we rely heavily on the tools available from the BioConductor project and recommend Causton et al. (2003) for an introductory guide to the basic approaches.

To provide some context to the techniques we describe, Chapter 11 introduces the way that microarray technology has been applied to problems in human health and disease, with a particular focus on cancer biology and infectious disease. We hope that these brief overviews of microarray technology in action will help the novice understand how the technique can be used and, broadly speaking, what type of biological questions can be addressed. By necessity this cannot be a comprehensive review, as we note above there are thousands of papers published every year that use the technology, but we have selected examples that show how relatively simple experiments can provide important biological insights. We end with a Chapter looking to the future of microarray technology, particularly to the areas where we feel there is scope for improvement in the field. Some of these areas are technical while others are analytical, but together the continued development of the technology suggests a continued future for array-based analyses in the research laboratory and, more significantly, in the diagnostics laboratory. Again, we emphasize that this is a young technology, barely out of the proverbial short trousers, and while there has been incredible progress in terms of reliability and reproducibility, the technology remains at heart a simple assay with a multitude of applications.

## REFERENCES

Allison DB, Cui X, Page GP, Sabripour M. (2006) Microarray data analysis: from disarray to consolidation and consensus. *Nat Rev Genet* **7**: 55–65.

Alwine JC, Kemp DJ, Stark GR. (1977) Method for detection of specific RNAs in agarose gels by transfer to diazobenzyloxymethyl-paper and hybridization with DNA probes. *Proc Natl Acad Sci U S A* **74:** 5350–5354.

Ball CA, Brazma A, Causton H, Chervitz S, Edgar R, Hingamp P, Matese JC, Parkinson H, Quackenbush J, Ringwald M, Sansone SA, Sherlock G, Spellman P, Stoeckert C, Tateno Y, Taylor R, White J, Winegarden N. (2004) Submission of microarray data to public repositories. *PLoS Biol* **2:** E317.

Brazma A, Hingamp P, Quackenbush J, Sherlock G, Spellman P, Stoeckert C, Aach J, Ansorge W, Ball CA, Causton HC, Gaasterland T, Glenisson P, Holstege FC, Kim IF, Markowitz V, Matese JC, Parkinson H, Robinson A, Sarkans U, Schulze-Kremer S, Stewart J, Taylor R, Vilo J, Vingron M. (2001) Minimum information about a microarray experiment (MIAME)-toward standards for microarray data. *Nat Genet* **29:** 365–371.

Casciano DA, Woodcock J. (2006) Empowering microarrays in the regulatory setting. *Nat Biotechnol* **24:** 1103.

Causton H, Quackenbush J, Brazma A. (2003) *Microarray Gene Expression Data Analysis: A Beginner's Guide*. Wiley Blackwell, Oxford, UK.

Chipping Forecast. (1999) *Nat Genet* **21:** Supplement 1.

Chipping Forecast II. (2002) *Nat Genet* **32:** Supplement 4.

Chipping Forecast III. (2005) *Nat Genet* **37:** Supplement 6.

Davidson EH. (2006) *The Regulatory Genome: Gene Regulatory Networks in Development and Evolution*. Academic Press, San Diego, USA.

Deshmukh SR, Purohit SG. (2007) *Microarray Data: Statistical Analysis Using R*. Alpha Science International Ltd., Oxford, UK.

Dyson FJ. (2001) *The Sun, The Genome, And The Internet: Tools Of Scientific Revolution*. Oxford University Press, Oxford, UK.

Fodor SP, Read JL, Pirrung MC, Stryer L, Lu AT, Solas D. (1991) Light-directed, spatially addressable parallel chemical synthesis. *Science* **251:** 767–773.

Fodor SP, Rava RP, Huang XC, Pease AC, Holmes CP, Adams CL. (1993) Multiplexed biochemical assays with biological chips. *Nature* **364:** 555–556.

Glonek GF, Solomon PJ. (2004) Factorial and time course designs for cDNA microarray experiments. *Biostatistics* **5:** 89–111.

Grunstein M, Hogness DS. (1975) Colony hybridization: a method for the isolation of cloned DNAs that contain a specific gene. *Proc Natl Acad Sci U S A* **72:** 3961–3965.

Jørstad TS, Langaas M, Bones AM. (2007) Understanding sample size: what determines the required number of microarrays for an experiment? *Trends Plant Sci* **12:** 46–50.

Kanji GK. (2006) *100 Statistical Tests*, 3rd edition. Sage Publications Ltd., London, UK.

Kerr MK. (2003) Experimental design to make the most of microarray studies. *Methods Mol Biol* **224:** 137–147.

Kimmel A, Oliver B. (2006) *DNA Microarrays Part A: Array Platforms and Wet-Bench Protocols*. *Methods in Enzymology* **Vol. 410:** Academic Press, San Diego.

Kimmel A, Oliver B. (2006) *DNA Microarrays Part B: Databases and Statistics*. *Methods in Enzymology* **Vol. 411:** Academic Press, San Diego.

Knudsen S. (2004) *Guide to Analysis of DNA Microarray Data*, 2nd edition. Wiley Blackwell, Oxford, UK.

Kornberg MJ. (2007) *Microarray Data Analysis: Methods and Applications*. Humana Press Inc., Totowa, USA.

Kreil DP, Russell RR, Russell S. (2006) Microarray oligonucleotide probes. *Methods Enzymol* **410:** 73–98.

Lee ML, Whitmore GA. (2002) Power and sample size for DNA microarray studies. *Stat Med* **21:** 3543–3570.

Lockhart DJ, Dong H, Byrne MC, Follettie MT, Gallo MV, Chee MS, Mittmann M, Wang C, Kobayashi M, Horton H, Brown EL. (1996) Expression monitoring by hybridization to high-density oligonucleotide arrays. *Nat Biotechnol* **14:** 1675–1680.

Noble D. (2006) *The Music of Life: Biology Beyond the Genome.* Oxford University Press, Oxford, UK.

Nygaard V, Hovig E. (2006) Options available for profiling small samples: a review of sample amplification technology when combined with microarray profiling. *Nucleic Acids Res* **34:** 996–1014.

Qin L, Rueda L, Ngom A. (2005) Spot detection and image segmentation in DNA microarray data. *Appl Bioinf* **4:** 1–11.

Reimers M, Carey VJ. (2006) Bioconductor: an open source framework for bioinformatics and computational biology. *Methods Enzymol* **411:** 119–134.

Schena M. (2000) *Microarray Biochip Technology.* Eaton Publishing, Natick, USA.

Schena M. (2007) *DNA Microarrays (Methods Express).* Scion Publishing Ltd., Bloxham, UK.

Schena M, Shalon D, Davis RW, Brown PO. (1995) Quantitative monitoring of gene expression patterns with a complimentary DNA microarray. *Science* **270:** 467–470.

Shi L, Reid LH, Jones WD, Shippy R, Warrington JA, Baker SC, Collins PJ, de Longueville F, Kawasaki ES, Lee KY, Luo Y, Sun YA, Willey JC, Setterquist RA, Fischer GM, Tong W, Dragan YP, Dix DJ, Frueh FW, Goodsaid FM, Herman D, Jensen RV, Johnson CD, Lobenhofer EK, Puri RK, Schrf U, Thierry-Mieg J, Wang C, Wilson M, Wolber PK, Zhang L, Amur S, Bao W, Barbacioru CC, Lucas AB, Bertholet V, Boysen C, Bromley B, Brown D, Brunner A, Canales R, Cao XM, Cebula TA, Chen JJ, Cheng J, Chu TM, Chudin E, Corson J, Corton JC, Croner LJ, Davies C, Davison TS, Delenstarr G, Deng X, Dorris D, Eklund AC, Fan XH, Fang H, Fulmer-Smentek S, Fuscoe JC, Gallagher K, Ge W, Guo L, Guo X, Hager J, Haje PK, Han J, Han T, Harbottle HC, Harris SC, Hatchwell E, Hauser CA, Hester S, Hong H, Hurban P, Jackson SA, Ji H, Knight CR, Kuo WP, LeClerc JE, Levy S, Li QZ, Liu C, Liu Y, Lombardi MJ, Ma Y, Magnuson SR, Maqsodi B, McDaniel T, Mei N, Myklebost O, Ning B, Novoradovskaya N, Orr MS, Osborn TW, Papallo A, Patterson TA, Perkins RG, Peters EH, Peterson R, Philips KL, Pine PS, Pusztai L, Qian F, Ren H, Rosen M, Rosenzweig BA, Samaha RR, Schena M, Schroth GP, Shchegrova S, Smith DD, Staedtler F, Su Z, Sun H, Szallasi Z, Tezak Z, Thierry-Mieg D, Thompson KL, Tikhonova I, Turpaz Y, Vallanat B, Van C, Walker SJ, Wang SJ, Wang Y, Wolfinger R, Wong A, Wu J, Xiao C, Xie Q, Xu J, Yang W, Zhang L, Zhong S, Zong Y, Slikker W. (2006) The microarray quality control (MAQC) project shows inter- and intraplatform reproducibility of gene expression measurements. *Nat Biotechnol* **24:** 1151–1161.

Simon R, Radmacher MD, Dobbin K. (2002) Design of studies using DNA microarrays. *Genet Epidemiol* **23:** 21–36.

Southern EM. (1975) Detection of specific sequences among DNA fragments separated by gel electrophoresis. *J Mol Biol* **98:** 503–517.

Speed T. (2003) *Statistical Analysis of Gene Expression Microarray Data.* Chapman and Hall/CRC, Boca Raton, USA.

Stekel D. (2003) *Microarray Bioinformatics.* Cambridge University Press, Cambridge, UK.

Subkhankulova T, Livesey FJ. (2006) Comparative evaluation of linear and exponential amplification techniques for expression profiling at the single-cell level. *Genome Biol* **7:** R18.

Timlin JA. (2006) Scanning microarrays: current methods and future directions. *Methods Enzymol* **411:** 79–98.

Watson JD, Crick FH. (1953) Molecular structure of nucleic acids; a structure for deoxyribose nucleic acid. *Nature* **171:** 737–738.

Wettenhall JM, Smyth GK. (2004) limmaGUI: a graphical user interface for linear modeling of microarray data. *Bioinformatics* **20:** 3705–3706.

Wit E, McClure J. (2004) *Statistics for Microarrays: Design, Analysis and Inference.* John Wiley & Sons, Hoboken, USA.

Yang YH, Speed T. (2002) Design issues for cDNA microarray experiments. *Nat Rev Genet* **3:** 579–588.

# The Basics of Experimental Design

The aim of the majority of microarray-based experiments is to quantify and compare gene expression on a large scale. Since the equipment, consumables and time involved are all expensive, and the amount of sample often limiting, microarray experiments require careful planning to optimize all aspects of the process. As with any scientific experiment, it is essential to have a clear objective from the outset and ensure that the experimental design optimizes the reliability of the data to best answer the specific questions posed. In this chapter we outline the fundamentals of experimental design and the factors that need to be considered when selecting controls and replication levels. This section is intended as a primer on design essentials: a more in-depth discussion of statistics and experimental considerations are provided in subsequent chapters. The experimental design and subsequent data analysis go hand in hand and it is recommended that statistically naive biologists consult an experienced statistician, preferably even before collecting samples,

Microarray Technology in Practice

to ensure that the planned strategy is appropriate. The planning stage and experimental execution in the laboratory are equally crucial and no amount of analysis can compensate for a poorly designed experiment. At worst, poor design will lead to nonrepresentative, misleading or entirely false results. The experimental design and analysis strategy may be quite different for different types of experiment and depend on a number of general variables. Therefore, at the outset a number of questions should be asked to provide a framework for subsequent design (Table 2.1). Yang and Speed (2002), Churchill (2002) and Bolstad et al. (2004) provide suggestions for planning microarray experiments and overview the major design issues involving cDNA microarrays and these texts are recommended as excellent starting points.

## 2.1 SOURCES OF VARIATION IN MICROARRAY GENE EXPRESSION MEASUREMENTS

Of course, the data obtained from all microarray experiments are subject to variation stemming from a number of sources and a good microarray experiment will be designed and executed so that sources of unwanted variation are reduced as much as possible. The typical sources of variation can be broadly divided into three major categories:

- **(i)** Variation within the biological sample.
- **(ii)** The performance of the technology itself.
- **(iii)** Variation in the spot signal measurements.

Fortunately, a comprehensive comparison of six different microarray technologies suggests that, with high quality microarrays and appropriate normalization methods, the majority of the variance in microarray experiments is biological rather than technical (Yauk et al., 2004). Thus good design can help focus on the biology of a system rather than the technicalities of an experimental method. There are a large number of processes that need to be performed between the printing of an array and data acquisition; each can have an impact on variation in the final processed data and it goes without saying that good laboratory practice along with quality control at multiple stages can help to minimize technical variation. The different types of variation encountered are dealt with in different ways and this needs to be taken into account when designing experiments. In the following chapters we explicitly deal with quality assessment for array printing (Chapter 3), RNA extraction and sample labeling (Chapter 4), hybridization (Chapter 5) and data preprocessing (Chapter 6). Suffice to say, a well-controlled set of protocols will reap rewards in terms of data reliability and appropriate effort should be directed in this direction (Altman, 2005).

It can be argued that variability is intrinsic to all biological processes and, at its most basic, can result from two different forces. Natural variation can result

**TABLE 2.1** Issues to Consider that Can Affect the Design of a Microarray Experiment: Adapted from Yang and Speed (2002)

| Scientific | Practical | Other |
|---|---|---|
| What are the experimental aims? | What types of RNA sample are available (mutant, wild type, treatment, control reference, time-course)? | Are all relevant experimental protocols in place (sample isolation, RNA extraction, labeling, etc.)? |
| How are specific biological questions prioritized within the framework of the experiment? | How much RNA can be obtained for each sample? | What controls are available on the array and how will they be used? |
| How will the analysis address the specific questions? | What types of microarray are available? | What subsequent validation methods are available and how many samples can they deal with? |
| | How many microarrays are available for the study? | How much can I afford to spend on the experiment? |

from inherent genetic differences between organisms or individual cells within an organism. However, even genetically identical organisms can be differentially affected by environmental factors. For whole organisms these factors can include nutritional status, metabolism, age, temperature, humidity, hormonal balance, infection, time of day, activity level and so on. Despite controlling the environmental conditions as much as possible, variation between different biological samples and between separate samples from the same subject are still inevitable. It is here that the analysis of replicate biological samples is necessary. Regardless of whether the original sample is a whole organism, extracted tissue, cultured cell line, subset of cells, single cells, etc., the RNA extraction procedure and any amplification techniques employed are likely to introduce further variation. Different extraction and amplification methods differ in the amount of variation and bias they are likely to introduce (Chapter 4). An added level of variation is in the labeling procedure, which will never behave identically, even with two aliquots of the same reaction mix. This is true even if such things as reagent batches, concentrations, volumes, temperature, the time and day, the experimenter and the laboratory environment are kept constant. To mitigate variability from these experimental sources, technical replication is appropriate. Added to this is the well-known dye bias effect, caused by differences in the incorporation or detection of the fluorescent dyes used in two channel experiments. This bias in labeling efficiency can be accounted for by dye swaps in the experimental design, a form of technical replication.

No two batches of printed arrays are exactly the same, no two arrays from the same batch are identical and neither are any two spots within an array. Similarly, even the carefully controlled in situ synthesized arrays available from commercial sources are subject to variability, though the industrialization of the process reduces this in comparison to 'home-made' arrays. Variation at the array production stage can occur for a number of reasons, for example, the probe concentration, position on the array, the particular printing pin used, the humidity and temperature during printing, etc. These variables can lead to differences in the absolute amount of probe that is deposited on the slide surface, the amount that remains tethered after processing, the morphology of the spot and the level of deviation from the expected spot location. All of these factors can have an impact on the amount of labeled target that can bind probe and on the efficiency of subsequent spotfinding and data extraction steps.

One of the largest sources of variation in the microarray experimental pipeline stems from the hybridization step. It is often apparent just by looking at the images from two separate hybridizations, that there can be considerable levels of unpredictable spatial variation in the signal. Automation of the hybridization process can help eliminate variation, nevertheless, uneven hybridization and washing cannot be totally eliminated. Scanning the hybridized array can introduce variation and so this should be minimized by using the same scanner for a set of experiments, regularly calibrating the scanner and optimizing the relevant settings (exposure or gain) and avoiding photobleaching (Chapter 5). Different spotfinding software will generate different spot measurement results and it is therefore essential to minimize variation by adopting the spotfinding method that is most suited to the array type, spot morphology, background levels and signal intensity range of each particular experiment. Obviously the same method should be employed for all arrays within an experiment.

As should be clear from the above, the total variation may be due to multiple small factors that can be difficult to precisely control. In addition, it is also possible that a degree of stochasticity may be an inherent feature of microarray studies that can never be totally eliminated. The only way to take such variation into account is to control as best as possible those parameters that can be controlled and reduce the impact of other effects with appropriate experimental design and replication. We deal in some detail with issues relating to statistical power and replication in Chapter 7, here we wish to emphasize the importance of considering the experimental design up front before committing sample to slide.

## 2.2 CONTROLS AND REPLICATES

### 2.2.1 Control Samples and Probes

The way an array is designed and printed can have an impact on the experimental data. In addition to the probes representing the genes under investigation, printing various control probes can aid spotfinding and normalization as well as provide a measure for array quality and consistency. The use of several negative control probes implies that defined positions on the array will give no signal after hybridization. Therefore, any signal associated with negative control spots is likely to be an indication of impurities in the labeled target, suboptimal hybridization from poorly blocked slides or hybridization and wash conditions that are not sufficiently stringent.

Positive control probes serve the opposite purpose: they are expected to give a signal after hybridization and therefore provide controls for labeling, hybridization and normalization. Ideally, multiple copies of each positive (and negative) control probe should be spread throughout the array and printed with each pin in the case of spotted arrays. This will enable any deviation from signal uniformity to be identified, indicating spatial variation, and will aid the quality control and normalization process. The positive controls may be internal controls, so-called housekeeping genes that are constitutively expressed in all the target samples. The selection of appropriate housekeeping genes is a contentious issue, since it has been reported that it is difficult to identify a set of genes that remain constitutively expressed at the same level under all experimental conditions (Thellin et al., 1999; Suzuki et al., 2000; Warrington et al., 2000). One possible method for identifying internal control genes is to mine data from previous microarray experiments, where they are available, and identify genes whose expression is constant under a variety of conditions (Lee et al., 2007). However, even if they are truly constitutive, such genes are typically expressed at high levels and may therefore be poorly representative of many genes of interest, which may be expressed at lower levels and therefore subject to intensity-dependent biases. A more suitable and potentially more controlled approach is to use exogenously added spike-in controls. Here, either gene sequences from an organism unrelated to the one under investigation or entirely synthetic sequences are used to generate probes. The most important features of a spike control probe are that it has no significant regions of homology with any sequences likely to be present in the sample under investigation and has the same general hybridization properties as the other probes on the array. Known amounts of RNA transcripts corresponding to the spike probes are added to the target RNA samples prior to the labeling reaction. By utilizing multiple different spike probes, corresponding RNAs can be added to the reverse transcription reaction at different concentrations and the resulting spread of signal intensities will be more representative of the full dynamic range of hybridization signals obtained with the transcripts in the biological sample. Such an approach generates an effective intensity calibration curve. Sets of external RNA spike-in controls are currently available from several sources (e.g. Affymetrix GeneChip Poly-A RNA Control Kit, GE Healthcare/Amersham Biosciences Codelink Whole Genome Controls, Stratagene SpotReport Alien Array Validation System), however, few probes corresponding to the controls are available on commercial microarrays as yet. One encouraging approach comes from the External RNA Control Consortium (ERCC), a collaborative group from academic, private and public organizations, which is currently developing a set of external RNA control transcripts that can be universally used to assess technical performance in gene expression assays. The ERCC is primarily focused on designing external controls for human, mouse and rat, the biomedically important species, but the reagents should be equally useful for other species. At present, candidate controls are being evaluated in both microarray and QRT-PCR gene expression platforms and the hope is that the set will be universally accepted and utilized by all microarray facilities and providers (External RNA Controls Consortium, 2005).

## 2.2.2 Probe Replication and Randomization of Array Layout

In order to apply statistical tests and to reduce the variability that is inherent in microarray experiments, experimental replication is essential. Replication can be divided into two basic classes: technical and biological. One obvious form of technical replication is within array probe replication. In addition to spotting multiple copies of control probes, it is advantageous to have duplicates, or preferably multiples, of all probes spotted on the same array, however, this may not be possible due to spotting density constraints. The precision of particular probe measurements will increase if the spot intensities of the replicate spots are averaged (Lee et al., 2000). It is, however, critically important to ensure that replicate spots are randomly distributed across the array and not spotted in adjacent locations in order to make inferences about the degree of variability across an array. Ideally each slide in an experiment will have the same probes but printed in a different randomized layout. In reality this is impractical with spotted arrays due to the limitations of current printing instruments and even if it were possible it would make tracking of spot identities a much more complex task (Churchill, 2002). However, there is one microarray platform that generates exactly this type of randomized layout: with Illumina BeadChips, oligonucleotide probe sequences are synthesized along with a unique addressing sequence and the hybrid oligonucleotides tethered to microbeads (Ferguson et al., 2000). Beads are randomly distributed among microwells fabricated on the array substrate such that each particular probe is represented at approximately 30 different locations on the array. Each array is provided with a deconvolution file, generated during the manufacturing process by detecting the locations of each unique address sequence. It should be noted that replicated spots are not entirely independent of each other and consequently they should be treated as correlated observations in the subsequent analysis. While the lack of independence of the measurements reduces their value for formal statistical tests, the platform is certainly the most robust system for within-array replication currently used.

## 2.2.3 Technical Replicates

Replicate slides hybridized with RNA from the same extraction are considered to be technical replicates. Such replication allows averaging across the replicates and more reliable expression estimates since averages are less variable than their component terms. There are characteristic, repeatable features of extractions, therefore technical replicates should result in a smaller degree of measurement variability than biological replicates. However, technical replicates are not sufficiently independent and therefore shared systematic biases will not be removed by averaging (Yang and Speed, 2002). While repeatedly measuring RNA from the same extraction will generate high confidence measurements for the particular sample by eliminating technical variation introduced during the experiment, it will not give any information about another sample from the same population. This would be akin to making conclusions about the height differences between males and females by measuring the same man and woman multiple times. Clearly, if we wish to extrapolate and make inferences about the whole population, it is necessary to measure several different males and several different females, that is we need biological replicates.

## 2.2.4 Biological Replicates

Biological replicates can be defined as hybridizations using RNA from different extractions, for example, independent preparations from the same source, or preparations from different sources, such as different organisms or different cultures of a cell line. As long as post-extraction steps are carried out separately for each RNA preparation, then the measurements will be independent. Of course true biological replicates utilize independent sample sources: strains or cells grown in parallel for example. Although the variation in expression measures for a particular gene will inevitably be greater than with technical replicates, biological replicates allow broader generalization of the experimental results since the bias introduced by each independent sample is almost eliminated by taking the mean of each measurement. Typically, biological replicates are used to validate generalization of conclusions and technical replicates are used to reduce the variability of these conclusions (Yang and Speed, 2002; Draghici et al., 2001).

## 2.2.5 Pooling of Samples

Often, the RNA available from individual samples can be limiting, for example if RNA is extracted from a specific cell type, from a small biopsy or from organisms where single individuals do not yield enough RNA for a labeling reaction. In such cases, one option for an expression study is to amplify the RNA, however this can cause unwanted distortions in gene expression measurements. An alternative strategy is to pool multiple RNA samples, a strategy that may also be considered in circumstances where the number of arrays is limiting or cost is an issue. However, pooling is only appropriate if the primary interest is in the mean expression levels within a population and not in the variation of that expression within the population. Pooling leads to a loss of degrees of freedom and a decrease in statistical power (Shih et al., 2004), consequently, a large number of individual samples contributing to the pool may be required to achieve a power comparable to that of nonpooled samples and the number of different pools should not be too small. The decision of whether to pool or not is ultimately a trade-off between the cost of RNA sampling and the cost of a microarray. An apparent disadvantage of pooling is that it may average out less significant changes in expression or hide information about biological variance that may be informative. In most cases this 'averaging out' is exactly the purpose of pooling. For example, in a comparison of gene expression differences between a wild type and a mutant genotype of *Drosophila*, differences between individual flies are not of interest. However, pooling is sometimes inappropriate for the type of question under investigation: for example, when studying the effect of a drug treatment on patients the gene expression in individual patients will be of particular interest and this information is lost by pooling. A number of publications discuss optimal pooling in more detail (Kendziorski et al., 2003, 2005; Peng et al., 2003; Zhang and Gant, 2005) and should be consulted by those considering this approach.

## 2.2.6 How Many Replicates are Required?

Microarray experiments should be appropriately designed such that they have a reasonable probability of answering the proposed hypothesis (statistical power) and also have a high probability that those genes declared significant are truly differentially expressed. Sample size, in this case the number of arrays, is a critical determinant of both statistical power and expected error rates. In a perfect world, experienced experimenters using a precise, well-calibrated system and hybridizing a pool of many RNA samples would theoretically need just a single array to measure differential expression, albeit without any way of supporting the data statistically. In reality, replication is needed to eliminate variability and evaluate the statistical significance of the conclusions. The question is then: how many replicates are needed? Obviously, the greater the number of replicates the more reliable the data, but the available budget and limited availability of RNA from different replicates may necessitate reducing replication to a minimum acceptable level. Optimal experimental design is therefore essential. Although it has been suggested that triplicates are sufficient for most forms of inference (Lee et al., 2000), it is in fact not possible to generalize. The actual number of replicates required will depend on the experiment, in particular on the noise present in the system and the level of certainty required for the conclusions. For example, if the extracted RNA has been amplified before labeling, then more replicates are likely to be needed than if the RNA is labeled directly. A framework for determining how many replicates are required for a given degree of certainty in gene expression levels is outlined by Wernisch (2002). Several other approaches have been proposed to estimate required sample size for microarray experiments (Pan et al., 2002; Lee and Whitmore, 2002; Wang and Chen, 2004), however, these methods calculate power based on an arbitrary level of gene expression difference being biologically relevant and constant across all genes. In reality, the level of variability is different for each probe on the array. A more recent method, the *PowerAtlas* (Page et al., 2006), uses information from pilot studies or public domain microarray data to estimate the variability in the expression level of each gene and takes this into account when calculating the sample sizes and statistical power for a proposed study.

The number of false positives that are considered acceptable in a microarray experiment may well depend on the type of validation that is to be subsequently performed. If a secondary screen of potential positives is quick and easy or amenable to high throughput, then more false positives may be tolerated since they can be efficiently eliminated in the validation studies. However, if the validation is time-consuming, costly or otherwise demanding, then a lower error rate should be accepted with the caveat that some of the true positives will be missed. The overall goal is to estimate the sample size required to be able to answer any hypothesis of interest with sufficiently high power and high probability of true positives with the minimum number of arrays. Of course, individual arrays can occasionally fail and it is therefore sensible to include extra arrays where budget and sample availability allow. To avoid introducing further bias, it is also important to hybridize the replicates, and indeed all samples in an experiment, to arrays selected at random from an array batch.

## 2.3 EXPERIMENTAL DESIGNS

Most microarray experiments typically fall into one of three broad categories depending on the objectives of the study. These are:

 **(i)**  class comparison,
 **(ii)**  class prediction,
**(iii)**  class discovery.

An example of class comparison is an experiment comparing individual samples with each other, for example comparing expression profiles of a male sample with a female, a mutant strain with a wild type, or a series of treatments (treatment 1 versus treatment 2 versus treatment 3, etc.). Class prediction, as the name suggests, involves using expression profiles from a class comparison study to determine which class a new sample belongs to. This is usually achieved by applying some type of multigene statistical model. For example, it may be desirable to predict whether a new cancer patient will respond well to a particular drug treatment. This type of study can be performed by first carrying out a class comparison to identify genes differentially expressed between tumors from patients that respond to the drug and those that are nonresponsive. Subsequently, a subset of genes that are predictive of response to the drug can be identified with a univariate statistical model and compared with the new patient's expression profile to indicate whether the new patient is likely to respond to the drug. With class discovery studies, as originally defined in a series of cancer studies (Golub et al., 1999), the objective is to partition a set of samples into particular categories based on their gene expression profiles. The categories or classes to which each sample is assigned are not known prior to the experiment. An example of such a study is the division of a set of apparently similar tumor samples into subclasses based on their expression profiles (Bittner et al., 2000). Time-course studies are another example that typically fall into the class discovery category. Any experiments that are designed to establish which groups of genes are coregulated can be considered as class discovery. We deal with the statistical methods and more advanced aspects of statistical design in Chapter 7.

### 2.3.1 Two-Channel Arrays: Dye Swap and Dye Balance

For single channel arrays, such as those available with the Affymetrix system, one sample is hybridized to each array. Consequently, there is no issue about how samples should be combined together for cohybridization on a single array. For two-channel microarrays, the experimental objectives need to be carefully considered before deciding which dye each sample should be labeled with and which samples should be combined together on each array. It is well known that if the same RNA sample is independently labeled with Cy3 and Cy5 dyes and these are then cohybridized together on an array, different signal intensities will be measured in each channel due to the different physical properties of the two dyes. While such systematic differences can, to a certain extent, be corrected during normalization, it is unlikely that the normalization process will be equally effective for every spot on every array. The end result is some residual dye bias in the data. An obvious way to reduce such dye bias is to perform dye-swap replications. For the first array, sample A is labeled with Cy3 and sample B with Cy5; for the second array the dyes are reversed and sample A is labeled with Cy5 and sample B with Cy3. Dye swapping is therefore a form of replication and can be performed for both technical replicates and biological replicates. As we note above, biological replicates have some advantages. Rather than labeling a technical replicate of each biological sample with each dye, which obviously doubles the number of arrays required, the experiment can instead be 'dye balanced', where half the samples from each class are labeled with one dye and half with the other dye. Labeling exactly half the samples from a class with one dye is preferable to labeling some other fraction since it provides a more accurate class comparison and is analytically straightforward. If a design is unbalanced, for example three samples labeled with Cy3 and four samples labeled with Cy5, then the dye bias can still be eliminated but a more complex weighted analysis is be required to adjust for the dye asymmetry. In certain designs, dye swap pairs are generally considered to be unnecessary, such as those involving a common reference sample (see below), since such designs are based on differences between slides and the systematic dye bias is removed during the analysis (Dobbin et al., 2003). Another approach that can be used to eliminate dye bias is to run a set of control arrays using the reference sample in both channels. This identifies the genes that display dye bias and these genes can be flagged as suspect if they show up as statistically significant in the experimental class comparisons (Dombkowski et al., 2004).

## 2.3.2 Direct and Indirect Comparisons

The ability to make a direct comparison between two samples on the same microarray slide can be an advantage of the dual-channel microarray system since it eliminates variation due to differences in individual arrays or to uneven hybridization. Often more than one design can be applied to an experiment and therefore a principal decision is whether the sample comparisons should be direct (within slides) or indirect (between slides). If the main aim of the experiment is to compare the gene expression profiles of two classes, A and B, then

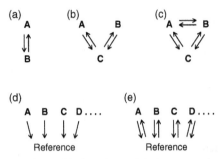

**FIGURE 2.1** Different experimental designs. Arrows join the two samples which are cohybridized on the same array, with the arrow-head representing the Cy3 labeled sample and the opposite end the Cy5 labeled sample. (a) Direct comparison between two samples, A and B, with a dye swap, making a total of two arrays. (b) Indirect comparison between samples A and B via reference sample C (4 arrays). (c) Block design, with all possible pairwise comparisons between three samples (6 arrays). (d) Reference design, with each sample of interest hybridized to the same reference sample (4 arrays). (e) Reference design with dye swap (8 arrays).

sample A and B can be labeled with Cy3 and Cy5 respectively and hybridized together on the same array. A second array should contain a dye swap replicate (Figure 2.1a). In an indirect experiment (Figure 2.1b), the samples A and B are not hybridized together, but sample A is hybridized to sample C and, on a separate array, sample B is hybridized to sample C. In this case sample C acts as a reference and the comparison between A and B is made indirectly via the reference. The main reason for directly comparing samples wherever possible is that it reduces the variance of any measurements. The efficiency of comparisons between two samples is determined by the length and number of paths connecting them (Kerr and Churchill, 2001; Yang and Speed, 2002). As long as there is a path of comparisons connecting two samples, they can be compared. However, the greater the number of intermediate steps the greater the statistical variance. Sample comparisons that are of the greatest interest should therefore be paired directly on the same array for the most reliable estimates of expression differences.

### 2.3.3 Common Reference Design

With just two or three samples, where all pairwise comparisons are of equal interest, it is feasible to make all pairwise comparisons directly (Figure 2.1c). However, with larger numbers of sample classes, it becomes increasingly impractical and expensive to pair every sample with every other sample and, of course, RNA may become limiting (e.g. for 6 RNA sources there are 15 pairwise comparisons requiring 5 samples of each RNA source, for 7 there are 21 requiring 6 samples, and so on; Yang and Speed, 2002). A commonly used solution to this problem is to use a reference design: where each sample of interest is hybridized to the same reference sample (Figure 2.1d). An advantage of the reference design is that there are never more than two steps in the path connecting any pair of samples. Consequently all comparisons are made with equal efficiency and, provided the amount of reference RNA is not limiting, the reference design can very easily be extended to accommodate more experimental samples at a later stage (Churchill, 2002). The simplicity of a reference design makes it very suitable for processing large numbers of samples. We shall see examples of large reference studies in biomedical studies in Chapter 11. With more complex designs, the risks associated with laboratory errors or sample handling increases. An obvious disadvantage of the reference design is that the reference RNA is hybridized on every array and therefore half of the data generated is of little or no interest. This makes the reference design inefficient and, perhaps more importantly, the statistical variance is four times higher than with simple direct comparisons.

An optimal reference sample should be easy to obtain in large quantities, be homogeneous and be stable over time. There are several types of reference samples in common use: a universal reference, a pooled reference and a biologically meaningful reference. A universal reference is any mix of RNA that is expected to hybridize to, and give a signal with, all or most of the elements on the array. For example, an RNA sample derived from an embryo at a specific developmental stage may be hybridized against a reference composed of a mix of all different developmental stages (e.g. Arbeitman et al., 2002). For some species such reference samples are commercially available, for example the Universal Human Reference RNA (Stratagene). Although each spot on the array should yield a measurable signal, a disadvantage of the reference approach is that the relative transcript levels for each gene may be quite different in the test and reference samples, making normalization more complex. In order to ensure that the signal intensities are similar between the test and reference samples, facilitating accurate normalization, a pooled reference can be generated by adding together an equal aliquot from each experimental sample to generate a pool. Rather than use a reference sample for which the data are less interesting, a biologically meaningful reference can be employed, that is where A versus reference and B versus reference are themselves of interest, rather than just A versus B being of interest. Examples of such studies are comparisons of tissues from different regions of a tumor with a healthy tissue as a reference, or different time points after a drug treatment compared with the untreated control as a reference. In reference design experiments, the general wisdom has been that the reference sample should be labeled with the same dye each time such that any dye bias remaining after normalization will affect all arrays equally and therefore not bias comparisons between samples (Dobbin and Simon, 2002). However, it has been subsequently reported that gene-specific bias exists in reference designs and that dye swap replicates should be incorporated where possible (Dombkowski et al., 2004).

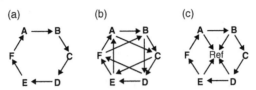

**FIGURE 2.2**   Different loop designs. (a) Basic loop. (b) Interwoven loop. (c) Combination of loop and reference design. Arrows join the cohybridized samples with the arrow-head representing the Cy3 labeled sample and the opposite end the Cy5 labeled sample.

## 2.3.4 Loop Designs

For large numbers of samples a reference design is the most efficient and for very small numbers a block design (all pairwise comparisons) may be feasible. However, when up to 8 conditions are being compared, it is more efficient to drop the reference sample and compare each sample with just one of the others. This has become known as a 'loop design' because in diagrammatic form (Figure 2.2a) it lines up all the conditions in a loop or daisy chain fashion (Kerr and Churchill, 2001). Without dye swaps, a loop design requires the same number of arrays as there are conditions, however, it is best to balance the dyes to negate dye bias by labeling an equal number of biological replicates with each dye. Loop designs become suboptimal when there are more than eight conditions and completely unmanageable for large experiments (Kerr and Churchill, 2001). A disadvantage of the design is that comparison of two samples far apart on the loop, that is with a greater number of intermediate steps, is inherently more variable than those that are more directly connected. If some comparisons within an experiment are of more interest than others, these should be designed to have the shortest connecting path and therefore the more precise measurements. Comparisons of lesser interest can be more distant and will be less precise. Another disadvantage is that the loop design is somewhat self-contained, making it hard to add in subsequent samples. In addition, it is a fragile design that is susceptible to collapse if there are one or more bad quality arrays, and can only be rescued by repeating the faulty arrays (Dobbin and Simon, 2002). Designs that interweave two or more loops (Figure 2.2b), or combine with a reference design (Figure 2.2c), help to create more than one pathway linking different samples and therefore improve the efficiency and reduce variance of the measurements. However, it should be borne in mine that the analysis of loops is more complex that with straightforward designs.

## 2.3.5 Experiments with Multiple Factors

The designs outlined above are suitable for investigating single factors, such as genotype, tissue type, treatment etc. Experiments where more than one factor is being investigated require more complex designs. Typically, factorial experiments can be used to analyze the expression profiles resulting from single factors or those arising from the combination of two or more factors. For example, Jin et al. (2001) studied the expression pattern of two different *Drosophila* strains at two different ages and in both sexes. Direct comparisons were made between the two age groups, but no direct comparisons were made between flies of different sex or different strain. This design reflects that fact that the primary goal of the experiment was to identify the effect of aging on gene expression with the effects of strain and sex being of secondary importance. Different error terms are used to test the age, sex, strain and sex-by-strain interaction effects. An alternative strategy is to use direct comparisons across all treatment factors in order to estimate the effect of each treatment with equal precision (Churchill and Oliver, 2001). The way that different sample pairings are arranged in a multifactor experiment can have an impact on the precision of different comparisons and this should obviously be taken into account during the design process.

## 2.3.6 Time-Course

Time-course experiments are those in which gene expression is monitored over time. For example, following a developmental process or monitoring a sample at several time points after induction of a treatment. As with all microarray experiments, numerous designs are possible (Figure 2.3) and the most suitable design for such experiments will depend on the specific questions the investigator wishes to address. The experimental aims should first be prioritized and then the best design determined by the comparisons that are of greatest interest and the number of time points being examined (Yang and Speed, 2002). Most

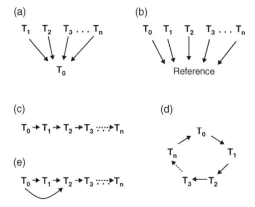

**FIGURE 2.3**   Examples of possible designs for time-course experiments. (a) Reference design with the starting point ($T_0$) as the reference. (b) Reference design with an independent common reference. (c) Sequential design. (d) Loop design. (e) Certain time points given preferential weighting. See text for details.

time-course experiments are aimed at identifying coregulated genes (gene class discovery). If the relative changes in gene expression between the different time points $(T_1, T_2, T_3, \ldots)$ and the starting point $(T_0)$ are of most interest, then comparing $T_1$ versus $T_0$, $T_2$ versus $T_0$, $T_3$ versus $T_0$ will be the most effective design, that is using $T_0$ as a common reference (Figure 2.3a). Alternatively, an unrelated reference sample can be used (Figure 2.3b). However, if the primary aim is to identify where sudden changes in gene expression occur, then a sequential design comparing neighboring time points (Figure 2.3c) or a loop design (Figure 2.3d) will be more appropriate. If not all time points are equally important, the experiment can be designed to weight time points differently depending on the objectives, with the more important samples using direct comparisons and others using indirect comparisons. For example, in Figure 2.3e, $T_0$ is compared directly with $T_1$ and $T_2$, but only indirectly with $T_3$. There are differing views on how best to design time-course experiments. Yang and Speed (2002) recommend a reference design, whereas Khanin and Wit (2005) recommend reducing large time-course experiments to a series of inter-woven loops or a combination of reference and loop design.

## 2.4 SUMMARY

Statistical quantification of evidence is preferable to qualitative descriptions. To provide a sound basis for statistical analysis, a microarray experiment needs careful planning to ensure simple and powerful interpretation. The most appropriate design will differ from experiment to experiment, but there are some points of general importance: the aims of the experiment should be the first consideration, direct comparisons should be made between samples of most interest, dye swapping or looping should be used to balance dyes and samples, and adequate numbers of biological replicates should be used. The design should be kept as simple as possible, but it is essential that there is an appropriate method available for analyzing the data produced. It is imperative that these considerations are taken into account before embarking on sample collection. In Chapter 7 we deal in depth with the statistical and analytical methods that are used to analyze each of the designs described in this section.

## REFERENCES

Altman N. (2005) Replication, variation and normalisation in microarray experiments. *Appl Bioinform* **4**: 33–44.

Arbeitman MN, Furlong EE, Imam F, Johnson E, Null BH, Baker BS, Krasnow MA, Scott MP, Davis RW, White KP. (2002) Gene expression during the life cycle of *Drosophila melanogaster*. *Science* **297**: 2270–2275.

Bittner M, Meltzer P, Chen Y, Jiang Y, Seftor E, Hendrix M, Radmacher M, Simon R, Yakhini Z, Ben-Dor A, Sampas N, Dougherty E, Wang E, Marincola F, Gooden C, Lueders J, Glatfelter A, Pollock P, Carpten J, Gillanders E, Leja D, Dietrich K, Beaudry C, Berens M, Alberts D, Sondak

V. (2000) Molecular classification of cutaneous malignant melanoma by gene expression pro-filing. *Nature* **406:** 536–540.

Bolstad BM, Collin F, Simpson KM, Irizarry RA, Speed TP. (2004) Experimental design and low-level analysis of microarray data. *Int Rev Neurobiol* **60:** 25–58.

Churchill GA. (2002) Fundamentals of experimental design for cDNA microarrays. *Nat Genet* **32:** S490–S495.

Churchill GA, Oliver B. (2001) Sex, flies and microarrays. *Nat Genet* **29:** 355–356.

Dobbin K, Simon R. (2002) Comparison of microarray designs for class comparison and class discovery. *Bioinformatics* **18:** 1438–1445.

Dobbin K, Shih JH, Simon R. (2003) Statistical design of reverse dye microarrays. *Bioinformatics* **19:** 803–810.

Dombkowski AA, Thibodeau BJ, Starcevic SL, Novak RF. (2004) Gene-specific dye bias in micro-array reference designs. *FEBS Lett* **560:** 120–124.

Draghici S, Kuklin A, Hoff B, Shams S. (2001) Experimental design, analysis of variance and slide quality assessment in gene expression arrays. *Curr Opin Drug Discov* **4:** 332–337.

External RNA Controls Consortium. (2005) Proposed methods for testing and selecting the ERCC external RNA controls. *BMC Genomics* **6:** 150.

Ferguson JA, Steemers FJ, Walt DR. (2000) High-density fiber-optic DNA random microsphere array. *Anal Chem* **72:** 5618–5624.

Golub TR, Slonim DK, Tamayo P, Huard C, Gaasenbeek M, Mesirov JP, Coller H, Loh ML, Downing JR, Caligiuri MA, Bloomfield CD, Lander ES. (1999) Molecular classification of cancer: class discovery and class prediction by gene expression monitoring. *Science* **286:** 531–537.

Jin W, Riley RM, Wolfinger RD, White KP, Passador-Gurgel G, Gibson G. (2001) The contributions of sex, genotype and age to transcriptional variance in *Drosophila melanogaster*. *Nat Genet* **29:** 389–395.

Khanin R, Wit E. (2005) Design of large time-course microarray experiments with two channels. *Appl Bioinform* **4:** 253–261.

Kendziorski CM, Zhang Y, Lan H, Attie AD. (2003) The efficiency of pooling mRNA in microarray experiments. *Biostatistics* **4:** 465–477.

Kendziorski C, Irizarry RA, Chen KS, Haag JD, Gould MN. (2005) On the utility of pooling biological samples in microarray experiments. *Proc Natl Acad Sci U S A* **102:** 4252–4257.

Kerr MK, Churchill GA. (2001) Experimental design for gene expression microarrays. *Biostatistics* **2:** 183–201.

Lee ML, Kuo FC, Whitmore GA, Sklar J. (2000) Importance of replication in microarray gene expression studies: statistical methods and evidence from repetitive cDNA hybridizations. *Proc Natl Acad Sci U S A* **97:** 9834–9839.

Lee ML, Whitmore GA. (2002) Power and sample size for DNA microarray studies. *Stat Med* **21:** 3543–3570.

Lee S, Jo M, Lee J, Koh SS, Kim S. (2007) Identification of novel universal housekeeping genes by statistical analysis of microarray data. *J Biochem Mol Biol* **40:** 226–231.

Page GP, Edwards JW, Gadbury GL, Yelisetti P, Wang J, Trivedi P, Allison DB. (2006) The Power-Atlas: a power and sample size atlas for microarray experimental design and research. *BMC Bioinform* **7:** 84.

Pan W, Lin J, Le CT. (2002) How many replicates of arrays are required to detect gene expression changes in microarray experiments? A mixture model approach. *Genome Biol* **3:** R0022.

Peng X, Wood CL, Blalock EM, Chen KC, Landfield PW, Stromberg AJ. (2003) Statistical implications of pooling RNA samples for microarray experiments. *BMC Bioinformatics* **4:** 26.

Shih JH, Michalowska AM, Dobbin K, Ye Y, Qiu TH, Green JE. (2004) Effects of pooling mRNA in microarray class comparisons. *Bioinformatics* **20:** 3318–3325.

Suzuki T, Higgins PJ, Crawford DR. (2000) Control selection for RNA quantitation. *Biotechniques* **29:** 332–337.

Thellin O, Zorzi W, Lakaye B, De Borman B, Coumans B, Hennen G, Grisar T, Igout A, Heinen E. (1999) Housekeeping genes as internal standards: use and limits. *J Biotechnol* **75:** 291–295.

Wang SJ, Chen JJ. (2004) Sample size for identifying differentially expressed genes in microarray experiments. *J Comput Biol* **11:** 714–726.

Warrington JA, Nair A, Mahadevappa M, Tsyganskaya M. (2000) Comparison of human adult and fetal expression and identification of 535 housekeeping/maintenance genes. *Physiol Genomics* **2:** 143–147.

Wernisch L. (2002) Can replication save noisy microarray data?. *Comp Func Genomics* **3:** 372–374.

Yang YH, Speed T. (2002) Design issues for cDNA microarray experiments. *Nat Rev Genet* **3:** 579–588.

Yauk CL, Berndt ML, Williams A, Douglas GR. (2004) Comprehensive comparison of six microarray technologies. *Nucleic Acids Res* **32:** e124.

Zhang SD, Gant TW. (2005) Effect of pooling samples on the efficiency of comparative studies using microarrays. *Bioinformatics* **21:** 4378–4383.

## Chapter 3

# Designing and Producing Microarrays

One of the critical factors for ensuring success in a microarray study is selecting an array type that will allow an informative analysis of the biology of interest. Unfortunately for the novice, there are several options available for the type of nucleic acid probe and the arraying technology, consequently it can be far from obvious what the best route to take is. At the most basic level, there are two ways to produce a microarray: either premade nucleic acids are deposited on a support via some type of robotic printer or the probe sequences are generated in situ by oligonucleotide synthesis on the array. While some of the current microarray technologies are suitable for use in individual or 'core' academic laboratories, others are only really applicable in an industrial setting. We outline the major technologies currently employed for array production and indicate some of the newer approaches that are beginning to find their way into users hands: a summary of these technologies is provided in Table 3.1.

First we need to consider the probe elements and what they need to achieve. We make no apologies for emphasizing the properties and design of probes since it is probably this area that is the major contributing factor to the success of a microarray analysis. The probe is the sensor that converts the abundance of nucleic acids in the sample into the digital readout we use to make inferences

**TABLE 3.1** Available Microarray Production Technologies

| Technology | Probe type | Elements/array | In-house |
|---|---|---|---|
| Contact printing | Any | ∼30 000 | Yes |
| Ink-jet printing | Any | ∼20 000 | Yes |
| Maskless photolithography | Oligonucleotide | 15 000 | Yes |
| Ink-jet synthesis | Oligonucleotide | 240 000 | No |
| Mask-based photolithography | Oligonucleotide | 6 200 000 | No |
| Maskless photolithography | Oligonucleotide | 2 100 000 | No |
| BeadArray | Oligonucleotide | ∼50 000 | No |
| Electrochemical | Oligonucleotide | ∼15 000 | No |

Contact and inkjet printers along with low-density maskless photolithographic synthesizers are available for in-house generation of custom arrays. The remaining high and very high-density platforms are only available from commercial providers.

about gene expression. Consequently, the fidelity with which a probe reports the presence of its cognate nucleic acid directly relates to the reliability of the experiment. We therefore need to consider the user selectable criteria that underlie the interaction between probe and target as well as the portion of the target molecule we design probes against. Probes can be either single stranded oligonucleotides, ranging from 25 to 70 or more bases in length, or double stranded DNA products, generally generated from PCR amplification of genomic DNA or of cDNA library clones. To some extent the choice of probe is dictated by the organism being studied since oligonucleotide probes require either a sequenced genome or, at the very least, a substantial amount of EST sequence. If genome sequence is not available, then the most frequent route is to prepare array probes from cDNA libraries. During the nineties, the use of cDNA arrays was prevalent, principally because synthetic oligonucleotides were expensive and genome sequence was limited for most organisms. As sequence information has accumulated, oligonucleotide synthesis costs reduced and in situ synthesis methods become more widely available, oligonucleotide probes are becoming increasingly prevalent and it is likely that for most large-scale studies oligonucleotide based arrays will be the method of choice.

## 3.1 PROBE SELECTION

The selection of probe sequences for a microarray design depends upon the use envisaged for the array, since different criteria apply when considering probes for gene expression studies, for genomic tiling or for genotyping. We consider tiling and other arrays in Chapter 10, focusing here on probe selection for gene expression. There are two issues to consider: what type of probe will be used and what portion of the target will be interrogated. Simply expressed, one wishes to design a set of probe sequences that have uniform properties: that is they all show the same thermodynamic behavior under the experimental conditions employed. At the same time the probes must be highly specific: that is they should only detect the sequence to which they are designed against. Specific probes can be designed against genes, transcripts or portions of transcripts. In reality, deciding on the target set and producing a good probe set can be difficult since there are several criteria that need to be balanced when designing probe sequences for an expression array: these are sensitivity, specificity, noise and bias (Kreil et al., 2006). Below we consider the theoretical aspects of probe design before looking at the more practical design and target selection issues likely to be encountered in the real world.

*Specificity* is the ability of a probe to detect the target mRNA that it is designed against while not hybridizing with any other sequence in the target population. A well-designed probe will maximize specificity by eliminating cross hybridization with related sequences. Specificity increases with length, up to probe lengths of approximately 100 bases. Thereafter, increasing length can reduce specificity since the chance of the probe containing a short random match to another sequence in the target population increases (Chou et al., 2004a). Specificity influences what is actually measured: a highly specific probe will only generate signals by binding a single target whereas a less specific probe will generate a mixed signal containing contributions from hybridization with the true target and some 'noise' due to cross-hybridization with related sequences in the sample. Clearly probes with reduced specificity will generate less accurate hybridization signals. There is however one application where specificity is less desirable, cross-species hybridization, where arrays designed for one species are hybridized with samples derived from a related species. This type of analysis is important in evolutionary or population studies where we know in advance that probe and sample will not perfectly match. In such cases, longer probes are frequently preferred (Bar-Or et al., 2007) and in general it is considered that targets with an overall sequence identity of around 70% are likely to cross hybridize with a long cDNA probe (Xu et al., 2001).

*Sensitivity*, which should also be maximized, is a measure of how little hybridization signal is lost between a probe and its target. In an ideal world, the hybridization signal detected by a particular probe will be a direct measure of the amount or abundance of its target in the sample population. As with specificity, a general trend is observed where longer probes are more sensitive since the binding energy between probe and target increases with length. This is fairly obvious if we simply consider that the more bases available for hybridization between probe and target, the stronger the potential signal will be. However, unlike specificity, which we can control by defining the probe sequence, sensitivity is more complex and depends upon the thermodynamic properties of the probe sequence, on the accessibility of the target sequence under hybridization conditions, on the surface chemistry of the microarray substrate and on the absolute concentration of the probe in an array element (Draghici et al., 2006).

*Noise* is defined as the random variation of a particular probe–target hybridization signal and is generally measured as the coefficient of variation of a set of replicate measurement (Kreil et al., 2006). Obviously noise should be reduced wherever possible and again it is found that noise reduces with probe length (Chou et al., 2004a). One way of effectively dealing with noise, especially when short probes are being considered, is to use multiple independent probes for a single target. This is essentially the approach taken by Affymetrix when designing their gene expression GeneChips.

Finally, *bias*, the systematic deviation of a measurement from the true hybridization signal due to probe-specific effects or technical limitations, should also be reduced. Bias can be more difficult to eliminate with our current state of knowledge, but it may be reduced by employing principled probe design. For example, by measuring a set of probes with random bias or by measuring bias for each probe experimentally and correcting for this in the analysis. A graphical representation of how these key parameters change with probe length is shown in Figure 3.1.

Balancing these factors is far from easy and inevitably the selection of good probe sequences involves trade-offs between each of these variables. In some cases it will be desirable to maximize the sensitivity of a probe set if, for example, it is known in advance that detection of low abundance transcripts is essential. Alternatively, a general probe set capable of being used for the analysis of several different species may be the goal, in which case relaxed specificity will be an important design criteria. In most cases, however, the aim is to generate a probe set that is capable of making specific measurements so that related transcripts may be distinguished with high reliability.

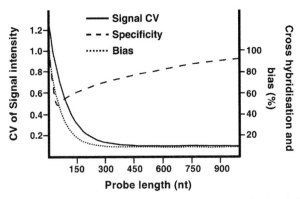

**FIGURE 3.1**   Composite graph showing the relationship between probe length in nucleotides (*X*-axis) and three key parameters: noise (variation in signal intensity from replicate probe elements on the left hand *Y*-axis), cross-hybridization and measurement bias (plotted on the right hand *Y*-axis). As probe length increases the noise and bias are reduced but the cross hybridization level increases. *Source*: Data from Chou et al. (2004a,b).

## 3.2 CDNA AND AMPLICON PROBES

With double stranded DNA probes, particularly those derived from the amplification of cDNA libraries, it is virtually impossible to control the key parameters outlined above. While individual probes are likely to be sensitive because
they are long and can hybridize with most or all of the target sequence, they are
far more prone to cross-hybridization with related sequences and thus lack
specificity. A further disadvantage is that cDNA clones are heterogeneous in
both size and sequence composition and consequently it can be difficult to
compare signals between different probe elements. As we note, however, the
lack of specificity can be an advantage if an array built with probes from one
organism is used to interrogate mRNA from a related species, since here reduced
hybridization stringency is an advantage. The forgoing may suggest there is no
place for cDNA arrays, this is, however, far from true and reinforces the view that
the type of array used is dependent on the biological question. If the goal is to
identify sets of differentially expressed or coordinately regulated genes, cDNA
arrays can be an effective tool and they have been widely used to explore human
disease, developmental biology and genome annotation (Alizadeh et al., 2000;
Furlong et al., 2001; Shyamsundar et al., 2005).

   If we wish to determine, for a particular species that has no genome
sequence, what sets of genes are specifically expressed in particular tissues or
in response to a particular environmental condition and what these genes are, the
cDNA array can be an effective tool for rapidly generating data. Normalized
cDNA libraries may be constructed from individual tissues and clones selected
from such libraries use to generate array probes. There are several methods for
normalizing cDNA libraries (e.g. Bonaldo et al., 1996; Carninci et al., 2000),
descriptions of which are beyond the scope of this review. Using normalized
libraries reduces the representation of cDNAs for abundantly expressed genes
and thus increases coverage without increasing array density. When arrays
generated with probes from normalized libraries are interrogated with RNA
samples from specific tissues or conditions, those probes showing specific
hybridization can be readily identified (e.g. Gracey et al., 2004). The clones
used to generate the probes can then be sequenced to identify the genes they
encode. An alternative approach is to first cluster cDNA libraries, via EST
sequencing, into unigene sets and array individual representatives of each cluster
(e.g. Vodkin et al., 2004). Here the benefits of generating a defined array where
sequences are known must be weighed against the initial up front effort required
for the EST sequencing project. In either case, inserts are prepared from cDNA
clones, generally via PCR amplification using vector-specific primers to drive
the PCR reaction. It is essential that quality control checks are preformed to
assess the PCR reactions and to quantify the amount of PCR product generated
(Knight, 2001). Much of the process can be automated if suitable robotics are
available and this can substantially reduce error and cross-contamination

(Burr et al., 2006). A note of caution: if this route is chosen, be extremely careful about the source of the cDNA libraries used (Bowtell, 1999). It is reported that as many as 30% of the clones in some unigene sets are misannotated or mislabeled (Halgren et al., 2001), which can lead to erroneous data interpretation. Consequently, it is recommended that gene sets are evaluated by end sequencing clones before using them to generate microarray probe sets (Taylor et al., 2001).

An alternative to using cDNA clones, that can circumvent the problems associated with size heterogeneity and help increase specificity, is to prepare specific PCR product probes rather than entire cDNA clones. There are two possibilities here; if genome sequence is available then specific primers are designed to amplify 3′ exons or fragments of genes from genomic DNA. If there is limited or poorly annotated genome sequence but good sequenced EST collections, then primers to amplify fragments from the 3′ end of cDNA clones can be designed. In both cases, standard primer design software, such as the popular Primer3 program (Rozen and Skaletsky, 2000), is used to select suitable PCR primers. In general, amplicons of 200–500 bp are generated, which fit reasonably well with the sensitivity and specificity constraints discussed above. Primers can be tailed with specific sequences to facilitate re-amplification since this can be more efficient than PCR amplification from cDNA clones carried in bacteria. Primers that incorporate specific sequences with utility in other genomics applications can be used. For example, preparing primers tailed with the recognition sequence for a bacteriophage RNA polymerase allows the generation of double-stranded RNA that can be used for high-throughput RNAi screening (Boutros et al., 2004). A typical example of the pipeline required to go from a clone set to a home-made cDNA microarray is given in White and Burtis (2000), however there are many basic texts in the literature that can guide the user (Sambrook and Russell, 2001).

## 3.3 OLIGONUCLEOTIDE PROBES

As we suggest above, specific oligonucleotides are widely held to be the probes of choice when genome sequence is available. Automated oligonucleotide synthesis chemistry is now well-established since its introduction in the mid 1980s (Caruthers et al., 1992) and the current widely employed phosphoramidite method allows the synthesis of comparatively large (around 100 nucleotides) oligomers at reasonable yield and purity. By carefully selecting probe sequences, specificity and sensitivity can be maximized, errors due to mislabeling clone libraries are eliminated and, most importantly, a good design can generate a set of probes with uniform characteristics. As we will describe below, oligonucleotide probes may be generated by conventional solid phase chemical synthesis and then deposited on the array or synthesized directly on the array substrate, so called in situ synthesis. In the latter case there are a variety of methods that have been developed for this process: mask-directed photolithography (Fodor et al., 1991), maskless photolithography (Singh-Gasson et al., 1999), ink-jet driven synthesis (Hughes et al., 2001) and electrochemical synthesis (Egeland and Southern, 2005). A slight variant on the standard array platforms is the Illumina BeadArray technology. Based on silica beads, each covered with a specific oligonucleotide probe, that randomly self assemble in microwells of a microfabricated substrate (Kuhn et al., 2004), the technology can be treated as an array composed of presynthesized probes. Each of these methods has its merits and drawbacks, the choice largely dictated by the requirements of the experiment or the local availability of equipment or expertise. Below we describe the basics underpinning the oligonucleotide synthesis methods before reviewing the principals of oligonucleotide design.

### 3.3.1 Oligonucleotide Synthesis

Oligonucleotides are synthesized by utilizing a chemical protection process, where reactive residues on a growing nucleotide chain are prevented from reacting by the presence of a protecting group. Controlled removal of the protecting group allows the addition of a single specific base, with cycles of deprotection and base addition generating the desired sequence. In standard solid phase synthesis, oligonucleiotides are built up $3'$–$5'$, beginning with a $3'$ priming residue attached to a solid support. The $5'$ end is protected by a dimethoxytrityl (DMT) group, which is removed in a detritylation step to generate a free hydroxyl group that allows the addition of a phosphopramidite nucleotide. The deprotection and base coupling reactions are not 100% efficient so any free hydroxyl groups remaining must be capped to prevent the synthesis of oligonucleotides with a missing base. The new bond is then oxidized and following a second capping step the cycle is ready to be repeated until the desired sequence is obtained. The key parameter we are concerned with is the coupling efficiency, the fraction of growing oligonucleotide chains that successfully incorporate a base at each cycle. Current solid phase chemistry achieves coupling efficiencies of at least 99% (Bellon and Wincott, 2000). While this seems impressive, in effect it means that when synthesizing a 50mer only 60% of the oligonucleotides will be full length and for a 100mer, less than 40% will be full length. This relationship is roughly sketched for 25, 50 and 70mer oligonucleotides in Figure 3.2. Since truncated oligonucleotides have been shown to reduce measurement specificity (Jobs et al., 2002) they should be eliminated if possible. When oligonucleotides are synthesized and then deposited on the array they can first be purified to ensure only full length probes are present. This may be achieved by HPLC purification or by adding a $5'$ amino group during the terminal synthesis round (Connolly, 1987) since only uncapped full-length sequences will be able to incorporate the last base. Amino-capped

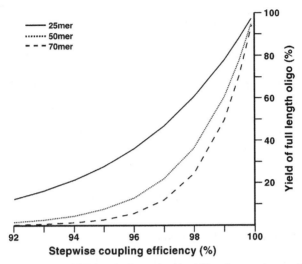

**FIGURE 3.2**  Dependency of full-length oligonucleotide yield on the stepwise coupling efficiency of oligonucleotide synthesis. Each curve represents the percentage yield at different efficiencies for 25, 50 and 70 nucleotide oligos.
*Source*: Data from Kreil et al. (2006).

oligonucleotides may be covalently attached to chemically modified array substrates and the unmodified truncated probes removed by washing (see below) (Kamisetty et al., 2007).

The alternative method for generating oligonucleotide arrays is to synthesis the probes in situ, building up the oligonucleotide sequences on the array substrate (Figure 3.3). In one method, conventional phosphoramadite synthesis is performed by delivering nanolitre volumes of the relevant chemicals to defined areas on the substrate with ink-jet printing heads (Hughes et al., 2001). The coupling efficiency of the synthesis process is estimated to be 98% and, although lower than the solid phase reaction described above, approximately 60% of 60mer oligonucleotides will be full length. A second method, light-directed oligonucleotide synthesis, uses alternative chemistry involving photolabile protecting groups rather than the DMT group described above. In this case there are two major technologies, the mask directed photolithography developed by Affymetrix corporation (Pease et al., 1994) and the maskless synthesis method that employs digital micromirror devices (DMD) (Singh-Gasson et al., 1999). With the former, a series of chrome/glass photolithographic masks control where high intensity light is delivered on a growing array and thus which oligonucleotide chains are deprotected. Unfortunately, due to the photolabile groups employed, the coupling efficiency is relatively low (92–94%; McGall et al., 1997). This effectively limits the size of oligonucleotides to 25mers or less since the yield of full-length probe for longer sequences is below 10%. In the case of

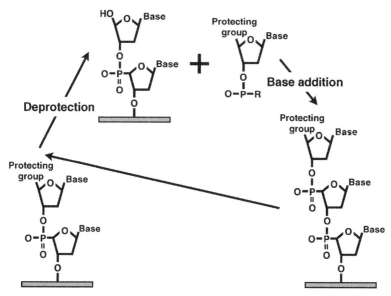

**FIGURE 3.3**   The basics of oligonucleotide synthesis. A growing oligonucleotide chain, attached to a solid substrate, is deprotected by light or chemical means and becomes available for the addition of the next protected base. Repeated cycles generate oligonucleotides of the desired sequence.

DMD directed synthesis, an array of precisely controlled mirrors directs high intensity light to particular array features and hence controls the deprotection step. In this case, improved photolabile groups can give synthesis yields up to 98%, allowing the synthesis of longer oligonucleotides (Nuwaysir et al., 2002). Thus, the DNA synthesis method dictates the type of probe that can be generated and while photolithographic methods restrict the range of available probe lengths, any drawbacks are more than compensated for by the very high densities that can be achieved with in situ synthesis technologies. For example, the latest Affymetrix arrays accommodate approximately 6 000 000 independent probe sequences on a single array, facilitating array designs that can control for loss of sensitivity or specificity by using of multiple independent probes for each target.

### 3.3.2 Designing Oligonucleotide Probes

For a variety of reasons, ranging from defining the target sequences that will be interrogated to understanding the thermodynamics of probe–target interactions under hybridization conditions, designing oligonucleotide probes is still a complex problem. For a single target transcript one can spend considerable effort in designing highly specific probes, however, when dealing with the tens of thousands of targets typical in complex eukaryotic genomes, there is clearly a need for an automated or semi-automated pipeline. Several tools, some of which are

**TABLE 3.2** A Selection of Popular Probe Design Tools and Their Sources

| Design tool | URL | References |
|---|---|---|
| ArrayOligoSelector | http://arrayoligosel.sourceforge.net/ | Bozdech et al. (2003) |
| BioSap | http://techdev.systemsbiology.net/biosap/ | NA |
| GoArrays | http://www.isima.fr/bioinfo/goarrays/ | Rimour et al. (2005) |
| OligoArray 2.1 | http://berry.engin.umich.edu/oligoarray2_1/ | Rouillard et al. (2003) |
| OligoPicker | http://pga.mgh.harvard.edu/oligopicker/ | Wang and Seed (2003) |
| OligoWiz | http://www.cbs.dtu.dk/services/OligoWiz/ | Wernersson and Nielsen (2005) |
| Oliz | http://www.utmem.edu/pharmacology/otherlinks/oliz.html | Chen and Sharp (2002) |
| Osprey | http://osprey.ucalgary.ca/ | Gordon and Sensen (2004) |
| Picky | http://www.complex.iastate.edu/download/Picky/ | Chou et al. (2004b) |
| PROBESEL | http://www.zaik.uni-koeln.de/bioinformatik/arraydesign.html | Kaderali and Schliep (2002) |
| ProbeSelect | http://ural.wustl.edu/software.html | Li and Stormo (2001) |

listed in Table 3.2, have been developed to assist in the process of large-scale probe design. While they differ in the precise algorithms they use or the exact approaches they take, by and large they all follow a common path: search for specific probes that are unique in the genome, that lack self-binding, that, as a set, are thermodynamically balanced and, most frequently, are close to the 3'end of their target genes. In general, a set of target transcripts and the genome sequence are provided, the user defines a set of parameters including probe length, number of probes per target, cross-hybridization threshold, etc., and the programs generate a list of probes for each target.

### 3.3.3 Target Selection

The first issue encountered when deciding to design a microarray is the source of target sequences: the set of transcripts to design probes against. While we have very high quality genome sequence for a range of organisms, including humans and many of the widely used experimental models, our understanding of gene structure and the transcript repertoire of these genomes is far from complete. Consequently target selection is never going to be complete or perfect. The use of tiling arrays, as we describe later in Chapter 10, is beginning to shed light on the complexity of genome transcription, however, at present we are generally restricted to using genome annotation to guide target selection. Here decisions have to be reached as to whether one wishes to try and interrogate every potential transcript isoform a particular genomic locus encodes or generate a smaller set of 'gene-specific' probes. With the former, a detailed understanding of intron–exon structures is required. In the latter case, the usual route is to compile a list of target sequences from the sequence databases or genome annotation projects.

As an example we consider the human genome: one could use the UniGene database from NCBI, the predicted annotations from Ensembl, or the highly curated gene model annotations from the Vertebrate Genome Annotation (VEGA) project. The choice of target sequence set obviously has consequences in terms of the number and complexity of probes. For example, as of November 2007 there are 122 036 UniGene entries, approximately 24 000 protein-coding gene models in Ensembl and 33 000 annotated genes in VEGA. Note that for the human genome there are almost 500 000 annotated exons in Ensemble representing a formidable challenge for any probe design algorithm. In Table 3.3 the current views of the genome and transcriptome for selected model organisms highlight the problem associated with selecting target sets in terms of balancing coverage with reliability.

The Human UniGene collection is built by examining all GenBank EST sequences for sequence homology. Sets of sequences that share virtually identical 3′ untranslated regions are grouped to form a UniGene cluster, which is then annotated with data relating to tissue expression and orthology. Almost 7 million sequences were used to build the 122 036 UniGene entries (Nov. 2007) and the collection is updated and revised regularly. It should be noted that as our understanding of gene structures and genome annotation improve, the UniGene collection changes with new entries appearing and some entries being retired. Consequently, gene annotations used to design a probe set will inevitable be different a year or two later. It is therefore important that array users revise their probe annotations regularly if they wish to infer gene expression changes that relate to the latest view of the genome. This can be relatively painlessly

**TABLE 3.3** Gene and Exon Statistics for Selected Model Organisms

| Organism | UniGenes | Predicted genes | Annotated genes | Annotated exons |
|---|---|---|---|---|
| Human | 122 036 | 23 686 | 18 825 | 499 606 |
| Mouse | 77 731 | 23 786 | 9 233 | 286 363 |
| Zebrafish | 56 768 | 18 825 | 9 741 | 119 200 |
| Chicken | 33 566 | 29 430 | 11 954 | 182 183 |
| Drosophila | 17 287 | – | 14 039 | 65 817 |
| Nematode | 21 657 | – | 20 140 | 141 560 |
| Arabidopsis | 29 918 | – | 27 029 | 167 212 |

UniGene data from NCBI, Predicted genes are from Ensemble. Annotated genes (protein coding not including pseudogenes or non-coding transcripts) and Exons are from VEGA or organism-specific databases.

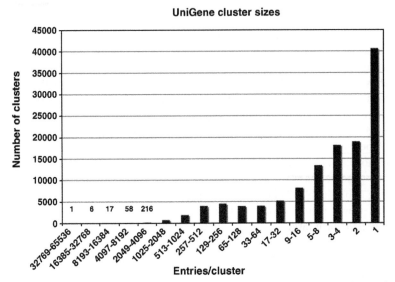

**FIGURE 3.4**   Graph showing the distribution of sequence contribution to UniGene clusters. Over 40 000 clusters are only represented by a single EST, indicating they should be treated with caution. For those few clusters containing tens of thousands of ESTs at the left hand end of the graph, the number of clusters are given above the *X*-axis.

accomplished by BLAST searching each probe sequence against the UniGene from time to time.

Each UniGene entry comprises a set of transcript sequences that appear to come from a single gene or expressed pseudogene locus and, since many of the clusters (almost 80 000) are defined by 4 or fewer sequences, these should be viewed with caution (Figure 3.4). There are certainly not 120 000 genes in the human genome (though this is not the place to enter into the 'what is a gene?' debate). In addition, since there may be exon sharing between UniGene entries, designing target specific probes can be complicated. It may therefore be more pragmatic to use annotated or predicted genes for designing probes, bearing in mind that some of the predictions may be wrong. Certainly if the aim is to generate printed arrays in-house, there will inevitably be compromises due to the relatively low array densities that can easily be achieved and it is therefore unlikely that probes against each UniGene entry will be made. The benefit of using annotated genes or predictions (note that we are not talking about purely computationally predicted genes here but those with some degree of support from orthology or EST sequence) is that they can be downloaded relatively easily from the relevant source database and are ready to plug straight into the oligo design software.

An alternative and more rigorous approach is to combine the UniGene set with various other databases to reduce the number of entries and improve the

TABLE **3.4** Sequence Sources Used in Designing the Affymetrix U133 Plus 2.0 GeneChip

| Sequence source | Clusters or sequences examined | Clusters or sequences used in design |
|---|---|---|
| UniGene | 108 944 | 38 572 |
| dbEST | 5 030 353 | 2 669 196 |
| GenBank | 112 688 | 49 135 |
| RefSeq | 18 599 | 13 692 |
| Total | 5 161 640 | 2 732 023 |

evidence for each target selected. This is not a trivial undertaking and requires good bioinformatics expertise, particularly familiarity in dealing with sequence databases and sequence clustering tools. A good example of a refined target selection pipeline is provided by the design process employed by Affymetrix to define the 47 401 transcript targets for the U133 Plus 2.0 Human Genome Array that interrogates 38 572 genes. Sequences were collected from dbEST, RefSeq and GeneBank and filtered to remove contaminants such as vector sequence, repeats and low quality sequence. The contributions from each of these sources is illustrated in Table 3.4. This collection was then aligned to the human genome assembly to orient sequences and detect features such as splice and polyadenylation sites, rejecting low quality sequences that were not consistent with being reliable transcripts. These data were combined with UniGene to generate an initial set of clusters. The clusters were then more rigorously clustered, partitioning alternative transcript isoforms into subclusters, using a dedicated and publicly available clustering tool (StackPACK; http://gene3.ciat.cgiar.org/stackpack/). This allowed identification of a core set of targets, including splice isoforms of particular genes that could readily be discriminated with unique probe sets. The resulting target collection provides a good compromise between the transcript complexity of the human genome and the reliably annotated gene features that can be accommodated on a single array. The complete design process is more fully described in technical documents available from the Affymetrix website (http://www.affymetrix.com/).

From the above, it is clear that target selection can be complex if the goal is to generate an array with as comprehensive a coverage as possible. For the commonly used mammalian models (mouse and human) it can be argued that the commercial providers provide the best route for obtaining comprehensive expression profiling arrays cost effectively. This is especially true when one considers that annotation of mammalian genomes is still very fluid and gene models are constantly being updated. Consequently the ability of some commercial providers to rapidly change probe sets (particularly true of those using ink-jet or maskless photolithography to synthesize probes in situ) can provide a degree of flexibility that is difficult for an academic lab to match. On the other hand, for those eukaryotes with smaller genomes, particularly the well annotated models such as *Drosophila* or *Caenorhabditis elegans*, or for prokaryotic genomes, where gene models are more clearly defined, the ability to select a fairly robust target set makes a relatively stable in-house array design less daunting.

### 3.3.4 Probe Design

As we describe above, there are a range of publicly available design tools that streamline probe design. To exemplify the design process, the typical steps required to design a probe set for an expression array are outlined. Note that individual steps may be different for particular software tools and anyone contemplating a probe design project is advised to thoroughly familiarize themselves with the documentation and background for the software selected. The obvious requirements for any design tool are to assess target specificity, usually taken care of with the BLAST tool (Altschul et al., 1997), and to calculate the thermodynamic properties of each probe. This is most commonly achieved by computing the probe melting temperature ($T_m$) with the Nearest-Neighbor Model developed by SantaLucia (SantaLucia, 1998). In this model, $\Delta H$ and $\Delta S$ are the dimer stacking and propagation energies, $R$ is the gas constant (1.9872 cal/K mol) and DNA is the DNA concentration (fixed at $10^{-6}$ M in OligoArray2.1). The calculation assumes a 1 M $Na^{2+}$ concentration. For full details see (SantaLucia, 1998; Rouillard et al., 2003)

$$T_m = \frac{\Delta H^\circ}{\Delta S^\circ + R \ln(\text{DNA}/4)} - 273.15$$

Probes with stable secondary structures at the hybridization temperature need to be eliminated and this is usually assessed with a secondary structure assessment tools such as mFOLD (Mathews et al., 1999) or its more modern derivative DINAMelt (Markham and Zuker, 2005). Additionally, some design tools allow undesirable sequences to be flagged in advance, known repeats for example, and some provide low complexity filters to eliminate simple sequence probes.

We take as an example the use of OligoArray2.1, since we find that this tools performs well in our hands, but emphasize that other software may perform just as well (Rouillard et al., 2003). The starting data are the list of target sequences in FASTA format, the genome sequence and a local installation of BLAST. To generate a suitable target set it is first necessary to remove highly related sequences from the initial target set as described above in the target selection section (i.e. duplicate genes with high sequence similarity). If sequences from a set of gene predictions are taken straight from a database then a preprocessing step using software tools such as nrdb90 (Holm and Sander, 1998) may be advisable to eliminate genes with extreme sequence similarity. A note of caution: from the outset, attention should be paid as to how samples will be labeled in subsequent analysis since this directs whether probes need to match the sense or antisense stand of the target transcript. For example, if samples will be labeled by direct label incorporation into 1st strand cDNA, labeled targets will be antisense to the transcript so the probe sequence must match the transcript sequence. Other labeling strategies sometimes require antisense probes, for example if one is directly labeling RNA. For most expression profiling applications, probes are generally biased towards the 3′ end due to the common cDNA synthesis-based sample labeling methods. In some cases this will be straightforward since 3′ ends are well defined because characterized full-length cDNAs are available or hundreds of ESTs have been sequenced. In other cases, where EST or cDNA support is less robust, then genomic features such as predicted polyadenylation sequences can be used. These issues should be resolved when selecting the target sequences and are more relevant to cases where targets are derived from a reduction of larger sequence databases. The issue is less likely to occur when one is using annotated or predicted transcripts since 3′ ends are defined in these cases.

The primary design run requires the selection of a number of parameters: oligonucleotide probe length range (min and max), oligonucleotide melting temperature ($T_m$, min and max), maximum distance from the 3′ end of the target, maximum number of probes per target sequence, temperature threshold for secondary structure prediction, temperature threshold for considering cross-hybridization, G + C content of the oligonucleotide (maximum and minimum percentage). In addition, one can specify if more than one probe per target sequence is required and the minimum distance between multiple probes. A list of forbidden sequences that should never appear in any probe, for example runs of nucleotides such as AAAAA or CCCCC can also be provided. These steps are best summarized by looking at the command line options for OligoArray2.1 as shown in Table 3.5.

The important user-defined parameters are probe length and thermodynamic properties such as $T_m$ and cross-hybridization threshold. The OligoArray2.1 length default is 45–47 nucleotides, which may be considered ideal if multiple probes per target are being selected. However, as we describe earlier, longer probes offer a better compromise between sensitivity and specificity and if the objective is to generate a single probe for each target, there is an argument that longer probes in the 65–70mer range are more effective (Bozdech et al., 2003; Kreil et al., 2006). Note that with some substrates there is evidence that 10–15 nucleotides of the probe sequence act as a 'spacer' between the array surface and the bases available for hybridization (Shchepinov et al., 1997) and consequently it may be sufficient to design 50mer probes and use spacer molecules to covalently attach probes to the array surface (Pack et al., 2007). In this respect it is believed that three-dimensional surface substrates (see below) are as effective as spacers and are now often used (Dorris et al., 2003). It is important to note that

**TABLE 3.5** Command Line Options for OligoArray2.1

| | |
|---|---|
| -I | The <u>input file</u> that contains sequences to process. |
| -d | The <u>Blast database</u> that will be used to compute oligo's specificity. |
| -o | The <u>output file</u> that will contain oligonucleotide data. |
| -r | The <u>file</u> that will contain sequences for which the design failed. |
| -R | The <u>log file</u> that will contain information generated during design. |
| -n | The maximum number of oligonucleotides expected per input sequences. *Default* = 1 |
| -l | The minimum oligonucleotide length. *Range* 15–75; *default* = 45 |
| -L | The maximum oligonucleotide length. *Range* 15–75; *default* = 47 |
| -D | The maximum distance between the 5′ end of the oligo and the 3′ end of the target. *Default* = 1500 |
| -t | The minimum oligonucleotide $T_m$. $<100 < maximum\ T_m$; *default* = 85 |
| -T | The maximum oligonucleotide $T_m$. $<100 > minimum\ T_m$; *default* = 90 |
| -s | Temperature for secondary structure prediction. Reject oligos folding into a stable secondary structure at this temp; *default* = 65 |
| -x | Threshold to start considering cross-hybridizations. Targets hybridizing with a $T_m$ above threshold are reported; *default* = 65 |
| -p | The minimum oligonucleotide GC content.$<100 < maximum\ GC$; *default* = 40 |
| -P | The maximum oligonucleotide GC content.$<100 > minimum\ GC$; *default* = 60 |
| -m | A list of prohibited sequences to mask in the input sequence. |
| -N | The number of sequences to process at the same time. Dependant upon processing power. $\leq 3$ *per processor*; *Default* = 1 |
| -g | The minimum distance between the 5′ end of two adjacent oligos when -*n* > 1. *To avoid overlaps* > -L. *Default* = 1.5 × *average oligo size* |

synthesis of long probes is not the only available option. As we indicated above, with some in situ synthesis platforms, due to the relatively low stepwise coupling efficiency of oligonucleotide synthesis, the effective length of probes is restricted. In this case, since short probes tend to have lower specificity and lower sensitivity, the solution is to synthesize multiple different probes for a single target. An example of how measurement bias for a particular target decreases when multiple independent short probes are used is shown in Figure 3.5. Generally such short probes are in the 25–36 nucleotide range and can be easier to design than longer probes. While the algorithms for designing short probes used by commercial manufacturers are proprietary, the same design tools that are described above can easily be used to design shorter probes. It should be noted that the accurate performance of short probes can be affected by base composition, particularly in the central 15 nucleotides (Naef and Magnasco, 2003; Wu et al., 2007), therefore particular attention should be paid to ensuring that, for example, runs of G or A residues are avoided. In addition, analysis methods that specifically take probe specific base composition into account are to be recommended (Seo and Hoffman, 2006; Irizarry et al., 2006b; see Chapter 6).

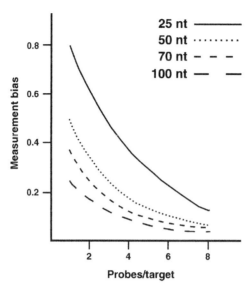

**FIGURE 3.5** Decreasing measurement bias for a given target as the number of independent probes increases.
*Source*: Data for four different probe lengths from (Chou et al., 2004a,b).

Affymetrix have considerable experience in the design and evaluation short probes, and while the exact details of their design criteria are proprietary, they appear to make extensive use of thermodynamic models of nucleic acid hybridization to predict probe binding affinity.

In addition to modeling approaches for predicting probe behavior, there is an increasing tendency, especially among commercial array designers, to incorporate empirical data derived from large-scale 'Latin Square' experiments to guide probe design (e.g. Wu et al., 2005). Given the increasing ease with which it is possible to generate arrays by in situ synthesis, it is now possible to pick several potential probes from a design run and synthesize a small number of arrays to test probe performance. By using such custom test arrays in conjunction with 'spike-in' experiments, where labeled targets at known concentrations are added to complex target mixtures, it is possible to collect data on probe performance, for example measuring the linearity of the hybridization response, the specificity and the sensitivity. Clearly attempting to perform spike-ins for every target on a 40 000 probe array is prohibitively expensive for most labs, however, the ability to assess performance of a selected subset of probes can substantially improve designs. Therefore, if the aim is to generate a relatively long-lasting array platform that will be used for thousands of hybridizations, effort employed up front in testing design criteria such as $T_m$, cross-hybridization threshold and probe length will pay dividends in terms of data quality once the final array is synthesized. A number of methods for assessing array performance have been described (for recent efforts see Shippy et al., 2006 and Tong et al., 2006) and some attempt at assaying probe sets is highly recommended for any new design.

As we have described, it is highly desirable that all of the probes on an array have uniform behavior at the hybridization temperature to be used. In practice this is currently achieved by constraining the $T_m$ of each probe within a narrow window. If the probes are constrained to a single length this can be problematic since it severely reduces the number of possible targets. It is therefore better practice to allow probe selection within a given size range (i.e. 65–69mer). This is obvious if one considers a 66mer oligonucleotide of a given sequence: adding a base will increase the $T_m$ while subtracting a base will decrease the $T_m$. In practice it is the behavior of the probe at the actual hybridization temperature that should be of greatest concern. Consequently it is probably best to think of probe design in terms of the free energy of the probe–target interaction at the hybridization temperature rather than $T_m$ and the hope is that the focus of design tools will move to more realistic thermodynamic considerations in the future. Some of the issues relating to solid-phase hybridization thermodynamics have recently been reviewed by Levicky and Horgan (2005).

The output of a run is a list of candidate probe sequences for each target: reporting any cross hybridization for each probe as well as numerical data such as free energy, enthalpy, entropy and $T_m$. In addition, a file containing rejected sequences, those targets for which an oligo satisfying the design parameters cannot be found, is also created. For these difficult targets one can either accept that a suitable probe cannot be found or perform an additional design run, relaxing one or some of the parameters. In any genome there will inevitably be difficult targets: members of related gene families, multiple splice variants of a given gene, short transcripts with limited design targets or transcripts with very unusual nucleotide composition. Consequently the user has to make pragmatic decisions balancing comprehensive coverage with uniformity. If the genome of the target organism is particularly intractable (i.e. some microbial genomes have very unusual nucleotide composition with extreme GC or AT contents) it may make more sense to design multiple short 25–30mer probes for each target.

An example of the OligoArray2.1 output for a specific and non-specific oligonucleotide is shown below. For a specific probe, the target sequence ID is listed followed by the position of the 5' end of the oligonucleotide on the target and the probe length. Next the thermodynamic parameters are listed: Free energy of duplex formation at 37 °C (kcal/mol), the enthalpy (kcal/mol), the entropy (cal/mol K) and the Tm (°C). There follows a list of targets matching the oligonucleotide, a single entry for specific probes, and the oligonucleotide sequence (5'–3', matching the target). For a non-specific oligonucleotide, a list of other targets with perfect/partial matches to the probe with their thermodynamic parameters is given. In this case, four additional targets have good matches to the probe sequence and this probe would not be used.

*Specific*

| YAL069W | 49 | 47 | −308.65 | −374.59 | −1014.59 | 85.38 |
|---|---|---|---|---|---|---|
| YAL069W | ACCACATGCCATACTCACCCTCACTTGTATACTGATTTTACGTACGC | | | | | |

*Non-specific*

| YAL069W | 126 | 47 | −302.33 | −366.4 | −985.60 | 87.54 |
|---|---|---|---|---|---|---|
| YAL069W | ACTTACCCTACTCTCAGATTCCACTTCACTCCATGGCCCATCTCTCA | | | | | |
| YJR162C | | | (−16.90 | −281.20 | −781.58 | 73.24 |
| acttaccctactctcacattccact- - - - -ccatcacccatctctca); | | | | | | |
| YLL065W | | | (−16.90 | −281.20 | −781.58 | 73.24 |
| acttaccctactctcacattccact- - - - -ccatcacccatctctca); | | | | | | |
| YFL063W | | | (−19.58 | −269.0 | −737.57 | 77.20 |
| -cttaccctactttcacattccact- - - - -ccatggcccatctctca); | | | | | | |
| YKL225W | | | (−18.77 | −281.90 | −778.13 | 75.58 |
| acttaccctactctcacattccact- - - - -ccatggcccag-tctca) | | | | | | |

## 3.3.5 Designing a *Drosophila* Probe Set

To exemplify a probe design project, we describe the recent development of a long oligonucleotide set for expression profiling in *Drosophila melanogaster* (Kreil et al., in preparation). The objective was to design a set of specific probes for each predicted gene in the fly genome, synthesize long oligonucleotides and distribute the probes to a number of printing labs (http://www.indac.net/). This consortium-based approach has a number of benefits: many laboratories use the same probe set thus increasing data comparability, the design process is in the hands of the user community and, not least, there are considerable cost savings compared to obtaining commercial oligonucleotide sets for printing in-house. For target selection we took advantage of the well-developed *Drosophila* genome annotation. We selected all predicted mRNAs, pseudogenes, tRNAs and miscellaneous RNAs from release 4 of the genome annotation, adding sequences for commonly used exogenous marker genes (GFP, GAL4, LacZ, FLP, P-transposase) and 21 *Arabidopsis thaliana* sequences, selected on the basis of lack of similarity with the fly genome, to use as exogenous spike-in controls. This set of sequences was processed with nrdb90 and CD-HI, merging sequences with more than 90% sequence similarity (e.g. splice variants of a single gene) to create a set of 15 829 non-redundant targets. Probes were restricted to a region within 1500 bases of the 3′ end of each target and designed to match the sense strand of the target. After some pilot runs examining oligonucleotides length and $T_m$

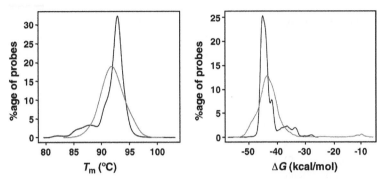

**FIGURE 3.6**   Distribution of probe melting temperatures ($T_m$) and binding free energies at 70 °C for the *Drosophila* INDAC set (dark curve) compared to sequences from a commercial supplier (light curve). Both the $T_m$ and $\Delta G$ distributions for the INDAC set are much tighter.

windows, we selected a set of parameters for the design run. Probe length was constrained between 65 and 69, the $T_m$ range was 90.6–95.6 °C, GC content was not constrained, 69 °C was used as the probe self-folding and cross-hybridization thresholds. All possible probes were screened for cross hybridization against both strands of the *Drosophila* genome sequence. For each target, probes passing these stringent filters were ranked by their proximity to the 3′ end of the target and the closest selected. For a small fraction of targets (<6%) with unusual sequence composition, no probes meeting the thresholds could be found and theses were subject to further design rounds with increasingly relaxed parameters. Distance from the 3′ end was increased to 2000 bases, recovering probes against some targets. For the remaining targets, cross-hybridization and self-folding thresholds were reduced and then the $T_m$ window reduced. After repeated design runs, acceptable probes could be found for 99.7% of the targets. The distribution of probe melting temperatures and probe–target binding energies shown in Figure 3.6, demonstrate the uniform properties of the set compared with a commercial *Drosophila* long-oligonucleotide probe set we evaluated. The $T_m$ distribution is very homogeneous, with only the difficult target sequences contributing to a small bulge to the left of the $T_m$ graph. The predicted hybridization behavior, as shown from the Free Energy graph, also has a very tight distribution, with the difficult targets contributing to the small number of probes with slightly higher $\Delta G$ values. The computational work was distributed among a cluster of Linux PCs using a Grid Engine, and from start to finish took approximately 3 months. This included target selection, initial parameter evaluation, re-runs for difficult targets and post design probe evaluation. The resulting probe oligonucleotides were synthesized in bulk by Illumina Inc. and aliquots distributed to 20 labs throughout the world where the set has been used for a variety of experiments (Choksi et al., 2006; Goldman and Arbeitman, 2007; Zeitouni et al., 2007).

### 3.3.6 Moving Forward

Probe design tools examine the behavior of the probe at the hybridization temperature, however, if arrays are to truly quantitative then the behavior of the target sequence must also be taken into consideration (Ratushna et al., 2005). It is clear that target sequences have the capability of forming secondary structures that may mask the sequence the probe has been designed against, rendering the probe less effective. A computational assessment of this is not trivial and becomes increasingly difficult as the number of targets being assessed increases. This is compounded by the common practice, at least for eukaryotic organisms, of synthesizing labeled targets by oligo-dT pimed reverse transcription, which results in targets of different lengths. For a more quantitative approach it is therefore highly desirable to synthesize full-length target sequences and explicitly account for target secondary structure effects in the probe design process. While easy enough to write, this is a very complex problem in practice. Modeling hybridization when probe and target are both in solution is a difficult enough problem, considering the situation when one of the sequences is tethered to a solid substrate, as in a microarray, is currently extremely challenging. Having said that, a number of groups are actively researching the area (Dimitrov and Zuker, 2004; SantaLucia and Hicks, 2004; Bishop et al., 2006) so we can expect that both the principled probe design and the quantitative assessment of hybridization signals will continue to improve. It is to be hoped that both probe design and analysis methods will improve in the future to specifically capture these complex variables. It should be noted that these complications are also relevant to both RT-PCR and Northern blotting techniques, consequently they are no more likely to quantitatively assess gene expression levels than a well designed microarray (Quackenbush and Irizarry, 2006).

### 3.4 PREPARING ARRAYS

There are many factors that govern the choice of microarray platform, including access to appropriate equipment for fabricating and processing arrays, the availability of pre-made arrays for the biological system under study and, of course, the cost of carrying out each study. Until comparatively recently, the availability of commercial microarrays was restricted to a limited range of key model organisms, often only human, mouse and rat. Thus any array-based exploration of more esoteric biological systems was firmly in the hands of home-made array technology. While some commercial platforms still remain focused on the biomedical market, the past few years have seen an increase in the number and availability of custom or boutique microarray providers and it is fair to say that it should be possible, albeit at a cost, to generate a microarray for virtually any organism or collection of sequences one can imagine. Indeed for the novice there is much to be said for taking advantage of a commercial operation or core academic facility that can assist with probe design and array fabrication. This is particularly the case for companies offering ultrahigh-density inkjet or photolithographic in situ synthesis platforms. The in situ synthesis platforms are particularly useful when assessing probe designs since a few test arrays can be synthesized and tested with relevant hybridizations. This approach facilitates the identification of suspect probes, which can then be redesigned as appropriate.

Platform choice is also, to some extent, dependent upon the scope of the envisaged experiments. For example, if a single or limited study aimed at identifying genes in a particular process for further detailed analysis is the primary objective, it would seem to make little sense investing the time and effort required to develop a well designed whole genome array. Here, the obvious option would be to purchase custom made or off-the-shelf arrays from a commercial provider or core lab. If, on the other hand, thousands of arrays are required, then the ability to fabricate and process arrays makes more sense from both a financial and a scientific perspective, since one can control the entire process and produce arrays comparatively cheaply (after an initial investment in equipment purchase and staff training of course). We stress however, that establishing a microarray production facility is not a trivial matter to be undertaken lightly. Aside from developing the necessary technical expertise to produce reliable and reproducible array data, the real costs, including equipment purchase and maintenance, staff recruitment and training, space provision and refurbishment as well as reagent costs, should all be factored in before deciding to embark down the home-made road (Auburn et al., 2005). We estimate that around a year is required to establish microarray production facility from scratch and almost certainly longer to be in a position where the generation of very high-quality array data is routine. This is supported by the surveys performed by the Association of Biomolecular Resource Facilities (ABRF) and their website, where surveys by the Microarray Research Group are available, provides a useful start-point for evaluating the experiences of other microarray facilities (http://www.abrf.org/).

### 3.4.1 Microarray Platforms

We have outlined the available array technologies and it is not our intention here to provide a detailed description of each platform. In subsequent chapters, we outline the important analytical differences between different technologies. Until recently there were relatively few objective comparative evaluations of the performance of different platforms and those published generally concluded that there was poor agreement between alternative technologies. More recently, the concerted activities of the Microarray Quality Control Project (MAQC) have provided a far more rigorous assessment of inter-platform performance, concluding that different platforms show excellent agreement with high correlation

**TABLE 3.6** Major Microarray Suppliers and Whether They Offer a Custom Array Service

| Company | Web | Custom arrays |
|---|---|---|
| Affymetrix | http://www.affymetrix.com/ | Yes[*] |
| Agilent Technologies | http://www.home.agilent.com/ | Yes[*] |
| Applied Biosystems | http://www.appliedbiosystems.com/ | No |
| Applied Microarrays | http://www.appliedmicroarrays.com/ | Yes[*] |
| CombiMatrix Corporation | http://www.combimatrix.com/ | Yes |
| Illumina | http://www.illumina.com/ | Yes |
| NimbleGen Systems | http://www.nimblegen.com/ | Yes[*] |
| Oxford Gene Technologies | http://www.ogt.co.uk/ | Yes[*] |
| Phalanx Biotech | http://www.phalanxbiotech.com/ | Yes |
| SuperArray Bioscience | http://www.superarray.com/ | Yes |

Those marked with an asterisk can provide high-density custom arrays.

coefficients and high concordance of gene lists obtained in comparative experiments (Patterson et al., 2006; Shi et al., 2006). Consequently, the user need not anguish over the platform choice and can use whatever is readily available or affordable; Table 3.6 list some of the major commercial array providers. For those interested in a more detailed description of the technologies and a potted history of the field, a recent review by Michael Heller provides an assessment of various microarray technologies as well as a comprehensive listing of reviews in the area (Heller, 2002). A word of caution: the forgoing is true of microarray platforms where careful probe design has been paramount, consequently it obvious that data obtained with recent oligonucleotides platforms is likely to be more robust, certainly at the quantitative level, than, for example that derived from an elderly cDNA array design. We focus mainly on a discussion of principals broadly relevant to array design and provide some guidance on the preparation of in-house spotted arrays.

### 3.4.2 Array Layouts

As with any experimental procedure, the reliability of microarray data is confounded by variability and technical artefacts (Chapters 2, 4 and 5). Some of this variability is due to the biology of the system being studied and can be accounted for by experimental replication. Other sources of variability are due to the microarray process itself and here there are two primary areas where artefacts can arise. Sample extraction and labeling can be variable and, as with biological variation, experimental replication can help capture this. The microarray itself can be a considerable source of variation for a variety of reasons, and aside from the probe design issues we describe above, this can be substantially reduced by carefully examining and optimizing the array printing and hybridization process. In terms of the microarray platform, there are several straightforward steps that can be taken to help reduce variability or bias, it is important to note that, contrary to the claims of some commercial manufacturers, these issues apply to all array platforms and not just homemade spotted arrays.

One clear and easily observed artefact is the spatial difference in signal intensities across an array. In the case of printed arrays this is often due to differences between individual printing pins or the print heads on non-contact printers and some of this variability can be reduced by so-called 'print-tip' normalization methods (Smyth and Speed, 2003, see Chapter 6 for a description of normalization techniques). However there are other sources of spatial variation, for example uneven hybridization across the slide or effects due to the site of probe introduction on an automatic slide processor and again this can be, to some extent, normalized out during the data processing. A common, and easily dealt with, artefact observable on early microarrays was due to the distribution of probes, their geographical layout on the array. In some cases, probes were arrayed in the same order as clone libraries were plated and as a result probes were distributed on the array according to clone size or chromosomal order. Arraying clones in strict plating order can also result in artefacts due to inconsistencies in probe concentrations between plates, further confounding other spatial effects and in some cases leading to spurious inferences about gene expression (Kluger et al., 2003). Early versions of the Affymetrix platform also had similar problems due to synthesizing the probe-pair sets for a particular target in a contiguous block. It is now recognized that spot layouts that are randomized with respect to probe source, genomic location and, in the case where multiple probes for a gene are used, individual target, can substantially reduce confounding spatial effects. Unfortunately, the control software for many commonly used robotic printers can be fairly inflexible and thus it is a tedious task to manually programme the arrayer to generate a random spot layout, a generic tool is available from our group that can help with this problem (Auburn et al., 2006).

### 3.4.3 Control Elements

The use of control elements to assess the quality and reproducibility of a particular hybridization is essential and there are several types of control that should be employed. These can be divided into 2 broad categories; array controls and sample controls. Array controls are spot elements used to assess background or non-specific sample interaction with the array and include non-printed 'spots' as well as spots printed with spotting buffer only. Measurements from these elements allow an estimation of the on and off spot background signals. Sample controls can include a variety of non-specific hybridization controls as well as probes that detect exogenous RNAs added to the biological sample. Non-specific controls can be a random oligonucleotide with no significant match to the transcriptome of interest or randomly sheared genomic DNA from an unrelated organism. The use of probes complementary to exogenously added RNAs, so called spike-in controls, is now regarded as an essential control for any well-designed array (Tong et al., 2006). Such probes provide a route for assessing the detection limit and linearity of the microarrays since the concentration of exogenous RNAs added to the sample can be carefully controlled. This allows an evaluation of hybridization responses over a wide dynamic range and can also provide useful data for assessing normalization strategies (Irizarry et al., 2006a). Probes designed against such control RNAs should have the same thermodynamic properties as the other specific probes on the array and this can be easily achieved by including the exogenous RNA sequences in with the targets during the probe design process. Identifying such sequences is not necessarily trivial; for our recent *Drosophila* array design we performed and exhaustive search comparing the *Arabidopsis* and *Drosophila* genomes to identify plant genes with the lowest sequence similarity to any sequence in the fly genome and generated probes and targets for 14 different sequences. These probes will obviously be useless as controls on any plant arrays. Fortunately, help is at hand for the array community since a group of manufacturers and academic microarray labs have joined together to form the External RNA Control Consortium (ERCC). The group are currently developing a fully tested and commonly agreed-upon set of universal controls that will be available to anyone generating an array (External RNA Controls Consortium, 2005). The development of a widely adopted standard set of controls will be enormously beneficial for assessing expression data and, more importantly, will provide a reference point for comparing experiments carried out on different array platforms.

To summarize, whether preparing your array in-house or commissioning a commercial array provider, users should ensure that all relevant controls are included in the design and, wherever possible, utilize standard controls. It goes without saying that all of the control elements employed must be randomly distributed throughout the array. If high quality data is the objective it is a good idea to use as many controls as possible. As always there is a trade-off between available spotting density, the number of different specific probes and the density of controls. We recommend a minimum of 10% of the array elements are dedicated to controls: 20% is even better! Of course if there is space, it is advisable to spot specific probes in duplicate (or more) on a single array.

### 3.4.4 Basics for Printing Arrays

Although a variety of arrayers are available for in-house microarray fabrication, including small-scale digital micromirror in situ synthesis devices and piezoelectric noncontact printers, it is fair to say that the majority of non-commercial array production utilizes contact printing robots. Pioneered by Patrick Brown's lab in Stanford (http://cmgm.stanford.edu/pbrown/mguide/index.html) the use of precision robotics to print high-density microarrays on relatively inexpensive glass microscope slide substrates revolutionized the genomics field (Schena et al., 1995). Although the plans for building a contact printer in-house are available on the web, as are the details for building your own inkjet printer or in situ synthesizer (Lausted et al., 2004; Fisher and Zhang, 2007), in the recent past most labs entered the field by purchasing a spotting robot from one of the commercial suppliers. While expensive, this has the benefit of not requiring in-house mechanical and electrical engineering skills to build, and more importantly maintain, the instrument. Over the past couple of years however, as the array boom settled and more reliable and varied commercial microarray supplies became available, there has been a decline in the number of arraying instruments on the market and currently there are only two main suppliers of contact printers and one inkjet printer manufacturer. Paradoxically then the choice of arrayer is now more limited, this is however balanced by the increased availability of in situ synthesis suppliers who can generate custom high-density arrays.

### 3.4.5 Substrates

For those who do chose to print their own arrays, once a probe set has been acquired they are faced with a choice of array substrates from a range of manufacturers. Many of these advertise miraculous properties in terms of increased sensitivity and low background: claims that need to be viewed with some scepticism. There are basically five major substrates in use, each with slightly different properties: amine, aldehyde, epoxy, 3D matrix and poly-L-lysine. Along with these, more esoteric substrates such as gold or nitrocellulose are also available, though less widely used. At the most basic level, substrates either form covalent attachments with the probe molecules or they bind probes via electrostatic interactions (Figure 3.7). The benefit of the former is that the probe-substrate attachment is very stable and can thus be treated far more robustly (stripping labeled samples and reusing the array for example) but on the downside, the probes often need to be modified with reactive amino-groups, increasing the costs of probe synthesis. With the latter, any DNA can be arrayed but the attachment is less well defined. Poly-L-lysine, the slide coating popular in the early days of home-made arrays, forms a non-covalent interaction with the probe molecules due to the positively charged lysine groups and behave similarly to amine coated slides. While some still use home-made poly-L-lysine coated slides, the improved performance of other surfaces coupled with increased slide consistency mean that many chose commercial slide providers. Most commercial suppliers use high quality glass, of uniform thickness and with excellent optical properties to eliminate slide surface variability. Since this helps reduce one area of confounding experimental noise, users are well advised to spend the extra few dollars on high quality substrates (most pre-coated substrates are

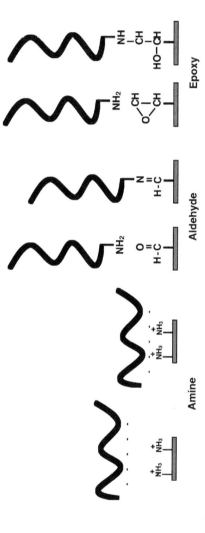

**FIGURE 3.7**  Basic types of binding substrate and their interactions with oligonucleotide probes. With amine surfaces, the negatively charged DNA interacts with the positively charged amine groups. With aldehyde surfaces, a covalent bond is formed between a primary amino group on a modified oligonucleotide and the primary aldehyde. With epoxy coatings, a covalent interaction between a primary amine on the probe and the epoxide group again makes a covalent attachment. It is also possible to make covalent attachments with the primary amines present on A, G and C residues, though this may reduce hybridization efficiency.

available for less than $20 each, comparatively little in comparison to the amount of data generated in a good experiment). The covalent attachment substrates, aldehyde and amine, are becoming increasingly popular since they can reduce variability in probe–surface interactions. This helps ensure that the time spent designing uniform probe sequences is not wasted by unpredictable effects of probe binding to the array substrate. More recently, the proprietary 3D matrices offered by a number of companies have become the substrate of choice. These are aqueous gels, usually carrying functional groups that will covalently attach amino-modified probes. They are believed to provide much better access for the labeled targets, since there is less steric hindrance from the slide surface preventing the target approaching the tethered probe. Whatever the chosen substrate, one thing is clear – evaluation experiments will be needed to ensure the slide and probe set perform well together.

Again, we emphasise we are not providing a lab manual for array printing but some useful pointers regarding areas that should be approached with caution. Anyone can make a good high-density microarray if enough care is taken to carefully assess the many variables that can affect the process. The objective is to obtain an ordered array of probes on the surface that will report specific hybridization to the targets and allow reliable extraction of the hybridization signal

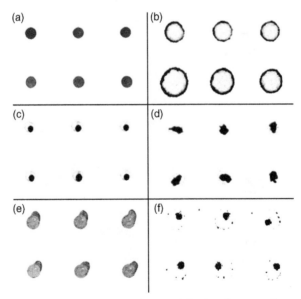

**FIGURE 3.8** Examples of poor spot morphologies. (a) Good quality spots of even size and signal distribution. (b) 'Rush-spots' frequently occur at the start of a print run due to excess liquid adhering to the printing pin but with some slide/buffer combinations can be caused by leaving the pin in contact with surface too long. (c and d) 'Fried eggs', where probes concentrate in the middle of the spot, either evenly or unevenly. (e) Spots deviate from ideal circle. (f) rapid solvent evaporation from spot leaves probe unevenly distributed.

while minimizing background and other non-specific effects. Array printing is affected by ambient temperature and humidity, the spotting buffer used to dissolve the probes, the precise mechanics of spot deposition in the case of contact printers and post printing array handling. Unfortunately it is not possible to give a precise recipe since environmental conditions vary from location to location. Even with the use of controlled clean rooms, environmental conditions, and as a consequence arraying protocols, will differ between any two labs. The best way to approach the process is to perform a series of pilot experiments assessing reproducibility, spot quality and specificity. Failure to do so can generate data that are difficult, if not impossible, to effectively analyze. Some horror spots are illustrated in Figure 3.8 and it is fairly straightforward to eliminate these types of printing defects by simply evaluating slide-buffer combinations under the typical temperature and humidity conditions found in the printing lab via a series of test hybridizations.

More insidiously, we have found dramatic effects of different spotting buffer/ slide chemistry combinations on hybridization signal specificity. This requires more careful analysis but we strongly recommend some specificity tests are performed otherwise the efforts aimed at designing specific probes may be

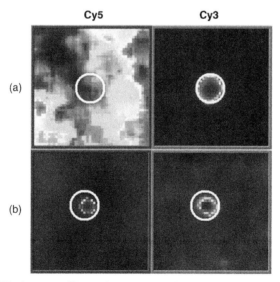

**FIGURE 3.9**  We show a specific spot from two arrays (a and b) where a 70mer oligonucleotide probe has been arrayed. The arrays were hybridized with a complex sample spiked with a Cy3 labeled RNA matching the arrayed probe (right hand side) and a Cy5 labeled probe that has no homology to the probe (left hand side). In slide buffer combination A, there is a signal in the Cy3 channel but the Cy5 channel is blank, even when the scanner gain is set at maximum. With slide buffer combination B there is a weaker Cy3 signal but in this case there is also appreciable hybridization of the Cy5 spike RNA.

wasted. In Figure 3.9 we provide an example of how slide/buffer interactions can generate misleading signals. Note that this is not a one-off oddity we have observed but an effect that can be reproducibly demonstrated. We therefore strongly recommend that any new array platform is evaluated for specificity and reproducibility via dedicated pilot studies.

From time to time, the arraying instrument should be calibrated to ensure it is working to specification and appropriate velocities and pin height parameters are as expected. In addition, the printing pins should be evaluated regularly to ensure that the set performs as expected in terms of spot morphology. Finally, the pin washing cycles and solutions should be checked to ensure there is no possibility of carry over from one source visit to the next. Such routine calibration and maintenance can be achieved using sonicated salmon or herring sperm DNA and hybridization with labeled random 9-mers or staining the array with SYBR dyes (George, 2006). This is particularly important if the QC steps below indicate a problem with the printing process.

### 3.4.6 Evaluating Arrays

Once a robust set of printing protocols have been established and a well-randomized array layout designed, printing in earnest can commence. With contact printing, the first few slides are usually employed to absorb the 'rush' of probe solution that occurs from a pin loading. Optimization of the printing process, controlling the way that pins enter and leave the source plates, pin-array contact time and the velocity with which the pins approach and leave the array surface, can reduce but not eliminate rush effects. Clearly, the number of slides lost to this is dependant upon the specifics of the particular printing set up. A random selection of slides, for example one each from the beginning, middle and end of the array run, should be set aside for quality control. Arrays can be assessed a number of ways, depending upon the probe composition. For example, test hybridizations with labeled random 9mers, a random primed genomic DNA sample or a complex RNA mixture. Alternatively a number of DNA-binding fluorescent dyes (e.g. SYBR_green or SYBR_555) or staining kits are available. After hybridization/staining control slides should be scanned and the images examined carefully, paying particular attention to spot morphology, even signal distribution and, with hybridization tests, low background.

Finally, one area that can pay dividends in terms of data quality is carefully evaluating the hybridization temperature for gene expression studies to optimize the sensitivity and specificity of the probe set. As we describe above, the probe design process is not an entirely empirical process and most of the models employ heuristics to select sequences that should have the desired sensitivity and specificity characteristics. It is not easy to assess the specificity and sensitivity of every probe on an array experimentally, but one can evaluate the overall performance of a set by performing some control hybridizations. The idea here is that performance can be improved by selecting the hybridization temperature that maximizes the detection of differential expression between two samples in a competitive hybridization (Kreil et al., in preparation).

## 3.5 SUMMARY

While the choice of array platforms or probe types may appear daunting, it is now becoming clear that when principled design rules are followed the results obtained with different platforms are likely to be very similar. Any user must simply decide whether they need to invest time and resources in developing their own arrays, utilize one of the commercial providers who can assist with array design and fabrication or use off-the-shelf commercial arrays. With the information we have provided in this chapter, users may evaluate the available options and decide what best fits with their needs. Certainly, if a few tens of arrays are needed there is little sense in designing and fabricating your own arrays unless one has access to a custom in situ synthesizer. On the other hand, if hundreds or thousands of arrays are needed then in-house array development may be the way forward in terms of cost. However, it is likely that, over the next couple of years, reliable microarrays will becoming relatively inexpensive commodity items, mirroring the way that DNA sequencing changed from a largely in-house technique to a commercial provision during the early part of the century. While there is some speculation that the next generation of high-throughput sequencing technologies will replace microarray-based analysis with counting molecules by sequencing, it is likely, in the medium term at least, that sequencing will remain expensive compared with array hybridization. In addition, much has been done over the past few years to iron out some of the technical difficulties with generating and analyzing array data, the same cannot yet be said for sequence-based methods. Thus we feel that arrays will be with us in some form for the foreseeable future – one can easily imagine that miniature, high density arrays will be routinely used to monitor gene expression much in the way the RT-PCR or Northern blotting is used today.

## REFERENCES

Alizadeh AA, Eisen MB, Davis RE, Ma C, Lossos IS, Rosenwald A, Boldrick JC, Sabet H, Tran T, Yu X, Powell JI, Yang L, Marti GE, Moore T, Hudson J, Lu L, Lewis DB, Tibshirani R, Sherlock G, Chan WC, Greiner TC, Weisenburger DD, Armitage JO, Warnke R, Levy R, Wilson W, Grever MR, Byrd JC, Botstein D, Brown PO, Staudt LM. (2000) Distinct types of diffuse large B-cell lymphoma identified by gene expression profiling. *Nature* **403:** 503–511.

Altschul SF, Madden TL, Schaffer AA, Zhang J, Zhang Z, Miller W, Lipman DJ. (1997) Gapped BLAST and PSI-BLAST: a new generation of protein database search programs. *Nucleic Acids Res* **25:** 3389–3402.

Auburn RP, Kreil DP, Meadows LA, Fischer B, Matilla SS, Russell S. (2005) Robotic spotting of cDNA and oligonucleotide microarrays. *Trends Biotechnol* **23:** 374–379.

Auburn RP, Russell RR, Fischer B, Meadows LA, Sevillano Matilla S, Russell S. (2006) SimArray: a user-friendly and user-configurable microarray design tool. *BMC Bioinform* **7:** 102.

Bar-Or C, Czosnek H, Koltai H. (2007) Cross-species microarray hybridizations: a developing tool for studying species diversity. *Trends Genet* **23:** 200–207.

Bellon L, Wincott F. (2000) Oligonucleotide synthesis. In: Kates S, Albericio F, editors. *Solid-Phase Synthesis: A Practical Guide.* Marcel Dekker Inc., New York, pp. 475–528.

Bishop J, Blair S, Chagovetz AM. (2006) A competitive kinetic model of nucleic acid surface hybridization in the presence of point mutants. *Biophys J* **90:** 831–840.

Bonaldo MF, Lennon G, Soares MB. (1996) Normalization and subtraction: two approaches to facilitate gene discovery. *Genome Res* **6:** 791–806.

Boutros M, Kiger AA, Armknecht S, Kerr K, Hild M, Koch B, Haas SA, Paro R, Perrimon N. (2004) Genome-wide RNAi analysis of growth and viability in *Drosophila* cells. *Science* **303:** 832–835.

Bowtell DD. (1999) Options available – from start to finish – for obtaining expression data by microarray. *Nat Genet* **21:** 25–32.

Bozdech Z, Zhu J, Joachimiak MP, Cohen FE, Pulliam B, DeRisi JL. (2003) Expression profiling of the schizont and trophozoite stages of *Plasmodium falciparum* with a long-oligonucleotide microarray. *Genome Biol* **4:** R9.

Burr A, Bogart K, Conaty J, Andrews J. (2006) Automated liquid handling and high-throughput preparation of polymerase chain reaction-amplified DNA for microarray fabrication. *Methods Enzymol* **410:** 99–120.

Carninci P, Shibata Y, Hayatsu N, Sugahara Y, Shibata K, Itoh M, Konno H, Okazaki Y, Muramatsu M, Hayashizaki Y. (2000) Normalization and subtraction of cap-trapper-selected cDNAs to prepare full-length cDNA libraries for rapid discovery of new genes. *Genome Res* **10:** 1617–1630.

Caruthers MH, Beaton G, Wu JV, Wiesler W. (1992) Chemical synthesis of deoxyoligonucleotides and deoxyoligonucleotide analogs. *Methods Enzymol* **211:** 3–20.

Chen H, Sharp BM. (2002) Oliz, a suite of Perl scripts that assist in the design of microarrays using 50mer oligonucleotides from the 3′ untranslated region. *BMC Bioinform* **3:** 27.

Choksi SP, Southall TD, Bossing T, Edoff K, de Wit E, Fischer BE, van Steensel B, Micklem G, Brand AH. (2006) Prospero acts as a binary switch between self-renewal and differentiation in *Drosophila* neural stem cells. *Dev Cell* **11:** 775–789.

Chou CC, Chen CH, Lee TT, Peck K. (2004) Optimization of probe length and the number of probes per gene for optimal microarray analysis of gene expression. *Nucleic Acids Res* **32:** e99.

Chou HH, Hsia AP, Mooney DL, Schnable PS. (2004) Picky: oligo microarray design for large genomes. *Bioinformatics* **20:** 2893–2902.

Connolly BA. (1987) The synthesis of oligonucleotides containing a primary amino group at the 5′-terminus. *Nucleic Acids Res* **15:** 3131–3139.

Dimitrov R, Zuker M. (2004) Prediction of hybridisation and melting for double-stranded nucleic acids. *Biophys J* **87:** 215–226.

Dorris DR, Nguyen A, Gieser L, Lockner R, Lublinsky A, Patterson M, Touma E, Sendera TJ, Elghanian R, Mazumder A. (2003) Oligodeoxyribonucleotide probe accessibility on a three-dimensional DNA microarray surface and the effect of hybridization time on the accuracy of expression ratios. *BMC Biotechnol* **3:** 6.

Draghici S, Khatri P, Eklund AC, Szallasi Z. (2006) Reliability and reproducibility issues in DNA microarray measurements. *Trends Genet* **22:** 101–109.

Egeland RD, Southern EM. (2005) Electrochemically directed synthesis of oligonucleotides for DNA microarray fabrication. *Nucleic Acids Res* **33:** e125.

External RNA Controls Consortium. (2005). Proposed methods for testing and selecting the ERCC external RNA controls. *BMC Genomics* **6:** 150.

Fisher W, Zhang M. (2007) A biochip fabrication system using inkjet technology. *IEEE Trans Autom Sci Eng* **4:** 488–500.

Fodor SP, Read JL, Pirrung MC, Stryer L, Lu AT, Solas D. (1991) Light-directed, spatially addressable parallel chemical synthesis. *Science* **251:** 767–773.

Furlong EE, Andersen EC, Null B, White KP, Scott MP. (2001) Patterns of gene expression during *Drosophila* mesoderm development. *Science* **293:** 1629–1633.

George RA. (2006) The printing process: tips on tips. *Methods Enzymol* **410:** 121–135.

Goldman TD, Arbeitman MN. (2007) Genomic and functional studies of *Drosophila* sex hierarchy regulated gene expression in adult head and nervous system tissues. *PLoS Genet* **3:** e216.

Gordon PM, Sensen CW. (2004) Osprey: a comprehensive tool employing novel methods for the design of oligonucleotides for DNA sequencing and microarrays. *Nucleic Acids Res* **32:** e133.

Gracey AY, Fraser EJ, Li W, Fang Y, Taylor RR, Rogers J, Brass A, Cossins AR. (2004) Coping with cold: an integrative, multitissue analysis of the transcriptome of a poikilothermic vertebrate. *Proc Natl Acad Sci U S A* **101:** 16970–16975.

Halgren RG, Fielden MR, Fong CJ, Zacharewski TR. (2001) Assessment of clone identity and sequence fidelity for 1189 IMAGE cDNA clones. *Nucleic Acids Res* **29:** 582–588.

Heller M. (2002) DNA Microarray technology: devices, systems and applications. *Annu Rev Biomed Eng* **4:** 129–153.

Holm L, Sander C. (1998) Removing near-neighbour redundancy from large protein sequence collections. *Bioinformatics* **14:** 423–429.

Hughes TR, Mao M, Jones AR, Burchard J, Marton MJ, Shannon KW, Lefkowitz SM, Ziman M, Schelter JM, Meyer MR, Kobayashi S, Davis C, Dai H, He YD, Stephaniants SB, Cavet G, Walker WL, West A, Coffey E, Shoemaker DD, Stoughton R, Blanchard AP, Friend SH, Linsley PS. (2001) Expression profiling using microarrays fabricated by an ink-jet oligonucleotide synthesizer. *Nat Biotechnol* **19:** 342–347.

Irizarry RA, Cope LM, Wu Z. (2006) Feature-level exploration of a published Affymetrix GeneChip control dataset. *Genome Biol* **7:** 404.

Irizarry RA, Wu Z, Jaffee HA. (2006) Comparison of Affymetrix GeneChip expression measures. *Bioinformatics* **22:** 789–794.

Jobs M, Fredriksson S, Brookes AJ, Landegren U. (2002) Effect of oligonucleotide truncation on single-nucleotide distinction by solid-phase hybridization. *Anal Chem* **74:** 199–202.

Kaderali L, Schliep A. (2002) Selecting signature oligonucleotides to identify organisms using DNA arrays. *Bioinformatics* **10:** 1340–1349.

Kamisetty NK, Pack SP, Nonogawa M, Devarayapalli KC, Watanabe S, Kodaki T, Makino K. (2007) Efficient preparation of amine-modified oligodeoxynucleotide using modified H-phosphonate chemistry for DNA microarray fabrication. *Anal Bioanal Chem* **387:** 2027–2035.

Kluger Y, Yu H, Qian J, Gerstein M. (2003) Relationship between gene co-expression and probe localization on microarray slides. *BMC Genomics* **4:** 49.

Knight J. (2001) When the chips are down. *Nature* **410:** 860–861.

Kreil DP, Russell RR, Russell S. (2006) Microarray oligonucleotide probes. *Methods Enzymol* **410:** 73–98.

Kuhn K, Baker SC, Chudin E, Lieu MH, Oeser S, Bennett H, Rigault P, Barker D, McDaniel TK, Chee MS. (2004) A novel, high-performance random array platform for quantitative gene expression profiling. *Genome Res* **14:** 2347–2356.

Lausted C, Dahl T, Warren C, King K, Smith K, Johnson M, Saleem R, Aitchison J, Hood L, Lasky SR. (2004) POSaM: a fast, flexible, open-source, inkjet oligonucleotide synthesizer and micro-arrayer. *Genome Biol* **5:** R58.

Levicky R, Horgan A. (2005) Physiochemical perspectives on DNA microarray and biosensor technologies. *Trends Biotechnol* **23:** 143–149.

Li F, Stormo GD. (2001) Selection of optimal DNA oligos for gene expression arrays. *Bioinformatics* **17:** 1067–1076.

Markham N, Zuker M. (2005) DINAMelt web server for nucleic acid melting prediction. *Nucleic Acids Res* **33:** W577–W581.

Mathews DJS, Zuker M, Turner D. (1999) Expanded sequence dependence of thermodynamic parameters improves prediction of RNA secondary structure. *J Mol Biol* **288:** 911–940.

McGall G, Barone A, Diggelmann M, Fodor S, Gentalen E, Ngo N. (1997) The efficiency of light-directed synthesis of DNA arrays on glass substrates. *J Am Chem Soc* **119:** 5081–5090.

Naef F, Magnasco MO. (2003) Solving the riddle of the bright mismatches: labeling and effective binding in oligonucleotide arrays. *Phys Rev E Stat Nonlin Soft Matter Phys* **68:** 011906.

Nuwaysir EF, Huang W, Albert TJ, Singh J, Nuwaysir K, Pitas A, Richmond T, Gorski T, Berg JP, Ballin J. (2002) Gene expression analysis using oligonucleotide arrays produced by maskless photolithography. *Genome Res* **12:** 1749–1755.

Pack SP, Kamisetty NK, Nonogawa M, Devarayapalli KC, Ohtani K, Yamada K, Yoshida Y, Kodaki T, Makino K. (2007) Direct immobilization of DNA oligomers onto the amine-functionalized glass surface for DNA microarray fabrication through the activation-free reaction of oxanine. *Nucleic Acids Res* **35:** e110.

Patterson TA, Lobenhofer EK, Fulmer-Smentek SB, Collins PJ, Chu TM, Bao W, Fang H, Kawasaki ES, Hager J, Tikhonova IR, Walker SJ, Zhang L, Hurban P, de Longueville F, Fuscoe JC, Tong W, Shi L, Wolfinger RD. (2006) Performance comparison of one-color and two-color platforms within the MicroArray Quality Control (MAQC) project. *Nat Biotechnol* **24:** 1140–1150.

Pease AC, Solas D, Sullivan EJ, Cronin MT, Holmes CP, Fodor SP. (1994) Light-generated oligo-nucleotide arrays for rapid DNA sequence analysis. *Proc Natl Acad Sci U S A* **91:** 5022–5026.

Quackenbush J, Irizarry RA. (2006) Response to shields: "MIAME, we have a problem". *Trends Genet* **22:** 471–472.

Ratushna VG, Weller JW, Gibas CJ. (2005) Secondary structure in the target as a confounding factor in synthetic oligomer microarray design. *BMC Genomics* **6:** 31.

Rimour S, Hill D, Militon C, Peyret P. (2005) GoArrays: highly dynamic and efficient microarray probe design. *Bioinformatics* **21:** 1094–1103.

Rouillard JM, Zuker M, Gulari E. (2003) OligoArray 2.0: design of oligonucleotide probes for DNA microarrays using a thermodynamic approach. *Nucleic Acids Res* **31:** 3057–3062.

Rozen S, Skaletsky H. (2000) Primer3 on the WWW for general users and for biologist programmers. *Methods Mol Biol* **132:** 365–386.

Sambrook J, Russell D. (2001) *Molecular Cloning: A laboratory Manual.* Cold Spring Harbor Laboratory Press, New York.

SantaLucia J. (1998) A unified view of polymer, dumbbell and oligonucleotide DNA nearest-neighbor thermodynamics. *Proc Natl Acad Sci U S A* **95:** 1460–1465.

SantaLucia J, Hicks D. (2004) The thermodynamics of DNA structural motifs. *Annu Rev Biomol Struct* **33:** 415–440.

Schena M, Shalon D, Davis RW, Brown PO. (1995) Quantitative monitoring of gene expression patterns with a complementary DNA microarray. *Science* **270:** 467–470.

Seo J, Hoffman EP. (2006) Probe set algorithms: is there a rational best bet?. *BMC Bioinformatics* **7:** 395.

Shchepinov MS, Case-Green SC, Southern EM. (1997) Steric factors influencing hybridisation of nucleic acids to oligonucleotide arrays. *Nucleic Acids Res* **25:** 1155–1161.

Shi L, Reid LH, Jones WD, Shippy R, Warrington JA, Baker SC, Collins PJ, de Longueville F, Kawasaki ES, Lee KY, Luo Y, Sun YA, Willey JC, Setterquist RA, Fischer GM, Tong W, Dragan YP, Dix DJ, Frueh FW, Goodsaid FM, Herman D, Jensen RV, Johnson CD, Lobenhofer EK, Puri RK, Schrf U, Thierry-Mieg J, Wang C, Wilson M, Wolber PK, Zhang L, Amur S, Bao W, Barbacioru CC, Lucas AB, Bertholet V, Boysen C, Bromley B, Brown D, Brunner A, Canales R, Cao XM, Cebula TA, Chen JJ, Cheng J, Chu TM, Chudin E, Corson J, Corton JC, Croner LJ, Davies C, Davison TS, Delenstarr G, Deng X, Dorris D, Eklund AC, Fan XH, Fang H, Fulmer-Smentek S, Fuscoe JC, Gallagher K, Ge W, Guo L, Guo X, Hager J, Haje PK, Han J, Han T, Harbottle HC, Harris SC, Hatchwell E, Hauser CA, Hester S, Hong H, Hurban P, Jackson SA, Ji H, Knight CR, Kuo WP, LeClerc JE, Levy S, Li QZ, Liu C, Liu Y, Lombardi MJ, Ma Y, Magnuson SR, Maqsodi B, McDaniel T, Mei N, Myklebost O, Ning B, Novoradovskaya N, Orr MS, Osborn TW, Papallo A, Patterson TA, Perkins RG, Peters EH, Peterson R, Philips KL, Pine PS, Pusztai L, Qian F, Ren H, Rosen M, Rosenzweig BA, Samaha RR, Schena M, Schroth GP, Shchegrova S, Smith DD, Staedtler F, Su Z, Sun H, Szallasi Z, Tezak Z, Thierry-Mieg D, Thompson KL, Tikhonova I, Turpaz Y, Vallanat B, Van C, Walker SJ, Wang SJ, Wang Y, Wolfinger R, Wong A, Wu J, Xiao C, Xie Q, Xu J, Yang W, Zhang L, Zhong S, Zong Y, Slikker W. (2006) The MicroArray Quality Control (MAQC) project shows inter- and intraplatform reproducibility of gene expression measurements. *Nat Biotechnol* **24:** 1151–1161.

Shippy R, Fulmer-Smentek S, Jensen RV, Jones WD, Wolber PK, Johnson CD, Pine PS, Boysen C, Guo X, Chudin E, Sun YA, Willey JC, Thierry-Mieg J, Thierry-Mieg D, Setterquist RA, Wilson M, Lucas AB, Novoradovskaya N, Papallo A, Turpaz Y, Baker SC, Warrington JA, Shi L, Herman D. (2006) Using RNA sample titrations to assess microarray platform performance and normalization techniques. *Nat Biotechnol* **24:** 1123–1131.

Shyamsundar R, Kim YH, Higgins JP, Montgomery K, Jorden M, Sethuraman A, van de Rijn M, Botstein D, Brown PO, Pollack JR. (2005) A DNA microarray survey of gene expression in normal human tissues. *Genome Biol* **6:** R22.

Singh-Gasson S, Green RD, Yue Y, Nelson C, Blattner F, Sussman MR, Cerrina F. (1999) Maskless fabrication of light-directed oligonucleotide microarrays using a digital micromirror array. *Nat Biotechnol* **17:** 974–978.

Smyth GK, Speed T. (2003) Normalization of cDNA microarray data. *Methods* **31:** 265–273.

Taylor E, Cogdell D, Coombes K, Hu L, Ramdas L, Tabor A, Hamilton S, Zhang W. (2001) Sequence verification as quality-control step for production of cDNA microarrays. *Biotechniques* **31:** 62–65.

Tong W, Lucas AB, Shippy R, Fan X, Fang H, Hong H, Orr MS, Chu TM, Guo X, Collins PJ, Sun YA, Wang SJ, Bao W, Wolfinger RD, Shchegrova S, Guo L, Warrington JA, Shi L. (2006) Evaluation of external RNA controls for the assessment of microarray performance. *Nat Biotechnol* **24:** 1132–1139.

Vodkin LO, Khanna A, Shealy R, Clough SJ, Gonzalez DO, Philip R, Zabala G, Thibaud-Nissen F, Sidarous M, Strömvik MV, Shoop E, Schmidt C, Retzel E, Erpelding J, Shoemaker RC, Rodriguez-Huete AM, Polacco JC, Coryell V, Keim P, Gong G, Liu L, Pardinas J, Schweitzer P. (2004) Microarrays for global expression constructed with a low redundancy set of 27,500 sequenced cDNAs representing an array of developmental stages and physiological conditions of the soybean plant. *BMC Genomics* **5:** 73.

Wang X, Seed B. (2003) Selection of oligonucleotide probes for protein coding sequences. *Bioinformatics* **19:** 796–802.

Wernersson R, Nielsen HB. (2005) OligoWiz 2.0 – integrating sequence feature annotation into the design of microarray probes. *Nucleic Acids Res* **33:** 611–615.

White K, Burtis K. (2000) Drosophila microarrays: from arrayer construction to hybridisation. In: Sullivan W, Ashburner M, Hawley R, editors. *Drosophila Protocols*. Cold Spring Harbor Laboratory Press, New York, pp. 487–507.

Wu C, Carta R, Zhang L. (2005) Sequence dependence of cross-hybridization on short oligo microarrays. *Nucleic Acids Res* **33:** e84.

Wu C, Zhao H, Baggerly K, Carta R, Zhang L. (2007) Short oligonucleotide probes containing G-stacks display abnormal binding affinity on Affymetrix microarrays. *Bioinformatics* **23:** 2566–2572.

Xu W, Bak S, Decker A, Paquette SM, Feyereisen R, Galbraith DW. (2001) Microarray-based analysis of gene expression in very large gene families: the cytochrome P450 gene superfamily of Arabidopsis thaliana. *Gene* **272:** 61–74.

Zeitouni B, Senatore S, Severac D, Aknin C, Semeriva M, Perrin L. (2007) Signalling pathways involved in adult heart formation revealed by gene expression profiling in *Drosophila*. *PLoS Genet* **3:** 1907–1921.

# Sample Collection and Labeling

The quality of data obtained from microarray experiments critically depends on the quality of the RNA that is used for labeling, since gene expression measurements are based on the assumption that an analyzed RNA sample closely resembles the in vivo transcript levels. Relatively standard molecular biology techniques are used to convert RNA populations into labeled pools suitable for hybridization and detection. A variety of protocols have been developed over the past few years to try and make this conversion as faithful as possible. Indeed it is now feasible to consider the analysis of single eukaryotic cells. However, no technique is perfect and inevitably there will be some alterations in the representation of molecules in the labeled pool compared with the initial sample population. Nevertheless, methods have improved and in terms of comparative analysis, where two or more samples are examined for differences in relative transcript abundances, the techniques are both robust and reliable when sufficient care is taken and relevant controls are performed. In this chapter we provide an overview of the commonly used methods and introduce some of the less well-known. We aim to show how different methods may be used to explore gene expression in diverse biological samples, ranging from large quantities of homogeneous cell populations to individual cells. While fluorescence is the most common label, users should be aware that other methods are available that may be more suitable for particular applications.

## 4.1 SAMPLE COLLECTION AND RNA EXTRACTION

Successful RNA isolation requires fast processing and careful handling of the tissue or cells to reduce degradation. Endogenous ribonucleases (RNases) are released from cellular compartments immediately after harvesting a tissue, consequently it is essential to inactivate these enzymes as soon as possible to prevent substantial RNA degradation. Several methods are employed to inactivate RNases, which are notoriously stable enzymes:

(i) Homogenization of samples immediately after harvesting in a chaotropic cell lysis solution (e.g. a guanidinium isothiocyanate).

(ii) Placing samples in a 'holding medium' such as RNA*later* (Ambion), an aqueous, nontoxic reagent that stabilizes and protects RNA in whole unfrozen tissue or cell samples. This method requires that the tissue samples be relatively small so that the solution can quickly permeate the tissue. Tissue samples treated with RNA*later* can be stored for a day at 37 °C, a week at 25 °C, or a month at 4 °C, making it ideal for certain nonlaboratory environments, such as field-work.

(iii) Flash-freezing samples in liquid nitrogen. Again, to minimize RNA degradation, it is important that the tissue be small enough to allow rapid and thorough freezing of the entire tissue.

The method of homogenization should be tailored to the cell or tissue type under study: cultured cells can be homogenized by vortexing in a cell lysis solution, whereas animal tissues, plant tissues, yeast and bacteria usually require more rigorous methods of disruption, such as treatment in a bead mill or tissue disruptor. Paraffin embedded histological specimens may require additional protease digestion. In addition to these tissue sources, the RNA may be extracted from selected subpopulations of cells by a number of methods, including manual dissection, laser capture microdissection (LCM) and fluorescence activated cell sorting (FACS), a specialized type of flow cytometry.

The RNA isolation method of choice will depend on the size and type of tissue and for the faint-hearted there are commercial suppliers of RNA extraction kits. The standard non-kit method, based on the technique developed by Chomczynski and Sacchi (1987), is to homogenize samples in a solution that combines phenol and guanidine thiocyanate in a monophase solution (available commercially as Trizol or TRI Reagent) to facilitate the immediate inhibition of

RNase activity. The addition of chloroform separates the homogenate into aqueous and organic phases and the RNA can then be precipitated from the aqueous phase with the addition of isopropanol. The entire protocol can be completed in approximately 1 h and is widely used because of the high RNA yields obtained even with difficult tissues.

There are many different RNA isolation methods commercially available, many of which are filter or column based or involve binding to magnetic beads. They vary in the quantity of RNA they are designed to extract, with anything from single cell to macro-scale kits now available. Due to the ease of handling, filter-based procedures are more suitable for working with multiple samples, but may lead to lower yields than when extracting RNA from each sample individually. A concern to be borne in mind is that any purification steps may lead to selective transcript loss, due to exclusion based on size, which could potentially result in a selection bias in subsequent microarray analysis. For example, in most of the column-based kits, only transcripts of about 200 nucleotides and longer are isolated.

Certain tissue types pose particular problems and require adaptations to the basic extraction protocols. Muscle tissue tends to be very fibrous and therefore difficult to homogenize. Spleen and thymus are nucleic acid-rich, making the lysate very viscous and consequently phase separation is problematic. Brain and plant tissue tend to be protein- and lipid-rich, again leading to phase separation problems. Spleen, thymus, liver and pancreas are particularly high in nucleases, requiring swift homogenization and inactivation of RNase activity to obtain good quality RNA. For best yield and purity, it is therefore important to use a method tailored to the tissue of interest. Consequently, we are unable to give specific advice on the most appropriate extraction method other than the general principles we have outlined. Researchers are advised to consult the literature for RNA extraction methods specific to the tissue and organism they are working with.

## 4.2 RNA QUALITY ASSESSMENT

The purity of the RNA sample can be easily assessed with a spectrophotometer and the NanoDrop (http://www.nanodrop.com/nd-1000-apps-microarray.html) is an ideal instrument in this respect since it can measure a range of RNA concentrations, from as little as 2 ng/$\mu$l up to 3000 ng/$\mu$l, from as little as 2 $\mu$l of sample. This means that only a small percentage of a typical sample is needed for the purity measurement. For pure RNA, the $A_{260}/A_{280}$ ratio should be 2.0, although above 1.8 is generally considered acceptable. A further purity indicator is the $A_{260}/A_{230}$ ratio, which should be around 2.2. An $A_{260}/A_{230}$ ratio lower than 1.8 indicates significant levels of organic contaminants such as carbohydrates, peptides, phenol and Trizol, which absorb light at 230 nm. The presence of these organic compounds may interfere with the subsequent molecular biology reactions and will ultimately affect the quality of the gene expression data (Naderi et al., 2004).

Even if the RNA is chemically uncontaminated, the optical density measurements do not give any indication of genomic DNA contamination or of the RNA integrity. DNA contamination is of concern if the RNA is to be reverse transcribed with random primers in downstream protocols but less of an issue if oligo(dT) primers are to be used, although the concentration measurements will be inaccurate. If genomic DNA contamination is a particular problem, for example when extracting RNA from tissues such as spleen that are high in DNA, the RNA should be additionally treated with a Ribonuclease-free DNase, which must be fully removed or inactivated before any subsequent steps.

With respect to integrity, it has been shown that if one RNA sample is intact and the other is degraded during isolation, then up to three-quarters of the differential gene expression measured will be solely due to differences in RNA integrity between the two samples (Auer et al., 2003). Therefore, before using RNA for expensive labeling and microarray hybridization it is wise to check its integrity. The cheapest method of obtaining information about the quality of the RNA is to electrophorese a small aliquot of the sample on an agarose gel (Figure 4.1). Intact total RNA run on a denaturing gel will have sharp ribosomal RNA bands (e.g. 28S and 18S rRNA in human samples). The 28S

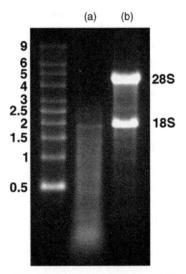

**FIGURE 4.1** Comparison of intact and degraded RNA. Degraded total RNA (a) and intact total RNA (b) were run along with molecular weight markers on a 1.5% denaturing agarose gel. The 18S and 28S ribosomal RNA bands are clearly visible in the intact RNA sample. The degraded RNA appears as a lower molecular weight smear. Figure based on http://www.ambion.com/techlib/append/supp/rna_gel.html. Reproduced with permission from Applied Biosystems.

rRNA band should be approximately twice as intense as the 18S rRNA band, though this is not true for all species. This 2:1 ratio (28S:18S rRNA) is a good indication that the RNA is intact. Partially degraded RNA will have a smeared appearance, will lack the sharp rRNA bands, or will have a lower 28S:18S ratio. Highly degraded RNA will appear as a very low molecular weight smear and genomic DNA contamination will be visible as a very high molecular weight band.

Although usually adequate, visual assessment of the 28S:18S rRNA ratio on an agarose gel can be misleading because the appearance of rRNA bands is affected by electrophoresis conditions, the amount of RNA loaded and saturation of ethidium bromide fluorescence. The detection threshold can be lowered significantly by using nucleic acid-binding dyes that are much more sensitive than ethidium bromide (e.g. SybrGold from Molecular Probes). A more expensive and widely used alternative to agarose gel electrophoresis is the Agilent 2100 Bioanalyzer, or the similar QIAcel system from Qiagen, which use a combination of microfluidics, capillary electrophoresis, and fluorescence to evaluate both RNA concentration and integrity (Figure 4.2). An additional advantage of the Bioanalyzer is that only small sample volumes are required,

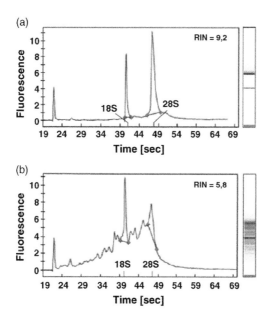

**FIGURE 4.2**   The Agilent 2100 bioanalyzer provides RNA quality control results in both gel-like image as well as electrophoretic data making it easy to detect even small degradation effects. In addition, an RNA Integrity Number (RIN) is provided for each total RNA sample allowing standardisation. Data are shown for an undegraded RNA sample (a) and for a degraded sample (b). Reproduced with permission, courtesy of Agilent Technologies, Inc. © Agilent Technologies, Inc. 2004.

allowing an assay of RNA quality with scarce samples. This method is particularly suitable for high-throughput.

It is important to treat all samples in the same way and to observe basic good laboratory practice in order to maintain integrity of the extracted RNA, thus reducing any bias in the expression analysis. These include, the use of RNase-free tips, tubes and solutions as well as gloved hands to reduce exposure to exogenous RNases, the use of an inert coprecipitant (such as linear polyacrylamide or DNase-treated glycogen) for quantitative recovery of low concentrations of RNA during precipitation, complete resuspension of any precipitated RNA pellets and careful storage. For short-term storage of RNA, $-20\ ^\circ$C is sufficient but for long-term storage $-80\ ^\circ$C is recommended. To avoid RNA damage from repeated freeze-thaw cycles, it is best to store larger RNA samples that will be used several times (for example reference RNA) in several smaller aliquots.

For some applications or labeling methods it may be desirable to purify polyA+ messenger RNA from the complex mix of total RNA. For example, if only mRNA is present, then random primers may be used rather than oligo(dT) primers in order to increase labeling efficiency (see below) or reduce 3' bias. In this case, standard polyA+ purification methods such as oligo-dT cellulose chromatography can be used.

## 4.3 CDNA PRODUCTION

### 4.3.1 Reverse Transcription

After successful extraction of good quality RNA, the next step in virtually all microarray expression experiments is to convert the RNA into a labeled form for hybridization. This most typically involves a reverse transcription step. In an ideal world, the reverse transcription should result in a cDNA population that reflects the original mRNA population in terms of transcript abundance and complexity. Furthermore, the reverse transcription reaction should be as efficient as possible to provide maximum sensitivity in the hybridization assay: consequently the reverse transcription protocol can have a large impact on the quality of the expression data.

Reverse transcriptase is a common name for an enzyme that functions as an RNA-dependent DNA polymerase, i.e. it uses an RNA template and copies it to make cDNA. With eukaryotic samples, the technique is usually initiated by mixing 23 base thymidine oligomers (oligo(dT)) with messenger RNA such that the oligomers anneal to the polyadenylate tail of the messenger RNA. Reverse transcriptase uses the oligo(dT) as a primer to synthesize so-called first strand cDNA. The used of an 'anchored' oligo(dT), with a random nucleotide at the 3' end of the primer, ensures that the primer initiates reverse transcription from the base just 5' to the poly-A tail of the transcript. During first strand synthesis the coding RNA, or sense strand, is copied into an antisense strand and whether it is appropriate to incorporate fluorescent dyes or other label into the antisense strand will depend on the type of probe present on the microarray. In protocols where purified polyA+ RNA is used as the template, random oligomers can be used to prime first strand synthesis. In this case, the length of the random primers can have an effect on the cDNA yield with some studies suggesting that 15-nucleotide-long random oligonucleotides (pentadecamers) can consistently yield at least twofold more cDNA than commonly used random hexamers (Stangegaard et al., 2006). It is important to bear in mind that transcripts with very short or even no polyA tails will be under-represented when oligo (dT) priming is used. This is of course important when analyzing RNA from prokaryotic sources since there is an absence of mRNA polyadenylation in such species. In such cases it is possible to enrich mRNA levels by depleting the rRNA via magnetic beads coated with sequences complementing the 16S and 23S rRNAs. Kits for rRNA depletion are available from Ambion (*MICROBExpress*) or Invitrogen (RiboMinus), the latter also offering similar systems for mouse, human and yeast.

## 4.3.2 Reverse Transcriptases

In addition to RNA quality and the priming strategy, enzyme efficiency is an important parameter for obtaining high yields of good quality cDNA. While all retroviruses encode a reverse transcriptase, the enzymes that are available commercially are derived from one of two sources, either by purification from the virus or expression of recombinant proteins in *Escherichia coli*. Moloney murine leukaemia virus (MMLV) encodes a single polypeptide enzyme whereas the Avian myeloblastosis virus (AMV) enzyme is composed of two peptide chains. Both enzymes have the same fundamental activities but differ in a number of characteristics, including their optimal temperature and pH. Most importantly, the MMLV enzyme has very weak RNase H activity (which degrades the RNA from RNA–DNA hybrids) compared to the AMV enzyme. This makes MMLV RT a clear choice when synthesizing cDNAs from long messenger RNAs and it is the enzyme most widely used in microarray experiments. Many commercially available RT enzymes are proprietary mutants that have been engineered and selected for improved characteristics (e.g. SuperScript III, StrataScript, PowerScript). In particular, they are more efficient at higher temperatures, facilitating more efficient copying of problematic templates such as those with extensive secondary structure. The engineered enzymes generally have increased half-life, for better stability, a reduced RNase H activity and have been optimized to give higher incorporation rate with dye-conjugated nucleotides, which are bulky and generally inhibit the reverse transcription activity of unmodified enzymes.

## 4.4 LABELING METHODS

### 4.4.1 Direct Labeling

In order to detect the cDNA or RNA species bound to an array probe it needs to be labeled. For ease of detection and safety, fluorescent dyes are preferable to radioactively labeled nucleotides. The simplest method of preparing labeled samples is to directly incorporate dyes into the cDNA during the reverse transcription step (Schena, 1995; Figure 4.3). This is achieved by using deoxynucleotide triphosphates (dNTPs) that have a dye molecule directly coupled to the base, with cyanine 3-dCTP (Cy3-dCTP) and cyanine 5-dCTP (Cy5-dCTP) the most popular. In general, the Cy-dCTP is added to the reverse transcription reaction in the place of a proportion of the unlabeled dCTP and is consequently directly incorporated into the antisense strand. The amount of incorporated dye

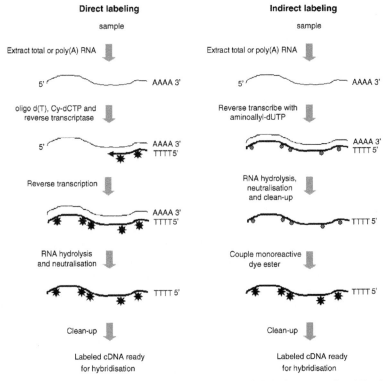

**FIGURE 4.3**   Schematic of direct and indirect labeling methods. In both cases an oligo d(T) primer is used to prime a reverse transcription reaction from either total RNA or purified poly(A) RNA in the presence of a modified nucleotide. Fluorescently labeled dCTP is typically incorporated during the reverse transcription reaction for direct labeling. Indirect labeling is a two step process: first aminoallyl dUTP is incorporated during the reverse transcription and then amine-reactive fluorescent dye is chemically attached to the aminoallyl group.

and the size of the resulting cDNA depend on the primers used (as described above), the ratio of labeled to unlabeled nucleotide (too little and the signal will be weak but too much and the close proximity of dye molecules to each other could lead to a reduction in signal due to fluorescence quenching), the temperature (secondary structure in the RNA will be less troublesome at 50 °C rather than at the typical 42 °C), duration of the reverse transcription reaction (longer reactions will, up to a point, yield more full-length transcripts and avoid losing representation of the 5′ ends).

The most commonly used dyes in two-channel microarray experiments are cyanine 3 (Cy3) and cyanine 5 (Cy5) (Figure 4.4). The cyanine dyes are both highly fluorescent and water-soluble, making them ideal for molecular biology reactions. Conjugates of the Alexa dyes, Alexa555 and Alexa647, the spectral analogues of Cy3 and Cy5, respectively, are reported to be more fluorescent and photo-stable than their Cy counterparts and are therefore the most commonly used alternatives. In addition, there is a seemingly endless list of other dyes available at many wavelengths across the spectrum, however, most of these tend to be less strongly fluorescent than the Cy dyes.

In order to drive the reverse transcription reaction, the nucleotides, including the labeled nucleotide, are added in excess and if allowed to hybridize to the array this would lead to high background signal. Therefore, a critical step in any labeling protocol is the removal of unincorporated nucleotides. This can be achieved in a number of ways, most typically by precipitation or with column-based methods. Glass filter or sieve-type columns retain the labeled cDNA products due to their larger size and allow the small mononucleotides to pass through: the bound labeled cDNA can then be eluted. The opposite approach is to use size exclusion chromatography, commonly with Sephadex columns, which retain the low molecular weight species allowing the larger labeled cDNA to pass through in the void volume. As discussed above for RNA extraction, particular purification methods may lead to sample loss and this should be taken into consideration when calculating the amount of labeled sample required. After clean-up, the labeled product usually requires reduction to a smaller volume, which is most easily achieved by evaporation of the sample in a vacuum dryer, passing through a Microcon membrane with the appropriate molecular weight cut-off, or when precipitation is used, the sample can simply be resuspended in a small volume.

The main advantages of direct labeling protocols are that they are quick and simple, requiring relatively few steps, and therefore are easy to scale up for high throughput. There are, however, some disadvantages. The typical amount of starting material for a direct labeling reaction is 25–100 $\mu$g total RNA (or approximately 0.5–2 $\mu$g mRNA). This is quite high and precludes experiments with samples from limited tissue sources. In addition, since the dye-coupled nucleotides are bulky, the incorporation efficiency of the reverse transcriptase is reduced. Of greater concern, Cy5 is a larger molecule than Cy3 and

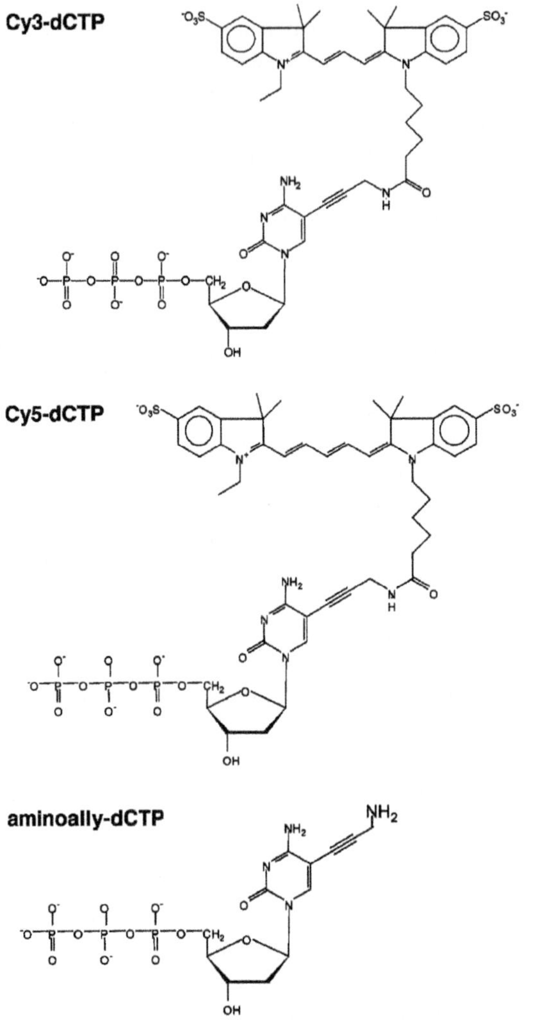

**FIGURE 4.4**   Molecular structure of dCTP modified with Cy3, Cy5 and aminoallyl moieties. Note that the Cy3-dCTP is significantly larger than aminoallyl-dCTP and that Cy5-dCTP is larger still. This affects their ease of incorporation by reverse transcriptase enzymes.

consequently Cy5-dNTPs are incorporated less efficiently by the reverse transcriptase. This can lead to a 'dye bias', where labeled cDNA products are not equivalent for the two dyes, potentially confounding subsequent analysis. One relatively straightforward way to deal with this problem is in the experimental design, where it is recommended that dye swaps are included, i.e. if a sample is compared with a control, then a Cy3-labeled sample is hybridized together with a Cy5-labeled control on one array and the dyes reversed on a replicate array (Cy5 sample, Cy3 control).

## 4.4.2 Indirect Labeling

One way of overcoming the problem of dye bias is to use an alternative two step, or 'indirect', method of labeling (Figure 4.3). Since the dye-coupled nucleotides are bulky and of different sizes, it is preferable to replace these molecules with something much smaller, such as an aminoallyl-dNTP (Figure 4.4). In the first step, the aminoallyl-dNTP is added to the nucleotide mix in the RT reaction to produce first strand cDNA: there is essentially no difference in the incorporation rate between experimental and control samples. In addition, since the aminoallyl-dNTP causes much less steric hindrance than the bulky Cy dyes, it is incorporated at a higher frequency and is also more likely to yield full-length product. In a second step, after first strand synthesis, an amine-reactive Cy dye is chemically coupled to the aminoallyl groups, thus labeling the cDNA. Since the substrate for this step is similar for both dyes, this method efficiently creates evenly labeled cDNA without any dye incorporation bias. The method is also more sensitive than direct labeling and consequently requires around fivefold less input RNA.

Although straightforward, the disadvantage of indirect labeling is that it requires more steps which increases the chance of distortions. In addition, despite the combination of aminoallyl nucleotide and amine-reactive dyes being cheaper than dye-coupled dNTP, the cost of the extra purification steps (once after cDNA synthesis and again after dye binding) makes the indirect method slightly more expensive than direct labeling. A word of caution: amine-reactive dyes will couple to any free amine groups in solution, such as the commonly used Tris-based buffers, it is therefore essential that cleanup of the aminoallyl cDNA product efficiently removes all traces of Tris before the dye coupling step.

Although indirect labeling can eliminate bias due to unequal nucleotide incorporation, it is known that Cy5 is potentially more labile than Cy3. In particular Cy5 appears to be much more sensitive to oxidation by environmental ozone. Therefore, if comparative hybridizations using Cy5 are planned then dye-swap controls should be incorporated into the experimental design unless steps are taken to regulate ozone levels with, for example, carbon filters (Branham et al., 2007).

## 4.4.3 Labeling with Klenow Polymerase

The importance of generating full-length cDNAs in the RT reaction will depend on the design of the array and the type of biological question being addressed. If, for example, the type of exon arrays described in Chapter 10 are being used then it is necessary to generate full-length labeled sample. If oligo(dT) priming of the

RT reaction is used, the cDNA synthesis needs to be efficient enough to generate a complete copy of the mRNA. This is not possible with direct incorporation and although the indirect labeling method helps to increase the processivity of the reverse transcriptase, the best chance of generating full-length cDNAs is to use no modified nucleotides. However, this obviously generates an unlabeled product. One of the simplest ways to obtain and label full-length cDNAs is via a multi-step reaction. First, the standard first strand reverse transcription is performed with unmodified dNTPs using conditions that favor full-length cDNA synthesis. Next, DNA polymerase, RNase H, *E. coli* DNA ligase and dNTPs are combined with the first strand cDNA for 'second strand' synthesis. The resulting double stranded cDNA can be used as a template for standard random primed labeling with a high concentration of an appropriate DNA polymerase, such as the Klenow exo-enzyme, generating a labeled representation of the entire double stranded cDNA. In this step either dye-coupled dNTPs are directly incorporated or an indirect labeling via aminoallyl dNTPs is used. A benefit of this approach is that there is a modest amplification, due to random priming and the strand displacement activity of Klenow (Lieu et al., 2005). In our laboratory we have determined that as little as 1–5 $\mu$g of total RNA starting material can be used, equating to 10–50-fold less input than with direct labeling (FlyChip, unpublished). As with first strand labeling, direct incorporation can lead to dye bias that should be controlled via dye-swapped experiments. Of course one disadvantage of this approach is that both sense and anti-sense strands are labeled, which may be unsuitable for experiments that require strand specificity.

### 4.4.4 Other Labeling Methods

Although fluorescent labeling with Cy dyes is widely used, various nonfluorescent detection methods have been reported. Standard microarray technology requires expensive scanners and reagents, in addition, the enzymatic treatments needed to introduce labels can skew the representation of the sample population. To address these issues, methods that circumvent reverse transcription and dye labeling have been developed. In one method (Sun et al., 2005), sample RNA is directly hybridized to the array and since the targets are not modified or subject to any molecular biology prior to hybridization it is potentially simpler, cheaper and less variable than cDNA synthesis based approaches. The sample RNAs hybridized to the complementary probes on the array are detected by incubating the array in a colloidal gold solution. The positively charged gold particles are electrostatically attracted to the negatively charged phosphate groups in the backbone of the target, resulting in precipitation of nanogold particles. The amount of precipitation is proportional to the amount of bound target RNA. Instead of an expensive confocal scanner, a relatively inexpensive flatbed scanner can used to detect the gold precipitate. It is important to remember that this, and all other methods that directly label RNA, can only be used with arrays containing appropriate probes such as double stranded DNA amplicons or antisense oligonucleotides, and not to the more commonly used sense oligonucleotide probes. Though of course in the case of the nanogold technique, double stranded probes cannot be used since they reassociate and thus would be detected.

A related method, employing nanogold and colourimetric silver staining detection has been applied to microarrays (Alexandre et al., 2001). This technique is reported to be of equal sensitivity to Cy dyes and considerably cheaper. The target cDNA in this case is labeled by incorporation of biotinylated nucleotides and, after hybridization to the array, streptavidin-nanogold particles are added which bind to the biotin. To enhance the signal, an additional silver precipitation step is included where, in the presence of a reducing agent, silver salts precipitate. Since the silver deposits strongly reflect light in the visible spectrum, light reflection-based readers can be used to detect the signal. A disadvantage is that silver solutions are sensitive to UV or sunlight and spontaneous conversion of silver solution into metallic grains can occur, leading to nonspecific silver precipitation. Careful control of parameters such as temperature, pH, the silver salt concentration and reducing agent concentration are necessary for optimal results with this technique.

A variation on the gold nanoparticle theme is a sensitive method that claims to require as little as 0.5 $\mu$g total RNA starting material (Huber et al., 2004). As with the method of Sun et al., the target does not require any labeling. Total RNA is bound directly to the array, followed by a second hybridization step in which the bound molecules are detected using oligo(dT)$_{20}$-modified gold nanoparticle probes that bind to poly-A tails and therefore detect only mRNAs. An extra silver enhancement step is added to increase the sensitivity. The method has been reported to be 1000 times more sensitive than fluorescent-based detection methodologies. As with other methods that have no reverse transcription step, the 3$'$ end bias of labeled products is avoided, making it a useful method for analyzing expression of splice variants. A disadvantage is that all different species of RNA will bind the probes, whereas only the poly-A$^+$ transcripts will be detected. It is therefore important that the amount of probe is not saturated by binding of poly-A$^-$ RNA and an initial poly-A$^+$ selection is probably advisable.

A further method which avoids reverse transcription uses a nonenzymatic method to label mRNA with Cy dyes (Gupta et al., 2003). This procedure involves a stable coupling between a monovalent platinum reagent, with Cy fluorophores synthetically attached via a linker arm, and the N7 of guanine residues in the RNA. Unlike alkylating reagents, this guanine modification does not cause depurination and resulting strand scission, and does not affect the ability of the modified single strands to hybridize. The platinum labeling reagent is mixed with total RNA and the reaction is complete within 15 min. The mRNA must then be purified with oligo-dT columns before hybridization. Since the labeling is with Cy dyes, standard detection methods are used.

Due to the potential hazards associated with working with radioactive materials, radioactive labeling methods have fallen out of favor and are generally being replaced by less hazardous methods. However, radioactivity has the advantage of being very sensitive, enabling smaller samples to be labeled without amplification of the RNA. It is also possible to utilize two different radionuclides on the same array, analogous to co-hybridizing with Cy3- and Cy5-labeled samples (Salin et al., 2002). In order to distinguish the signal from the two different nuclides they must be of significantly different energies, for example $^3$H together with $^{35}$S or $^{33}$P. A single image of the radioactive disintegration events is captured with a high-resolution digital autoradiography system and an algorithm used to filter the data into two subimages, one for each nuclide. In addition to superior signal to noise ratio, this technique has been shown to be between 2- and 100-fold more sensitive than fluorescent labeling, although multiple exposures of the same array may be necessary to extract the full dynamic range.

## 4.5 SIGNAL AMPLIFICATION

### 4.5.1 End Labeling

An important consideration for methods that rely on incorporation of modified nucleotides into cDNA during reverse transcription, is that the incorporation is sequence dependent. In practice this means that the observed signal is not directly proportional to the abundance of the transcript in the population. An ideal solution to the problem is to end-label each transcript such that one quantum of labeling equates with one copy of the transcript. By including known standards on the microarray and using spike-in controls at defined concentrations, it should be possible to calculate actual numbers of bound cDNA molecules per probe. The drawback of end-labeling is that one fluorophore molecule per transcript molecule is, in practice, undetectable with current scanner technology. Consequently methods have been developed to amplify end-label signals. One well-developed example is the use of dendrimer technology (Shchepinov et al., 1997, 1999; Stears et al., 2000) (Figure 4.5). Dendrimers are branched oligonucleotides with multiple arms that offer the possibility of attaching a fluorophore to each arm, thereby increasing the signal obtained from each molecule. In a relatively straightforward approach, an oligo(dT) primer containing a specific capture sequence extension is used to prime first strand cDNA synthesis. For dual channel experiments two (or more) different capture sequences may be used. The resulting cDNAs hybridized to an array have capture sequences at their 5′ end that are subsequently detected by hybridization with fluorescently labeled dendrimers coupled to the complement of the capture sequence. In principle, each cDNA molecule binds a single dendrimer and, since the number of fluorophores conjugated to the dendrimer is known, absolute quantification of transcript numbers is possible. Current tests indicate that

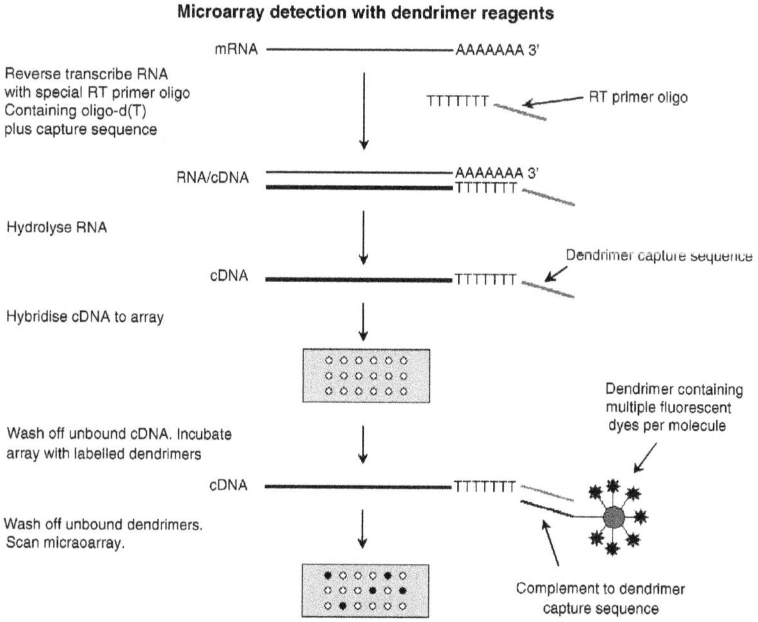

**Microarray detection with dendrimer reagents**

FIGURE 4.5  Schematic of the dendrimer microarray detection method. Dendrimers are branched oligonucleotides with a fluorophore attached to each arm. First strand synthesis is primed by a special oligo d(T) primer containing a specific capture sequence extension. For two channel experiments, a different capture sequence is used for the second channel. The resulting cDNAs are hybridized to an array and then incubated with fluorescent dendrimers that are coupled to oligonucleotides complementary to the capture sequence.

the signal obtained from 1 to 2 $\mu$g of total RNA starting material is equivalent to that obtained with 40–50 $\mu$g of RNA in a standard direct labeling protocol, so this method shows great promise as a sensitive and, more importantly, quantitative detection system.

## 4.5.2 Tyramide Signal Amplification

Tyramide signal amplification (TSA) uses a signal amplification technique originally developed to improve the sensitivity of immunohistochemistry for fluorescence microscopy (Wiedorn et al., 1999). The method has since been adapted for use in amplification of microarray signals and claims to require 20–100 times less RNA than direct cDNA labeling (Karsten et al., 2002). In this method, the RT reaction incorporates either biotin- or fluorescein-conjugated nucleotides into the cDNA that are detected after hybridization in sequential steps with specific antibodies. The microarray is first incubated with anti-fluorescein conjugated to horseradish peroxidase (HRP). This antibody–enzyme conjugate specifically binds to the hybridized fluorescein-labeled cDNA.

The HRP catalyzes the local deposition of Cy3-labeled tyramide. The reaction is quick (less than 10 min) and results in the deposition of numerous Cy3 labels immediately adjacent to the immobilized HRP. During this enzymatic process, the amount of tyramide relative to fluorescein label is greatly amplified. Prior to the next step, the residual HRP must be inactivated. In the second step, streptavidin–HRP binds to the hybridized biotinylated cDNA. The HRP portion of the enzyme conjugate catalyzes the local deposition of Cy5-labeled tyramide, again with amplification. Unfortunately, in some hands the TSA method fared relatively poorly compared with other labeling methods, suffering from high variation and noise (Badiee, 2003), and it remains to be seen whether the technique will see wider adoption in the array field.

## 4.6 RNA AMPLIFICATION

An obvious drawback for all the labeling procedures described above is that they require relatively large amounts of starting RNA and are consequently unsuitable for studies with limited biological material. Even the signal amplification techniques require at least 1 $\mu$g of total RNA and, since an average metazoan cell contains approximately 10 pg of total RNA, the minimum number of cells required is of the order of $1 \times 10^5$. In the case of in vivo studies with metazoans, where most tissues are generally a heterogeneous mix of different cell types, isolation of large amounts of a specific cell type may be difficult. This can be important since expression profiling of a tissue comprising different cell types makes analysis difficult with the data tending to reflect gene expression in the dominating cell type, drowning out potentially interesting signals from less abundant cells (Szaniszlo et al., 2004). It is therefore highly desirable to be able to expression profile specific cells of interest or, in the extreme case, individual cells. A variety of technologies facilitate the precise selection of cells of interest from a tissue (e.g. LCM, biopsy, immunomagnetic selection or FACS sorting) however these techniques generally yield too few cells for conventional RNA labeling methods. The only realistic way that such small samples can currently be used for microarray expression analysis is to amplify the RNA.

The most important issue relevant to sample amplification for microarray analysis is conservation of the relative transcript abundance of the starting sample. Any methods that distort this distribution will lead to erroneous and misleading results. Since increasing levels of amplification are needed as input sample sizes decrease, there is an increasing potential for biases as fewer cells are analyzed. Various RNA amplification methods exist, each with different levels of sensitivity, fidelity and reproducibility, and it is essential that these factors are taken into consideration when designing an experiment involving extremely low input RNA levels. A list of commercially available amplification kits is provided in Table 4.1 and we summarize the most common methods below.

**TABLE 4.1** Summary of Commercially Available RNA Amplification Kits

| Manufacturer | Range of input total RNA[a] | Kit | Amplified product[b] | Amplification method[c] |
|---|---|---|---|---|
| Agilent Technologies | 50 ng–5 μg | Low RNA Input Linear Amplification Kit | Antisense cRNA, or sense cDNA | Linear 1 round |
| Ambion | 100 ng–2 μg | MessageAmp aRNA Kit | Antisense cRNA | Linear 1 round |
| Arcturus | 2–10 μg | RiboAmp Plus RNA Amplification Kit | Antisense cRNA | Linear 1 round |
| Arcturus | 500 pg–5 ng | RiboAmp HS Plus RNA Amplification Kit | Antisense cRNA | Linear 2 round |
| Clontech | 100 ng–5 μg | BD SMART mRNA Amplification | Sense RNA | Template switch plus linear 1 round |
| Clontech | 10 ng–1 μg | BD Atlas SMART mRNA Amplification | ds cDNA | Template switch plus PCR |
| Epicentre Biotechnologies | 25–500 ng | TargetAmp 1-Round aRNA Amplification Kit 103 | Antisense cRNA | Linear 1 round |
| Epicentre Biotechnologies | 10–500 pg | TargetAmp 2-Round aRNA Amplification Kit 2.0 | Antisense cRNA | Linear 2 round |
| Invitrogen | 100 ng–5 μg | SuperScript RNA Amplification System | Antisense cRNA | Linear 1 round |
| NuGen | 500 pg–50 ng | WT-Ovation Pico RNA Amplification System | Antisense cDNA | Linear isothermal amplification |
| Roche Applied Science | 50 ng–1 μg | Microarray Target Amplification Kit | Antisense cRNA | Exponential (5–10 PCR cycles) plus linear 1 round |
| Sigma | 10–100 ng | TransPlex Whole Transcriptome Amplification Kit | ds cDNA | Exponential |
| Telechem | 50 ng–1 μg | ArrayIt MiniAmp mRNA Amplification Kit | Antisense cRNA | Exponential (5–10 PCR cycles) plus linear 1 round |

[a] Optimum input range recommended by manufacturer. Some recommend a minimum input amount which is lower than the optimum input range, suggesting that the amplification will be of reduced fidelity.

[b] Many companies offer alternative or modified kits whereby aminoallyl UTP can be incorporated into the amplified product.

[c] Kits using a single round of linear amplification can usually be extended to second round.

### 4.6.1 Linear Amplification

#### 4.6.1.1 Eberwine Method

The first RNA amplification protocols were developed before the microarray era and were devised to amplify mRNA from limited numbers of brain cells for the construction of specific cDNA libraries (Van Gelder et al., 1990). One protocol, commonly referred to as the Eberwine method, aims to 'linearly' amplify the mRNA and forms the basis of many, more recently developed, techniques.

The general principle involves a reverse transcription reaction primed with a modified oligo(dT) primer containing a T7 RNA polymerase promoter site (Figure 4.6). First strand cDNA is converted to double stranded cDNA using the conventional methods described above and the resulting dsDNA used as the template in an in vitro transcription (IVT) reaction. T7 RNA polymerase binds to its promoter site in the modified oligo(dT) primer and transcribes multiple copies of antisense RNA (aRNA). In order to label the aRNA, aminoallyl nucleotides can be incorporated during the IVT step, followed by dye coupling as described above for indirect labeling. Approximately 1 $\mu$g of input total RNA is required for one round of amplification by this method and 1000–2000-fold amplification is typical. However, the products of the first round of amplification can be used as the starting point for second and subsequent rounds of amplification, which are initiated by random priming. A second round of amplification can result in up to $10^6$-fold amplification of the original material (Phillips and Eberwine, 1998) and is suitable for submicrogram quantities of input RNA. Such a T7-IVT amplification step is integral to the sample preparation methods

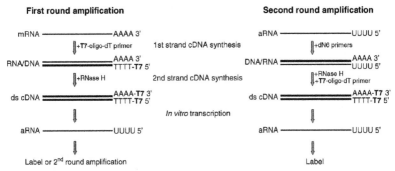

**FIGURE 4.6**   Schematic of linear mRNA amplification procedure, based on the well established 'Eberwine' protocol (Van Gelder et al., 1990). First strand cDNA synthesis is primed by an oligo d(T) primer containing a T7 polymerase binding site. RNase H treatment digests the mRNA strand from the mRNA/cDNA hybrid to leave small fragments of mRNA which are used to prime second strand cDNA synthesis. An in vitro transcription reaction using T7 RNA polymerase then produces multiple copies of antisense RNA. Labeled nucleotides can be incorporated directly during the RT reaction to generate labeled antisense RNA. Alternatively, the amplified RNA can be used as a template for a second and further rounds of amplification by priming the first strand cDNA synthesis with random primers.

for the Affymetrix GeneChip platform, where biotin-conjugated nucleotides are incorporated during the IVT step to generate biotinylated-cRNA. This is cleaned up and fragmented before hybridization to the arrays; the hybrids are detected by streptavidin-coupled phycoerythrin binding to the biotin. If the starting amount of total RNA is between 10 and 100 ng, then the product of the first round IVT reaction (without biotin) can be used as the template for a further round of RT and IVT.

### 4.6.1.2 *Template Switching*

The classical Eberwine method is 3′ end biased due to the use of oligo(dT) primers to initiate cDNA synthesis from the 3′ end of the mRNA: there is no

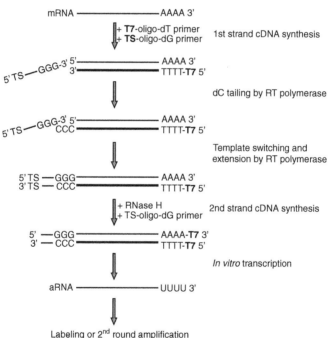

**FIGURE 4.7**  Linear amplification with template switching ensures full-length amplification (based on Wang et al., 2000). As with the classical Eberwine protocol, first strand cDNA synthesis is primed by an oligo d(T) primer containing a T7 polymerase binding site. A template switch (TS) oligo d(G) primer is also added to the mix to exploit the template switching effect of the reverse transcriptase enzyme. Template switching refers to the fact that the RT enzyme automatically incorporates nontemplate dCTPs at the 3′ end of the transcript, then switches template and continues replication to the end of the TS oligo d(G) primer, resulting in full-length cDNA. Second strand cDNA synthesis is initiated by the oligo d(G) primer. An in vitro transcription reaction using T7 RNA polymerase then produces multiple copies of antisense RNA.

realistic way to ensure that the cDNA extends to the very 5' end of each transcript. This problem has been addressed by exploiting the terminal transferase property of reverse transcriptases in a modification to the original Eberwine protocol known as template switching (Wang et al., 2000) (Figure 4.7). In this process, the reverse transcriptase enzyme adds additional (primarily cytosine) nontemplate residues to the 3'end of the cDNA. In addition to the T7-oligo(dT) primer, the reverse transcription reaction contains a second, oligo(dG) primer that has a defined sequence attached to its 3' end. This primer pairs with the C residues added by the reverse transcriptase of the newly synthesized cDNA and the enzyme is able to switch strands and continue replicating to the end of the oligo(dG) primer sequence. Since the C residues are only added to the 3' end of the cDNA when it has reached the 5' end of the mRNA, the method ensures that only full-length cDNAs are generated. For second strand synthesis, a primer with bases complementary to the specific sequence attached to the oligo(dG) primer is used. As with the classical protocol, antisense RNA is transcribed from the dsDNA template by T7 RNA polymerase.

## 4.6.2 Exponential (PCR-Based) Amplification

### 4.6.2.1 Basic PCR-Based Amplification

A drawback of the so-called 'linear' amplification methods is that they are time-consuming multistep procedures that are susceptible to failure or misrepresentation at any step. In contrast, exponential (PCR-based) amplification methods are both quick and easy, offering the most cost-effective and easiest way to scale up for processing large numbers of samples simultaneously. The general scheme of a PCR-based RNA amplification procedure (Figure 4.8) is to first reverse transcribe the mRNA with an oligo(dT) primer and then to add an oligo(dA) tail to the 3' end of the cDNA using a terminal transferase enzyme. Addition of the poly(dA) stretch means that both ends of the RNA–DNA molecule have the same sequence and so a single oligo(dT) or oligo(dT)-adaptor primer can be used in subsequent conventional PCR amplification cycles. It is critical that the number of PCR cycles is optimized to avoid reaching saturation. This can be done via pilot studies where the PCR products are quantified after each cycle and a plot of cycle number against product amount generated. After a certain number of cycles the reaction will reach saturation and plateau. For the real sample, it is essential to terminate the reaction at least one cycle prior to saturation of abundantly expressed genes (Endege et al., 1999). This ensures that the majority of transcripts are in the exponential phase of amplification and should maintain the original transcript distribution. Unlike the aRNA generated by linear amplification methods, the dsDNA products of PCR amplification are suitable for hybridizing to arrays with probes of either orientation.

A major concern with exponential amplification is that it can skew the original relative transcript abundances because the PCR reaction is biased

**PCR-based amplification**

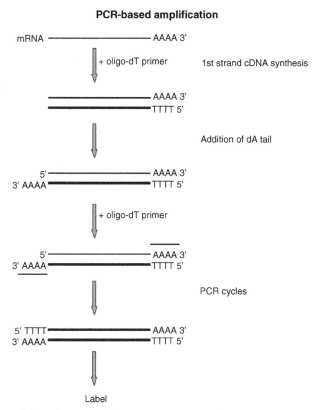

**FIGURE 4.8**   Schematic representation of a typical exponential RNA amplification method. First strand cDNA synthesis is primed by an oligo d(T) primer. A terminal transferase enzyme is then used to create an oligo d(A) tail at the 3′ end of the cDNA. This enables the same oligo d(T) (or oligo d(T)-adaptor) primer to be used in the subsequent PCR reaction. The end product of the amplification is double stranded cDNA.

against long cDNAs or sequences with high GC content, leading to nonrepresentative amplification. Various modifications of the basic protocol have been published that optimize the reaction in order to minimize biases. Iscove et al. (2002) restrict the extent of the initial reverse transcription to only a few hundred bases at the 3′ end of the RNAs by limiting the reaction duration and reagent concentrations. In principal this creates cDNAs of equal length and therefore makes them all equally likely to be amplified in the subsequent PCR. By comparing results from amplified and unamplified samples, they suggest that the exponential approach preserves abundance relationships after as much as $10^{11}$-fold amplification. The amount of input RNA required is reduced to the picogram level: in the single cell range. Clearly this method is specifically designed not to produce full-length transcripts and will therefore only be compatible with arrays with probes designed close to the 3′ end.

#### 4.6.2.2 *Template Switching*

In order to amplify only full-length cDNAs, the basic PCR-based amplification can be combined with the template switch mechanism described above, as exemplified in the SMART PCR approach (Petalidis et al., 2003). In this case, the T7-oligo(dT) primer is designed to have an additional nucleotide sequence, which is the same as that in the oligo(dG) primer, enabling primers with this nucleotide sequence to amplify the full-length product from both ends in a PCR reaction. An alternative strategy (Stirewalt et al., 2004) couples the use of template switching, single strand amplification and IVT. The template switch concept is employed to generate dsDNA, but unlike the SMART protocol, which simultaneously amplifies from both ends of the transcript, this method linearly amplifies from only the 5′ end of the cDNA template, producing large amounts of sense DNA strands. A separate reaction is used to synthesize the second strand cDNA and the resulting dsDNA serves as a template for IVT. Biotinylated nucleotides can be introduced during this step for hybridization to Affymetrix arrays.

### 4.6.3 Single Primer Amplification

A further variation, which mixes linear amplification and PCR, involves using a single primer during the amplification cycles (Smith et al., 2003). As before, the mRNA is primed by a modified oligo(dT) primer to generate first strand cDNA. The complement to this primer is incorporated into the 3′ end of the cDNA following second strand synthesis. A primer equivalent to the modified part of the oligo(dT) primer is then used to direct single primer, semi-linear *Taq* polymerase amplification where only the antisense cDNA strand is amplified. The single primer amplification concept has also been used in isothermal amplification of cDNA (Dafforn et al., 2004) (Figure 4.9). This technology uses a chimeric RNA–DNA primer with a known RNA sequence tag on the 5′-end of an oligo-dT cDNA primer. Incorporation of the chimeric RNA–DNA primer means that, after first and second strand synthesis, the resulting double stranded cDNA contains a short RNA–DNA heteroduplex segment of known sequence at the 5′ end of the first strand. It is this heteroduplex, the distinctive aspect of this technology, which provides the basis for the linear amplification. Amplification is initiated by incubation with a mix of RNaseH, DNA polymerase and a second chimeric primer containing a stretch of ribonucleotides at its 5′ end joined to a stretch of deoxyribonucleotides at its 3′ end (SPIA). RNaseH only digests RNA that is hybridized to DNA, so it digests the short RNA segment leaving the complementary cDNA portion exposed. Exposure of this complementary sequence allows the SPIA primer to hybridize to its complementary sequence and recruit the polymerase to initiate synthesis of another cDNA strand, which displaces the original strand. A continuous cycling of RNaseH degradation, primer binding with extension, and displacement of a portion of the previous

**Isothermal linear amplification procedure**

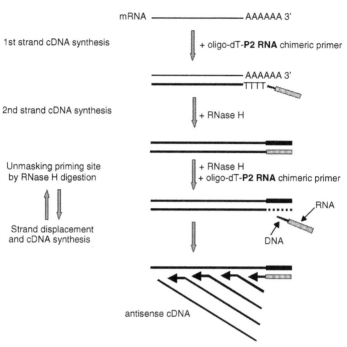

mRNA ————————— AAAAAA 3'

1st strand cDNA synthesis     + oligo-dT-**P2 RNA** chimeric primer

————————— AAAAAA 3'
————————— TTTT

2nd strand cDNA synthesis     + RNase H

Unmasking priming site        + RNase H
by RNase H digestion          + oligo-dT-**P2 RNA** chimeric primer

Strand displacement           RNA
and cDNA synthesis            DNA

antisense cDNA

**FIGURE 4.9**   Schematic of the isothermal linear amplification procedure, based on the method of Dafforn et al. (2004). This method makes use of a special chimeric primer which is part RNA, part DNA. In the first step, mRNA is reverse transcribed with the chimeric primer. Second strand synthesis results in an RNA–DNA heteroduplex at one end. The RNA portion is digested with RNase H which unmasks the priming site. The RNA–DNA primer binds to the exposed single stranded segment and is extended by a strand-displacing DNA polymerase to synthesize cDNA. Once the extension is initiated, the RNase H again digests the RNA segment of the RNA–DNA hetroduplex enabling a new primer molecule to bind and extend. This cycle of umasking, strand displacement and extension leads to a continuous isothermal generation of single stranded antisense cDNA copies.

cDNA enables rapid linear production of thousands of replicates from each cDNA template. The starting material for this method can be as little as 5 ng total RNA.

## 4.6.4 Sense Strand-Specific Amplification Methods

Exponential amplification results in a double stranded DNA product that, when labeled, is suitable for hybridizing to oligo arrays containing either sense or antisense sequences. The linear methods described above tend to amplify just one strand, mostly the antisense. While this is of little concern when using amplicon arrays, oligo arrays are strand specific so care must be taken. Most oligo arrays contain sense strand sequences, and so the antisense amplification product needs to contain the label or tag. This means that either the modified nucleotides must be incorporated at the IVT or single primer linear amplification stage, or further steps must be added to the protocol in order to incorporate label into the correct strand. Additional processes mean additional time and cost and also a greater chance that the labeled end-product is no longer a faithful representation of the transcript distribution in the original sample. To avoid this issue, some amplification protocols have been specifically designed to generate sense strand as their end-product. The amplified product can then be treated as unamplified RNA, for example for direct labeling in an RT reaction. In a variation of the template switching method, Rajeevan et al. (2003) incorporated a primer containing the SP6 RNA polymerase binding site at the 3′ end of the first strand cDNA product. This SP6 promoter is therefore at the 5′ end of the sense strand and an IVT reaction with SP6 polymerase will generate multiple copies of sense strand RNA.

In a procedure involving two rounds of IVT, mRNA was first amplified linearly with oligo(dT)-T7-primed reverse transcription and IVT by T7 RNA polymerase. In a second RT reaction, random 9-mer primers attached to a T3 RNA polymerase binding site were used to generate sense DNA from the antisense RNA template. T3 primers and T3 RNA polymerase were then used in a second IVT reaction to generate sense-strand amplified RNA (Kaposi-Novak et al., 2004). A further amplification method for generating labeled antisense involves the classical Eberwine procedure to produce amplified antisense RNA followed by an RT reaction to transcribe the aRNA into sense cDNA. The sense cDNA is mixed with random primers and fluorescently labeled nucleotides and provides the template for Klenow labeling to produce labeled antisense cDNA ready for hybridization to sense arrays (Schlingemann et al., 2005).

Using sequence-specific 'terminal continuation' (TC) primers, an IVT amplification method has been adapted to produce either amplified sense or antisense RNA (Che and Ginsberg, 2004). TC is based on the ability of the reverse transcriptase enzyme to add nucleotides nonspecifically at the end of the mRNA template. To generate sense RNA, an oligo(dT) primer and a TC primer, containing oligo(dC) and a T7 promoter sequence in the sense orientation, are added to the RT reaction for first strand cDNA synthesis. The TC primer anneals with the stretch of Gs and provides a binding site for priming second strand cDNA synthesis. In IVT reactions, the transcription is driven from the T7 polymerase site at the 5′ end and therefore produces amplified sense RNA. To generate the antisense strand, an oligo(dT)-T7 primer is used in the place of oligo(dT), and an oligo(dC) primer is used in the place of the oligo(dC)-T7 TC primer. This produces IVT products primed from the opposite end.

## 4.6.5 Degraded RNA

Formalin fixed, paraffin-embedded histological samples are, by nature, partially degraded and contain low amounts of total RNA. A limitation of the amplification procedures involving oligo(dT) in the reverse transcription, is that they are not very effective with degraded samples since any breakages in the RNA will most likely lead to only a small amount of 3' sequence still retaining the poly-A tail. This is clearly a problem for detection with oligo arrays where the probe is designed to more 5' sequences (e.g. exon arrays), since there will be no detectable signal. A method for amplifying degraded samples has been published (Xiang et al., 2003) which is similar to the method of Kaposi-Novak et al., described above. Rather than using oligo(dT) primers, total RNA is primed using random 9mers with a T3 RNA polymerase promoter sequence at the 5' end (T3N9). After conventional second strand synthesis, antisense RNA is transcribed from the double stranded DNA template with T3 RNA polymerase and should not be 3' biased. The same T3N9 primer can be used to drive first strand cDNA synthesis in subsequent rounds of amplification and, when tested on artificially damaged RNA, up to four rounds of amplification were shown to give reproducible results. The number of differentially expressed genes detected with this random primed T3N9 method was higher than when an oligo(dT)-T7 primer was used. However, it should be noted that the first reverse transcription reaction will drive cDNA synthesis not only from damaged and intact mRNA but also from any other poly-A$^-$ transcripts. For organisms which produce polyadenylated transcripts, such a protocol should only be considered when damaged RNA is unavoidable and the available arrays contain probes that are not close to the 3' end of the transcript.

## 4.7 AMPLIFICATION METHODS COMPARED

### 4.7.1 Amplification Efficiency

The amplification efficiency achieved by a given method is usually measured by dividing the final yield of amplified product by the starting amount of mRNA to give an amplification factor or fold. Such a method is usually only an estimate, since the actual amount of mRNA in the sample is, in most cases, not known but is generally assumed to be 1–3% of the total RNA. When evaluating protocols, the typical procedure involves a comparison of unamplified sample against a dilution that has been amplified. The results are likely to be an overestimate of the amplification efficiency achievable for real samples, where the RNA may be less pure or of a different complexity. In order to make a direct comparison between multiple different methods, the use of a standard RNA source is required. Nevertheless, several studies have compared one or two different methods using the same total RNA samples (e.g. Puskás et al., 2002; Klur et al., 2004). Currently, the typical fold amplifications reported for linear amplification are $\sim 10^3$-fold after one round and $\sim 10^6$ after a second round. In comparison, a PCR-based method has reported a factor of $3 \times 10^{11}$ (Iscove et al., 2002) and methods combining both linear and PCR amplification are around $10^6$–$10^7$, i.e. higher than with two rounds of linear amplification, but lower than purely exponential methods (Ohtsuka et al., 2004).

### 4.7.2 Reproducibility

If the microarray results obtained with an amplified sample are to be trusted, it is important to assess the reproducibility of the method. In practice this usually involves dividing a sample into aliquots and amplifying each in independent reactions: after labeling and hybridization, the correlation of the expression profiles between the samples is measured. It is commonly observed that the reproducibility of replicates from amplified samples is actually higher than that from hybridizations of unamplified samples. Although still high, the reproducibility of the amplification has been shown to decrease as the input RNA is decreased (Wang et al., 2000; Baugh et al., 2001). A comparison of reproducibility between different protocols has been less conclusive, with some studies showing that linear amplification is slightly more reproducible than PCR-based methods, whilst others have concluded the opposite (see, Subkhankulova and Livesey, 2006, for an example).

### 4.7.3 Fidelity

Even if a protocol is reproducible, it is of little use if it is reproducibly wrong! It is therefore critical that any amplification procedure maintains the relative abundances of the original transcripts, otherwise any quantitative measures of gene expression levels may be highly suspect and misleading. It is difficult to know which criteria are best for assessing fidelity, but the commonly used methods for verifying whether genes are differentially expressed are quantitative or real-time polymerase chain reaction (RT-PCR), Taqman assays or even high-throughput sequencing, comparing results from the amplified and unamplified samples. Although considered the 'gold standard' by some, real time PCR will also be less reliable as the input RNA amount decreases since more cycles of amplification are needed. Of course any PCR reaction suffers from the same potential biases as exponential RNA amplification.

From the above discussion, it should be clear that it is not possible to provide a particular recommendation as to which protocol should be used. Techniques vary from lab to lab and it is clear that the methods chosen should reflect the experimental question being addressed. If only a few cells are available then exponential amplification is probably the method of choice but when thousands of cells can be obtained perhaps an IVT-based method. All we can do is urge users to take the utmost care and optimize each stage of an amplification reaction, the effort will pay off in terms of final data quality.

## 4.8 QUALITY CONTROL FOR FLUORESCENTLY LABELED SAMPLES

Microarrays are an expensive commodity and it is therefore important to ensure that the labeled product is as optimal as possible before hybridizing it to an array. Even if the RNA from the starting material was of good quality, and has been amplified effectively, it still may not yield reliable microarray expression data if the labeling was substandard. As with quality control of the extracted RNA, labeled cDNA can be evaluated by gel electrophoresis, spectrophotometry and the Agilent Bioanalyzer. The NanoDrop spectrophotometer has the ability to measure both the cDNA and dye concentration (http://www.nanodrop.com/nd-1000-apps-microarray.html). If the sample is electrophoresed on an agarose gel, the size distribution can be used as an indicator of integrity. Since the cDNA is already labeled with a dye, a phosphorimager with the appropriate emission filter is required to visualize the labeled target and it is essential to use a loading dye that does not autofluoresce at this wavelength. Any residual unincorporated dye-coupled nucleotides will appear as a low molecular weight band and the labeled target should give a smear of higher molecular weight. The same will be true when using the Agilent Bioanalyzer. The exact size distribution and label intensity that classifies a sample as good will, of course, depend on the platform being used. It is therefore difficult to make this assay absolutely quantitative, but over time it will be possible to decide which characteristics of the analyzed samples correlate with good signal and dynamic range after hybridization. These criteria can then be used as a screening tool to avoid wasting valuable arrays. An additional method that has been used as a quality control screen for cDNA synthesis, is to amplify a control gene by PCR using both $3'$ and $5'$ primer sets. Measuring the ratio of $3':5'$ products will give an indication of the percentage of full-length cDNA (Degenkolbe et al., 2005). All the above measures will avoid the costly mistake of wasting expensive microarrays on bad samples (Grissom et al., 2005) or even worse, wasting time due to inappropriate interpretation of spurious data.

## 4.9 SUMMARY

As we have described, there are plenty of potential pitfalls when preparing and labeling complex samples for hybridization to microarrays. However, careful attention with each reaction and at every purification stage means many problems can be mitigated. Of course, careful experimental design can identify potential problems and 'normalize' out any labeling or amplification biases, ensuring that high quality comparative gene expression data is delivered. While only the bravest would claim that arrays are quantitative, they are currently at best semi-quantitative, it is clear that comparisons between samples can deliver very precise measures of differences in transcript abundance between samples. Even relatively small fold changes involving genes expressed at relatively low levels, which may be of considerable biological relevance, can be reliably detected. It goes without saying that sample treatment goes hand in hand with high-quality microarray probe design, as described in Chapter 3, if robust experiments are to be reproducibly performed.

## REFERENCES

Alexandre I, Hamels S, Dufour S, Collet J, Zammatteo N, De Longueville F, Gala JL, Remacle J. (2001) Colorimetric silver detection of DNA microarrays. *Anal Biochem* **295:** 1–8.

Auer H, Lyianarachchi S, Newsom D, Klisovic MI, Marcucci G, Kornacker K. (2003) Chipping away at the chip bias: RNA degradation in microarray analysis. *Nat Genet* **35:** 292–293.

Baugh LR, Hill AA, Brown EL, Hunter CP. (2001) Quantitative analysis of mRNA amplification by in vitro transcription. *Nucleic Acids Res* **29:** E29.

Branham WS, Melvin CD, Han T, Desai VG, Moland CL, Scully AT, Fuscoe JC. (2007) Elimination of laboratory ozone leads to a dramatic improvement in the reproducibility of microarray gene expression measurements. *BMC Biotechnol* **7:** 8.

Che S, Ginsberg SD. (2004) Amplification of RNA transcripts using terminal continuation. *Lab Invest* **84:** 131–137.

Chomczynski P, Sacchi N. (1987) Single-step method of RNA isolation by acid guanidinium thiocyanate-phenol-chloroform extraction. *Anal Biochem* **162:** 156–159.

Dafforn A, Chen P, Deng G, Herrler M, Iglehart D, Koritala S, Lato S, Pillarisetty S, Purohit R, Wang M, Wang S, Kurn N. (2004) Linear mRNA amplification from as little as 5 ng total RNA for global gene expression analysis. *Biotechniques* **37:** 854–857.

Degenkolbe T, Hannah MA, Freund S, Hincha DK, Heyer AG, Köhl KI. (2005) A quality-controlled microarray method for gene expression profiling. *Anal Biochem* **346:** 217–224.

Endege WO, Steinmann KE, Boardman LA, Thibodeau SN, Schlegel R. (1999) Representative cDNA libraries and their utility in gene expression profiling. *Biotechniques* **26:** 542–550.

Grissom SF, Lobenhofer EK, Tucker CJ. (2005) A qualitative assessment of direct-labeled cDNA products prior to microarray analysis. *BMC Genomics* **6:** 36.

Gupta V, Cherkassky A, Chatis P, Joseph R, Johnson AL, Broadbent J, Erickson T, DiMeo J. (2003) Directly labeled mRNA produces highly precise and unbiased differential gene expression data. *Nucleic Acids Res* **31:** e13.

Huber M, Wei TF, Muller UR, Lefebvre PA, Marla SS, Bao YP. (2004) Gold nanoparticle probe-based gene expression analysis with unamplified total human RNA. *Nucleic Acids Res* **32:** e137.

Iscove NN, Barbara M, Gu M, Gibson M, Modi C, Winegarden N. (2002) Representation is faithfully preserved in global cDNA amplified exponentially from sub-picogram quantities of mRNA. *Nat Biotechnol* **20:** 940–943.

Kaposi-Novak P, Lee JS, Mikaelyan A, Patel V, Thorgeirsson SS. (2004) Oligonucleotide microarray analysis of aminoallyl-labeled cDNA targets from linear RNA amplification. *Biotechniques* **37:** 582–588.

Karsten SL, Van Deerli VM, Sabatti C, Gill LH, Geschwind DH. (2002) An evaluation of tyramide signal amplification and archived fixed and frozen tissue in microarray gene expression analysis. *Nucleic Acids Res* **30:** e4.

Klur S, Toy K, Williams MP, Certa U. (2004) Evaluation of procedures for amplification of small-size samples for hybridization on microarrays. *Genomics* **83**: 508–517.

Lieu PT, Jozsi P, Gilles P, Peterson T. (2005) Development of a DNA-labeling system for array-based comparative genomic hybridization. *J Biomol Technol* **16**: 104–111.

Naderi A, Ahmed AA, Barbosa-Morais NL, Aparicio S, Brenton JD, Caldas C. (2004) Expression microarray reproducibility is improved by optimising purification steps in RNA amplification and labeling. *BMC Genomics* **5**: 9.

Ohtsuka S, Iwase K, Kato M, Seki N, Shimizu-Yabe A, Miyauchi O, Sakao E, Kanazawa M, Yamamoto S, Kohno Y, Takiguchi M. (2004) An mRNA amplification procedure with directional cDNA cloning and strand-specific cRNA synthesis for comprehensive gene expression analysis. *Genomics* **84**: 715–729.

Petalidis L, Bhattacharyya S, Morris GA, Collins VP, Freeman TC, Lyons PA. (2003) Global amplification of mRNA by template-switching PCR: linearity and application to microarray analysis. *Nucleic Acids Res* **31**: e142.

Phillips J, Eberwine JH. (1998) Antisense RNA amplification: a linear amplification method for analyzing the mRNA population from single living cells. *Methods* **10**: 283–288.

Puskás LG, Zvara A, Hackler L, Van Hummelen P. (2002) RNA amplification results in reproducible microarray data with slight ratio bias. *Biotechniques* **32**: 1330–1340.

Rajeevan MS, Dimulescu IM, Vernon SD, Verma M, Unger ER. (2003) Global amplification of sense RNA: a novel method to replicate and archive mRNA for gene expression analysis. *Genomics* **82**: 491–497.

Salin H, Vujasinovic T, Mazurie A, Maitrejean S, Menini C, Mallet J, Dumas S. (2002) A novel sensitive microarray approach for differential screening using probes labeled with two different radioelements. *Nucleic Acids Res* **30**: e17.

Schlingemann J, Thuerigen O, Ittrich C, Toedt G, Kramer H, Hahn M, Lichter P. (2005) Effective transcriptome amplification for expression profiling on sense-oriented oligonucleotide microarrays. *Nucleic Acids Res* **33**: e29.

Shchepinov MS, Udalova IA, Bridgman AJ, Southern EM. (1997) Oligonucleotide dendrimers: synthesis and use as polylabeled DNA probes. *Nucleic Acids Res* **25**: 4447–4454.

Shchepinov MS, Mir KU, Elder JK, Frank-Kamenetskii MD, Southern EM. (1999) Oligonucleotide dendrimers: stable nano-structures. *Nucleic Acids Res* **27**: 3035–3041.

Smith L, Underhill P, Pritchard C, Tymowska-Lalanne Z, Abdul-Hussein S, Hilton H, Winchester L, Williams D, Freeman T, Webb S, Greenfield A. (2003) Single primer amplification (SPA) of cDNA for microarray expression analysis. *Nucleic Acids Res* **31**: e9.

Stangegaard M, Dufva IH, Dufva M. (2006) Reverse transcription using random pentadecamer primers increases yield and quality of resulting cDNA. *Biotechniques* **40**: 649–657.

Stears RL, Getts RC, Gullans SR. (2000) A novel, sensitive detection system for high-density microarrays using dendrimer technology. *Physiol Genomics* **3**: 93–99.

Stirewalt DL, Pogosova-Agadjanyan EL, Khalid N, Hare DR, Ladne PA, Sala-Torra O, Zhao LP, Radich JP. (2004) Single-stranded linear amplification protocol results in reproducible and reliable microarray data from nanogram amounts of starting. *RNA Genomics* **83**: 321–331.

Subkhankulova T, Livesey F. (2006) Comparative evaluation of linear and exponential amplification techniques for expression profiling at the single cell level. *Genome Biol* **7**: R18.

Sun Y, Fan WH, McCann MP, Golovlev V. (2005) Microarray gene expression analysis free of reverse transcription and dye labeling. *Anal Biochem* **345**: 312–319.

Szaniszlo P, Wang N, Sinha M, Reece LM, Van Hook JW, Luxon BA, Leary JF. (2004) Getting the right cells to the array: gene expression microarray analysis of cell mixtures and sorted cells. *Cytometry A* **59:** 191–202.

Van Gelder RN, von Zastrow ME, Yool A, Dement WC, Barchas JD, Eberwine JH. (1990) Amplified RNA synthesized from limited quantities of heterogeneous cDNA. *Proc Natl Acad Sci U S A* **87:** 1663–1667.

Wang E, Miller LD, Ohnmacht GA, Liu ET, Marincola FM. (2000) High-fidelity mRNA amplification for gene profiling. *Nat Biotechnol* **18:** 457–459.

Wiedorn KH, Kühl H, Galle J, Caselitz J, Vollmer E. (1999) Comparison of in-situ hybridization, direct and indirect in-situ PCR as well as tyramide signal amplification for the detection of HPV. *Histochem Cell Biol* **111:** 89–95.

Xiang CC, Chen M, Ma L, Phan QN, Inman JM, Kozhich OA, Brownstein MJ. (2003) A new strategy to amplify degraded RNA from small tissue samples for microarray studies. *Nucleic Acids Res* **31:** e53.

## FURTHER READING

Schena M, Shalon D, Davis RW, Brown PO. (1995) Quantitative monitoring of gene expression patterns with a complementary DNA microarray. *Science* **270:** 467–470.

Badiee A, Eiken HG, Steen VM, Løvlie R. (2003) Evaluation of five different cDNA labeling methods for microarrays using spike controls. *BMC Biotechnol* **3:** 23.

# Hybridization and Scanning

**Microarray Technology in Practice**

During the hybridization reaction, labeled targets interact with the tethered probes due to sequence complementarity. It is therefore critical to ensure that the hybridization conditions are appropriate, otherwise the measured signals will not be a true reflection of the levels of each RNA in the sample. Assuming that the probes on the array have been well designed so that they will not significantly cross hybridize to other target sequences (Chapter 3), then the chosen hybridization conditions must ensure that nonspecific hybridization is minimized while the specific signal is maximized. After washing, another step where optimization can pay dividends in terms of data quality, the arrays are scanned to acquire an image of the hybridization. The images are the primary raw data of a microarray experiment and careful attention to the scanning parameters and the subsequent image processing steps of spotfinding, intensity extraction and background estimation are essential if good quality data are to be obtained. In this chapter we review all of these stages with an emphasis on spotted arrays for gene expression. It is important to note however, that the methods and concepts are generally applicable to any array platform or application.

## 5.1 HYBRIDIZATION

The hybridization procedure involves several steps: first, the arrays are blocked to minimize background. Second, the labeled target is added to the array at a specific temperature to allow complementary sequences to anneal. Finally, the arrays are washed to remove unbound or weakly hybridizing material. It is necessary to evaluate and optimize each of the steps to obtain reliable data from relatively background-free arrays.

### 5.1.1 Blocking

Before hybridization, the array is treated to prevent nonspecific interactions between the nucleic acid in the labeled sample and the array surface. The treatment should effectively block the areas of the microarray surface that remain reactive after printing to eliminate fluorescent background from the hybridization reaction. Many blocking methods have been described and the particular method employed is mainly determined by the chemistry of the slide coating (see Chapter 3 for an overview of slide types). For example, poly-L-lysine arrays need to have exposed amines blocked prior to hybridization to prevent binding of labeled material and this is most often achieved with a mixture of succinic anhydride, 1,2-methyl pyrrolidinone and sodium borate. During the blocking, succinic anhydride reacts with and caps the amines before the excess DNA from the printed probes leaches from the spot area and binds nearby exposed lysines. After blocking, double stranded DNA arrays are boiled to denature the double stranded molecules and thus enhance their availability for hybridization. This step is obviously unnecessary for oligonucleotide arrays, since the molecules are already single stranded. Given the range of available substrates, a discussion of the different methods is not feasible especially since some commercial substrates require specific treatments. Our advice is to follow the manufacturer's recommendations in the first instance, evaluating the overall background and signal uniformity of scanned arrays carefully, before trying alternative methods.

## 5.1.2 Hybridization

Hybridization depends on the ability of the labeled target to anneal to a complementary probe strand tethered to the array. This occurs just below the melting temperature ($T_m$) of the target–probe duplex. The $T_m$ is the temperature at which half of the duplexes in a population will be dissociated. As we describe in Chapter 3, for the best data all probe sequences should be designed so that they have a similar $T_m$ for a single hybridization condition to be close to optimal for all probes on the array. Outside of the actual sequence, there are a number of factors that affect the $T_m$ and therefore influence the hybridization of the target to the probe (Relogio et al., 2002). The main variables are temperature, pH, monovalent cation concentration and the presence of organic solvents. If the hybridization temperature is too low then nonspecific probe–target binding will occur (specificity will decrease). In contrast, if the temperature is too high, hybridization of the specific targets will be reduced with a concomitant loss of signal (sensitivity will decrease).

The composition of the hybridization solution can be adjusted to increase the efficiency of the hybridization process and there are a range of solutions in common use. Again caution must be exercised since some chemicals may not be compatible with particular surface chemistries and consequently users are advised to start with slide manufacturer's recommendations in the first instance. In general, hybridization solutions contain a high concentration of salts, detergents, accelerants, and buffering agents. The most common components of solutions are:

(i) Sodium chloride and sodium citrate (SSC) as a source of monovalent cations, these interact mainly with the phosphate groups of the nucleic acids decreasing the electrostatic interactions between the two strands.

(ii) Formamide and dithiothreitol (DTT) are organic solvents that reduce the thermal stability of the hydrogen bonds formed between probe and target and the inclusion of such reagents allows specific hybridization at lower temperatures than in purely aqueous solutions.

(iii) Dextran sulfate is often added because it becomes strongly hydrated, effectively increasing the probe concentration thus increasing the rate of hybridization (Ku et al., 2004).

(iv) EDTA is divalent cation chelator that removes free $Ca^{2+}$ and $Mg^{2+}$ from the hybridization solution since these ions stabilize duplex DNA and can reduce specificity.

**(v)** Further components, which are typically added in order to decrease the chance of nonspecific binding of the target, include sonicated salmon sperm DNA, polyA, Denhardt's solution and tRNA to act as a carrier RNA.

Reaction time is another critical variable, since the true level of each RNA in the sample cannot be measured unless the hybridization of probe and target pairs has reached equilibrium (Dorris et al., 2003). This depends upon the relative concentration of the reacting species and since, to a first approximation, probes are all at a constant concentration and in excess with respect to the targets, the rate is determined by the concentration of the target in the complex sample. The hybridization reaction follows first order kinetics and, in short, high abundance transcripts hybridize with their probes quickly whereas low abundance sequences take longer. The relationship between concentration and hybridization is fairly well understood when both probe and target are in solution, however in an array experiment (or indeed any filter-based hybridization assay such as a Northern blot) where one of the molecules is tethered, things are less clear although there have been some empirical studies that indicate similar kinetics (Vernier et al., 1996). There is also an interesting observation that perfect match hybrids take longer to reach equilibrium than nonspecific binding (Dai et al., 2002) and this can, to a certain extent, be used to increase specificity. Pragmatically, hybridizations are typically carried out overnight to ensure that the hybridization is approaching equilibrium, but for higher throughput, hybridization times can be reduced (see below) with the caveat that attention must be paid to the relative contributions of perfect and nonspecific hybridization.

After the hybridization process, the arrays must be washed in order to remove unbound target and any target loosely bound to imperfectly matched sequences. As with Southern blotting, hybridization stringency can be controlled by washing (Korkola et al., 2003; Zhang et al., 2005). For standard arrays, initial washes should be carried out at, or close to, the required stringency conditions (temperature and salt concentration) defined by the probe $T_m$. This is generally at the same stringency as hybridization takes place. The stringent washes are usually followed with a final low stringency wash. Although the hybridization step is more efficient in small volumes (since sample concentration can be maximized) the washes benefit from much larger volumes, which more effectively dilute impurities and contaminants that may contribute to the background. As we suggested in Chapter 3, it is sometimes the case that arrays designed for one species are hybridized with samples from a related species. In this case both hybridization temperature and, more importantly, the washing conditions are altered. For example, one can reduce stringency by increasing the salt concentration of the washing solution and lowering the wash temperature. For good quality arrays, it is essential that both hybridization and washing is uniform across the array and that the surface is evenly dried before scanning. Any nonuniformity is likely to result in high and uneven background.

A good array design should provide control elements that allow aspects of the hybridization and washing conditions to be assessed. At the very least, arrays should contain negative control spots, that is probes for which no signal is expected, as well as spotting buffer and blank elements. As with all stages in the microarray process, pilot experiments should be carried out to ensure that the conditions are optimized to maximize signals from the positive spots and minimize signals from negative controls. Such pilots can be performed using spike-in controls; for example, the experiments we describe for slide-spotting buffer evaluation in Chapter 3. Commercial array providers go to some lengths to optimize hybridization and washing conditions for their platforms so when using such arrays the manufacturer's recommendations are obviously an ideal starting point. However, it is unwise to proceed with a set of experiments without first ensuring that the conditions in your own laboratory generate data that is within the quoted specifications and most commercial suppliers provide control samples that allow such evaluations.

## 5.1.3 Manual or Automated Hybridization

Once the hybridization conditions for a particular platform have been optimized to achieve a high signal to noise ratio, it is essential that these conditions are reliable and reproducible. Not least, establishing highly reproducible conditions can minimize the number of technical replicates needed for good quality data and obviously fewer replicates means reduced cost. Manual hybridization is the simplest and cheapest to implement. This generally involves placing a drop of hybridization solution containing the labeled probe onto the printed surface of the array and then carefully placing a glass coverslip on top so that the solution spreads evenly between the array surface and the coverslip, avoiding the introduction of air bubbles. Arrays can then be placed in an oven for incubation at the required temperature. However, this method can result in evaporation and produce high background, particularly at the edges of the array. To overcome this, many manufacturers provide hybridization cassettes where the array and coverslip can be sealed in a watertight humidity chamber and the whole chamber then submerged in a waterbath. The temperature distribution is likely to be more even across the array and a reservoir inside the cassette chamber ensures that the arrays are well humidified. While straightforward, manual hybridization is less suitable when many arrays need to be processed simultaneously, although methods exist to hybridize two arrays back to back, halving some of the effort required for manual hybridization (Ting et al., 2003). Not only is the manual procedure labor intensive, but also it can be prone to inter- and intraslide variability that is hard to precisely control. Another potential drawback of the manual approach is that there tends to be relatively limited sample diffusion. Microarray hybridizations are carried out with small volumes to increase the concentration of the labeled sample. However, the rate at which sample molecules can find their cognate probe sequences is restricted by the diffusion rate in the small space between the array and the coverslip. In a typical micro-array hybridization the diffusion constant is very small and it may consequently take molecules a long time to travel over the entirety of the slide. Lifter slips (e.g. LifterSlip™, Erie Scientific), rather than coverslips, have a raised edge design that provides better slide–coverslip separation. These facilitate more even dispersal of the hybridization solution but while this aids hybridization uniformity it does not greatly improve the diffusion properties.

A number of technological innovations designed to enhance the diffusion kinetics have been published, including continuous and discontinuous rotating microchambers (Vanderhoeven et al., 2005a), surface acoustic waves on a piezoelectric substrate (Wixforth, 2005), a shear driven system (Vanderhoeven et al., 2005b) and various other microfluidic devices (Yuen et al., 2002; McQuain et al., 2004; Peytavi et al., 2005). These all result in faster and improved mixing, in principal facilitating shorter hybridization times as well as increased spot intensities by improving sensitivity.

A number of commercial manufacturers have developed automated hybridization and washing instruments that practically implement some of these agitation technologies to facilitate higher throughput microarray processing (e.g. Advalytix, Biomicrosystems, GeneMachines, GE Healthcare, Genomic Solutions, Tecan). While such instruments have different specifications in terms of the number of slides processed and the technology employed, in general they aim to provide consistency and reproducibility with precise temperature control, reduced sample evaporation, uniform sample mixing and some form of agitation to improve diffusion kinetics. Automating the hybridization process removes some of the human factor, reduces hands-on time and is a step forward in the path to minimizing variation (Yauk et al., 2005). The use of a hybridization instrument can improve data quality by increasing reproducibility making it easier to compare data from different experiments, performed on different days or by different individuals and can even improve consistency between laboratories. Of course this will only be true if the instrument is properly calibrated and maintained. In addition, users should be aware that it is highly unlikely that protocols developed and validated for manual hybridization will be directly transferable to an automated instrument or that procedures for one instrument will be transferable to another. As ever, evaluation is essential, but in our view the potential benefits, especially if many arrays will be processed, indicate that the payoff in terms of data quality justify investing in an instrument.

## 5.2 DATA ACQUISITION

### 5.2.1 Scanning

Once the labeled sample has been hybridized to the array it must be scanned. Clearly the objective is to measure the amount of labeled target that is bound to the immobilized probe using the reasonable assumption that the intensity of fluorescent light varies with the strength of hybridization. A microarray scanner uses a light source to excite the fluorophores present on the sample molecules and then detects the emitted light. The fluorescence information is captured in the form of a digital image, most typically separate 16-bit tiff (tagged image format) files for each wavelength scanned. Each image is composed of a matrix of pixels where each pixel represents the fluorescence intensity of a small area of the array. It is important to emphasize again that this digital image represents the 'raw data' from which the fluorescent signal of each array element will subsequently be quantified and the expression level of each target inferred.

## 5.2.2 Scanner Purchasing Advice

The microarray scanner market is constantly changing, especially as ultrahigh density arrays are becoming more widely available and there are many instruments to choose from. Scanners are expensive pieces of equipment and while cost may be a major deciding factor in any purchase, there are other key considerations that should be taken into account when deciding on an instrument. As with most pieces of equipment, an inexpensive model may suffice initially, but it should be borne in mind that even a low-end scanner should at least have the potential for upgrading so that the machine does not become obsolete too soon. The availability of adequate service and support contracts should be taken into account since the delicate optics are not really user serviceable. The majority of scanners on the market are designed for the standard format of 1 in. × 3 in. glass slides, which are typical for most gene expression arrays. However, some scanners are more flexible and accommodate other array formats such as microtitre plates. The maximum scan area should at least be large enough to scan the complete area of any arrays you plan to use: it is surprising how many instruments are restricted in this respect. The scanner must also be sufficiently sensitive to detect the expected signal intensities, for example, if the signal intensities are expected to be particularly low, or if a high dynamic range is anticipated, it would be prudent to establish this prior to obtaining a particular instrument. The resolution of an instrument may become an issue since, as we allude to above, very high density arrays have higher spot densities and concomitantly smaller feature sizes, requiring a higher resolution than standard spotted arrays to differentiate individual spots. In some cases, for example Affymetrix GeneChip technology, the scanner choice is restricted due to the closed nature of the system.

One important feature is the scanner control software, which should be as flexible and user-friendly as possible. While most scanners come with similar basic software, some have additional features. For example, it may be advantageous to have at least a simple graphical user interface that facilitates scan area selection. More importantly, it is helpful to be able to perform a preview scan in real time, during which scan parameters can be adjusted and optimized before performing the final high resolution scan. Control of individual channel excitation and light acquisition levels will also provide a greater degree of control.

The level of throughput and automation required will also be a purchasing consideration. With the simplest scanners each slide must be manually loaded and the scanning initiated. This is perfectly adequate for relatively small numbers of slides where the operator is on hand to save each scan before feeding in the next array, but it is not optimal for high throughput. Optional autoloaders, in the form of stacks or carousels with capacities up to 48 arrays, are available for several instruments. In such cases, barcode readers may be advantageous for keeping track of each array. Some scanner software is flexible enough to allow multiple scans of the same array, which may be very useful for extending the dynamic range (see below). The time taken to scan an individual array may also need to be taken into consideration. For example, some scanners can take over half an hour to acquire two channels from an array. If high throughput is required or very rapid image acquisition desirable, for example, in a clinical setting, a faster instrument may be needed. The number of separate channels that can be acquired may also be a potential issue. Since two-color microarrays are the most common format, most scanners are equipped to read Cy3 and Cy5. However, some assays may utilize additional channels, in which case the scanner will need additional emission filters to distinguish the signals and/or the ability to excite at multiple wavelengths. If these are anticipated future needs, then it would be prudent to obtain an instrument that at least has an upgrade option.

## 5.2.3 Fluorescence Detectors

Scanners fall into two general categories, laser and charge-coupled device (CCD)-based, according to the type of excitation and detection technology they utilize (Figure 5.1).

### 5.2.3.1 *Laser Scanners*

These use narrowband laser illumination to excite the fluorophores and capture the resulting fluorescence with a photomultiplier tube (PMT) detector. In general, a single-wavelength laser beam of a few microns in diameter is scanned back and forth across the array to excite an area representing a single pixel. Light that is emitted from the excited fluorophore travels back through the excitation lens and is focused on the PMT detector. The PMT amplifies the signal from each photon of light, which is converted into a digital value used to create the image representing signal intensity at each pixel position. Most laser scanners acquire images for each channel consecutively and as a consequence each channel must be registered before accurate comparisons can be made. Any slight movement in the position of the array or difference in the starting point of the scanning head can lead to misregistration (see below).

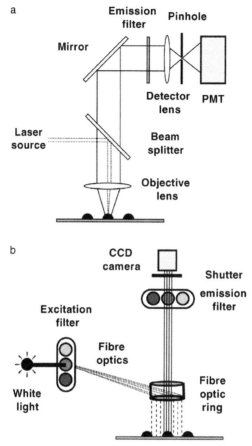

**FIGURE 5.1**   Microarray scanners: (a) laser scanner. The laser light (dotted lines) travels via a beam splitter through an objective lens where it is focused at the array surface. Emitted light (solid lines) is reflected from a mirror, through an emission filter and focused through a pinhole into the photomultiplier tube. (b) CCD scanner. High intensity white light (dotted line) is passed through an excitation filter to produce light of a defined wavelength. It travels via fibre optic guides through a fibre optic ring to illuminate an area of the slide surface. Emitted light (solid line) travels through an emission filter where it is collected by the CCD camera. The shutter controls the light admitted into the camera.

## 5.2.3.2 CCD Scanners

The excitation source for a CCD detector is broad-spectrum white light, such as that obtained from a xenon or mercury lamp. The excitation wavelength is selected by filtering the white light into a narrower wavelength range. The filtered light illuminates a large area of the array and the fluorescent emission from the entire field of view is collected by a stationary CCD. An imaging aperture or shutter is opened for a predetermined exposure time to allow the

CCD to collect sufficient light from the array to create a representative image. A CCD chip is an array of semiconductor devices, or camera pixels. Each camera pixel stores an electrical charge generated by light from the emission filter. This charge is proportional to the intensity of the light, or number of photons, that reaches the semiconductor. Electronic circuitry on the camera converts pixel electron counts from the CCD chip into a digital signal that represents the intensity of each pixel. A CCD array chip that is large enough to capture the whole area of a microscope slide in a single image would be ridiculously expensive, therefore CCD scanners tend to capture multiple images from different areas of the array and then stitch them together to create a single image. A potential disadvantage of this approach is that any imprecision in the stitching, or photobleaching due to repeated exposure of overlapping regions, will result in inaccurate and uneven signal quantification. An advantage for multichannel image scanning is that data from all wavelengths are simultaneously collected and therefore there are no misregistration issues.

### 5.2.4 Sensitivity, Dynamic Range

Typically, laser scanners deliver more excitation photons to the sample than CCD-based instruments: in effect more emission photons per pixel can be collected in a given amount of time. This is because the broadband excitation source is less efficient than a laser and, in addition, CCD detectors are much less sensitive than PMTs. Therefore CCD cameras must integrate for a longer period to capture the same amount of fluorescence signal as a laser scanner, typically seconds or minutes compared to the microseconds for laser illumination. Mitigating this decrease in speed, the CCD captures a larger area of the slide in a single exposure and thus requires fewer acquisition events. For optimum system performance, the detector must have a good linear range. This is the range of input signal intensities over which the detector can accurately measure change, so that a given degree of change in input signal generates the same degree of change in output signal. PMTs typically have a range of four orders of magnitude. The signal from a PMT is converted to a 16-bit digital value, which means that intensity values can be divided into $2^{16} = 65\ 535$ steps, covering about 4.5 orders of magnitude. The magnitude of signal output from a PMT is adjusted by changing the voltage applied to it, which determines the amplification of the signal, known as the gain. PMTs have an optimum working range over which the signal response is most linear. If the gain is set below this range, not all photons will be converted into a signal. Conversely, if the gain is too high, noise becomes more significant. The linear range of a CCD detector is typically about 3.5 orders of magnitude because 12-bit digital resolution is employed ($2^{12} = 4096$). The signal intensity range of a CCD is adjusted by changing the exposure time. In contrast to a PMT, a CCD array response is generally linear with increasing integration time. However, signal generated by random electrons flowing through the device in the absence of light (dark current), increases proportionally with exposure and so leads to an increase in background signal. Since the linear range of the fluorescent dyes used in microarray studies is usually no more than 2.5 orders of magnitude, then although the linear range for a CCD camera is narrower than that of a PMT, it is sufficient to capture the full range of signals on a typical microarray. Whatever type of scanner is being used, optimal dynamic range is usually achieved by adjusting the settings (PMT gain or exposure time) so that the brightest spots are just below the saturation level of the detector.

## 5.2.5 Pixel Size

In laser scanners, the size of a pixel is determined by how far the scanning beam moves each time it collects a data point, the distance between data points is known as the pixel resolution. A second important component of a scanner's resolution is its optical resolution: the size of the area measured for each pixel. Pixel resolution is often finer than the optical resolution (beam diameter) of the scanner. The light intensity of a laser is not constant across the diameter of the beam but has a Gaussian distribution with a peak of intensity in the centre, falling away to either side. Consequently, when the laser beam is larger than the pixel, its fluorescence signal will contain a significant contribution from pixels adjacent to the pixel of interest. The downstream effect of this effect is to mask noise, which may make images look prettier or smoother due to multipixel averaging, but it can no longer be considered truly 'raw' data. For example, if a scanner has a beam diameter of 10 $\mu$m there is no benefit derived from pixel resolutions lower than 10 $\mu$m since images obtained at lower pixel resolutions will contain more pixels but the information content will be the same, because no features of the microarray smaller than the beam diameter can be revealed by the small pixels. It is generally recommended that a pixel size no more than about one tenth the size of each feature on the microarray is used. For example, microarrays with spots of 100 $\mu$m should be scanned with a pixel size of 10 $\mu$m or less.

## 5.2.6 Confocal Versus Nonconfocal Optics

Confocal imaging was originally developed for taking thin optical sections of thick samples without the need for mechanical sectioning. This is achieved by creating a very narrow focal depth so that any signal from beyond the focal plane is ignored. Repeated scanning at different depths generates multiple images that can be reconstructed to create a three dimensional image of the sample. While popular in the early days of microarrays and still widely available, confocal optics is now considered superfluous for scanning microarrays. This is mainly due to the fact that most of the background or interfering signals come from nonspecific binding to the slide surface, which is in the same focal plane as the specific spot signal. Additionally, slide surfaces can have slight fluctuations in thickness so the narrow focal plane of confocal optics can result in different areas being in and out of focus with a consequent loss of signal. In this respect microarray substrates with optically flat surfaces are available and therefore able to minimize these inaccuracies. Some studies indicate that the use of a confocal scanner design can result in increased noise levels as well as sizable signal reduction, ultimately producing an image with a lower signal to noise ratio (Cheung et al., 1999).

### 5.2.7 Sequential Versus Simultaneous Scanning and Channel Crosstalk

The most popular microarray experimental format cohybridises two samples, each labeled with a different fluorophore, to the same array. The most commonly used fluorophores are Cyanine 3 (Cy3), with peak absorption at 550 nm and emission at 570 nm, and Cyanine 5 (Cy5), with peak absorption at 649 nm and emission at 670 nm (Figure 5.2). One of the major reasons for using these particular fluorophores is that their absorption and emission spectra are sufficiently different from one another. Arrays with more than one fluorophore can either be scanned at each wavelength simultaneously or sequentially. The advantages of simultaneous scanning are that the imaging time is reduced and there is no need for image registration. Two images will be produced, one for each wavelength, but the pixels in each image will be perfectly aligned. The downside is that the two fluorophores must be spectrally separable to avoid crosstalk. Crosstalk occurs when light emitted from one channel is detected in another channel and usually occurs because emission spectra tend to be rather broad with a significant amount of light emitted far from the peak emission wavelength. The emission spectrum of Cy3 overlaps slightly with the absorbance wavelength of Cy5, so fluorescent signal from the Cy3 channel can potentially be detected in

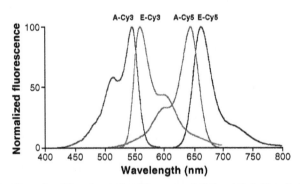

**FIGURE 5.2**   Absorption (A) and emission (E) spectra of Cy3 and Cy5. Note that although the emission spectra are well separated, there is a slight overlap between the emission spectrum of Cy3 and the absorbance wavelength of Cy5.

the Cy5 channel. A common way to avoid this crosstalk is to use emission filters that block any light outside of the desired wavelengths, however, narrowing the emission bandwidth inevitably leads to a loss of signal. The problem of crosstalk is reduced by sequential scanning. In this case one channel is imaged first followed by the second and any subsequent channels. This generates two or more separate images that need to be aligned or 'registered' for subsequent analysis. Imperfect image registration means that pixels used to measure the signals in the separate channels may not be in exactly corresponding positions and as a consequence there is a loss in accuracy. Misregistration makes analytical methods that compare channels at the level of individual pixels much more difficult.

## 5.2.8 Multichannel or Single Channel

Some laboratories and commercial array providers, such as CodeLink or Affymetrix, use just one sample and one fluorophore on each microarray. While this means two arrays are required for a single comparison, and a range of internal controls are necessary in order to make comparisons between chips, it eliminates crosstalk and image registration issues. Many platforms use a dual-channel approach, comparing two different samples directly on a single microarray, cutting down on the number of arrays needed (and hence cost) and potentially increasing the statistical power of an analysis. In theory, there is no limit to the number of different channels that can be scanned simultaneously as long as each channel is spectrally separated by appropriate combinations of excitation and emission filters. This makes it possible to compare multiple samples directly on the same array; for example, using additional channels for spot quality controls (Hessner et al., 2003a, 2003b; Waukau et al., 2003) or as a reference channel (Maratou et al., 2004). In reality it is not so easy to find fluorophores with significantly different spectra that also fulfill the other biochemical properties that are required, such as photostability, efficiency of incorporation during labeling, etc. In theory, three-color loop design experiments should be more efficient than two-color loop designs in terms of reduced number of arrays required for comparing the same number of samples. However, it has been shown that the improved efficiency of the design is somewhat offset by a reduced dynamic range and increased variability in a three-color experimental system (Woo et al., 2005). Four color combinations have been reported for microarray hybridizations, using Alexa 488, Alexa 594, Cy3 and Cy5, but some crosstalk was detected, in the worst case a 13% crosstalk from Cy5 when scanning at the Alexa 594 settings (Staal et al., 2005). Such crosstalk can obviously significantly influence differential gene expression analyses, especially if the signals for Cy5 and Alexa 594 differ drastically within a spot. Although there have been other reports of using three different dyes to label three different targets and hybridizing them simultaneously (Forster et al., 2005), the dual and single channel systems remain the most popular formats for array studies.

## 5.2.9 Photobleaching

Photobleaching is the destruction of a fluorophore by high intensity light, resulting in reduced signal intensity and consequent underestimation of expression level. Microarray scanners use intense light sources and the greater the applied light intensity, either by the use of high laser power or increasing the duration of illumination (laser dwell or CCD exposure times), the greater the photobleaching effect. It is therefore advisable to minimize photobleaching as much as possible by limiting both light source power and pixel dwell times. Repeated scans will result in successively reduced signals across the array. Scanning a selected area, for example while adjusting and optimizing scan settings during a preview scan, will result in more photobleaching in that area. This can be a problem if the whole array is subsequently scanned because of the significant intra-array variation in signal intensity that will inevitably occur. Another issue that should be remembered is that dyes are not equally photostable and therefore differ considerably in their susceptibility to photobleaching. This becomes an issue in multichannel experiments where photobleaching in one channel, but not the other, can lead to inaccurate signal ratio measurements.

## 5.2.10 Calibration

Scanners are precision instruments and since they are used as analytical tools their performance must be consistent in order to compare results over time. Compared to the large amount of variance coming from biological samples and upstream technical processing of the microarrays, scanners contribute very little variation. Nevertheless ensuring that the scanner is used at the optimum settings will aid the separation of experimental error and true biological variation. To ensure that the various optical, mechanical and electrical components of the scanner are working reproducibly over time it is advisable to calibrate the scanner against a known standard. The standard should be fluorescent materials that absorb and emit light at the same wavelengths as the excitation and emission bands of the scanner. The standard should be photostable so that it always gives the same output signal even after prolonged use. If these criteria are met then any difference in calibration signal from the original measurements will indicate a change in scanner performance. Calibration slides are commercially available from a number of sources, including Paragon, Invitrogen and Full Moon BioSystems (FMB). As an example, the FMB slide contains two separate array blocks of Cy3 and Cy5 dilution series. Each block consists of 28 sets of twofold dye dilutions along with three sets of blanks and one set of position markers. Each column contains 12 repeats of each sample. This, or other similar calibration slides, can be used to quantitatively analyze the dynamic range and detection limit of a scanner, detect and analyze variations in performances among different microarray scanner, detect channel crosstalk, verify laser alignment and determine scanning uniformity.

Microarray-based measurements assume a linear relationship between fluorescence intensity and the dye concentration. In reality, however, the calibration curve can be nonlinear (Shi et al., 2005). By scanning a calibration slide under different PMT gains, Shi et al. (2005) evaluated the calibration characteristics of Cy5 and Cy3, demonstrating that:

**(i)** The calibration curve for the same dye under the same PMT gain is nonlinear at both the high and low intensity ends.

**(ii)** The degree of calibration curve nonlinearity depends on the PMT gain.

**(iii)** The Cy5 and Cy3 PMTs behave differently even under the same gain.

**(iv)** The background intensity for the Cy3 channel is higher than that for the Cy5 channel.

The impact of such scanner characteristics on the accuracy and reproducibility of measured mRNA abundance was demonstrated, with the conclusion that it is preferable to scan microarray slides at fixed, optimal gain settings under which the linearity between concentration and intensity is maximized. A method for calculating ratios based on concentrations estimated from the calibration curves was proposed for correcting ratio bias. Such studies clearly demonstrate the utility of taking the time to properly calibrate scanners.

### 5.2.11 Multiple Scans and Data Merging

Since the relationship between fluorescence intensity and dye concentration is nonlinear, scanner gain settings with maximal linearity should be used. This creates a dilemma: with laser scanners, higher PMT settings are generally recommended due to improved signal-to-noise ratios with low-intensity spots. However, this is offset by saturation of high-intensity spots because spots containing saturated pixels are either discarded from subsequent analysis, thus wasting valuable data, or have less reliability. The alternative is to use a lower PMT setting, avoiding saturation, but potentially losing spots with low signal since these are now below the level of detection. To overcome this dilemma several dynamic range extension methods have been developed (Romualdi et al., 2003; Dudley et al., 2002; Dodd et al., 2004; Lyng et al., 2004; Bengtsson et al., 2004). In such approaches the same array is scanned several times at different PMT gains and the linear ranges from each scan combined to generate an extended linear scale.

### 5.3 FINDING SPOTS

The raw microarray images produced by the scanner need to be processed so that the fluorescence intensity associated with each arrayed spot in each channel can be determined, generating a table of values linking intensity values to each array feature. This process can be viewed as a series of key stages:

(i) Grid placement or addressing: where preliminary spot locations within the image are determined.
(ii) Spot segmentation: using the available grid information, spots are individually segmented into two classes, foreground and background.
(iii) Pixel intensities are extracted from the areas defined during the segmentation process.

In principle, the spotfinding process should be simple – we can all look at a microarray image and see where the spots are – in practice it is a difficult task and certainly not adequately solved at present. An ideal array will have perfectly round spots of identical size arranged in an exactly defined grid pattern. Signals will be bright and homogeneously distributed within each spot. There will be no, or low and homogeneous, background. Such arrays can be simulated in the computer but not even the best manufacturing process will produce such perfection. In reality the position, shape and size of spots, as well as the background fluorescence, can fluctuate significantly across the array, making the accurate identification of spots challenging. Such fluctuations can result from the array production process, from postprint processing or during hybridization. In addition, environmental artefacts such as dust may be introduced at any stage during the process and also cause problems, typically producing high fluorescence. All of these issues present a challenge to the spotfinding software.

### 5.3.1 Grid Placement/Spot Addressing

The detailed layout of a microarray is determined by the arrayer configuration, therefore, when the array is printed or synthesized the two-dimensional coordinates of each spot are in principle known with a precision of a few microns. Most spotted microarrays are arranged as a series of subgrids within an overall metagrid. Each subgrid is generated by a single printing pin and the metagrid is defined by the number and distribution of different pins in the printhead. For example, a single 18 240 element metagrid could be composed of 12 rows and 4 columns of subgrids (48 pins), each with 20 rows and 19 columns of spots (380 spots/pin). Since the actual spot positions may fluctuate slightly from the model of the array generated by the printer, the scanned image must be 'addressed' or matched to the model. In order to place the spotfinding grid accurately a number of parameters need to be determined and software tools approach the problem differently. Most typically, these parameters are: the overall position of the array within the scanned image, the distance between the rows and columns of each subgrid, translation of individual subgrids (resulting from slight variations in print-tip positions), distance between the rows and columns of spots *within* each subgrid, small translations of individual spots.

Other parameters that may be taken into account by the gridding software are misregulation of the channels, rotation of the whole array within the image and skew in the array. Clearly the gridding/addressing process must be accurate in order for the downstream spot segmentation steps to be reliable. The gridding step is usually automated (see Figure 5.3), since this is more efficient than manual gridding and ensures that if the gridding process is repeated then the outcome will be the same. However, good software should allow for additional, quick manual adjustment of the grid points if, as often occurs, the automatic method incorrectly identifies some spot positions.

## 5.3.2 Spot Detection and Image Segmentation

After positioning the grids the next step is segmentation, a general term in image analysis used to describe techniques that distinguish objects of interest from the rest of an image. In the case of microarrays, this means distinguishing pixels which are part of a spot from the background. The commonly used premise is that regions where probe is located have different brightness characteristics in comparison to nonspot regions without probe. It sounds simple to separate the foreground spot signal from the background but in reality it is a complex issue due to the imperfect nature of microarray images. For example, the commonly observed doughnut artefact, where a spot is devoid of hybridization in its centre presents a challenge. Are the nonhybridizing pixels included when the final spot intensity is calculated or can these pixels be identified and removed while leaving the 'good' pixels? Ideally, these and other imperfections should be automatically recognized in the image analysis with the estimated intensities adjusted to take account of them. Segmentation is an active research area and

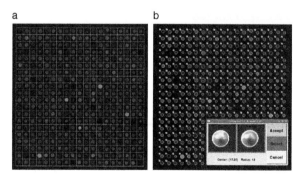

**FIGURE 5.3**  Example of automated gridding and spotfinding by the Dapple software (Buhler et al., 2000). (a) For this particular array, a grid with 20 rows and 19 columns is automatically centred over the spots, but the grid position can also be manually adjusted. (b) In the next step each spot within the grid is segmented, with the foreground pixels bordered by a white circle. The inset shows one particular spot in two channels with an automated accept or reject status. The status can be manually overridden.

new segmentation algorithms are appearing all the time. It is consequently not feasible to describe all of them in detail, rather we provide several different examples that illustrate the basic principles underpinning current spotfinding approaches. The field has been well reviewed (Qin et al., 2005).

### 5.3.3 Fixed Circle Algorithm

This segmentation algorithm was one of the first applied to microarray image analysis, it relies on the assumption that all the spots on the array are circular and with a constant radius. After gridding, a circular mask of a fixed radius is placed over each spot location. Everything within the mask is counted as spot foreground and everything else as background. While the algorithm is widely available as an option in some spotfinding applications, due to the variability reasons we have described it is somewhat crude and requires considerable manual intervention.

### 5.3.4 Adaptive Circle

As the name suggests, adaptive circle algorithms allow for some flexibility compared with the traditional fixed circle. The approach still assumes that all spots are circular, but the radius for each spot is estimated separately. Some spotfinding software allows the user to manually adjust the radius for each spot, which is extremely time consuming and laborious considering that there are typically thousands of spots on each microarray, however automated versions exist. In the Dapple software for example, the radius of each spot is estimated using edge detection (Buhler et al., 2000). The outline of each spot is enhanced using the second-difference approximation of Laplacian and then the radius of a circle matching the given enhanced edges is identified with matched filtering. Changing the radius of the circle facilitates more accurate intensity quantitation since, in principle, nonhybridizing pixels out the spot can be excluded by reducing the radius while increasing radius can accommodate all the hybridizing pixels in larger spots.

### 5.3.5 Seeded Region Growing

The seeded region-growing algorithm (Adams and Bischof, 1994) was first used for microarray segmentation in the Spot software package. The algorithm segments each spot by iteratively growing separate regions with respect to a set of predefined seed points that provide a starting point for the segmentation. During each iteration, the algorithm includes the most homogeneous pixels from the immediate neighborhood in the segmented regions. The algorithm aims to make the final segmented regions as homogeneous as possible given the connectivity constraint. The region originating from the foreground seeds is considered as the spot foreground, and the region originating from the background seeds as the background. The benefit of this approach over the adaptive circle method is that noncircular spots can be more accurately defined.

### 5.3.6 Mann–Whitney

The Mann–Whitney test is a nonparametric statistical test for assessing the statistical significance of the difference between two distributions. A segmentation algorithm for accurately separating foreground and background based on this test is implemented in the BASICA tool and the widely used QuantArray software (Chen et al., 1997; Hua et al., 2004). First, a circular target mask is selected, which encloses all possible foreground pixels and separates them from the known background. Second, a set of random pixels from the background are compared against a selected number of pixels with the lowest intensity within the foreground target mask using the Mann–Whitney test. If the difference between the two sets is not significant, then the algorithm discards a predetermined number of pixels from the target area and selects new pixels from the target area. This process is repeated until the two sets differ significantly from each other. The pixels remaining within the target mask after the final iteration are considered to be the spot foreground. Again spots with irregular shapes are more accurately defined and, in addition, this approach is able to discard nonhybridizing pixels from within doughnut spots.

### 5.3.7 *k*-Means Segmentation

The $k$-means segmentation algorithm is based on traditional $k$-means clustering and was first adapted for use on microarray images by Bozinov and Rahnenführer (2002). The segmentation result is derived using information from both channels simultaneously. That is, for each spatial location, the intensities from both channels are combined as one feature vector. The number of cluster centers, $k$, is set as two, since the segmentation needs to divide the image into two classes (foreground and background). The pixels with the minimum and maximum intensities are selected as the initial cluster centers. All data points are then assigned to their nearest cluster center according to Euclidean distance. New cluster centers are then calculated and the algorithm iteratively repeated until the cluster centers stay unaltered. As with the Mann–Whitney test, this method can deal with irregular and doughnut spots but has the added advantage that both channels are dealt with simultaneously. This improves comparative experiments, particularly in cases where there is expression in one channel but not the other since the spot area can be accurately defined for very low intensity spots.

### 5.3.8 Hybrid *k*-Means

The hybrid $k$-means algorithm (Rahnenführer and Bozinov, 2004) is an extended version of the original $k$-means segmentation approach and combines the advantages of both spot shape methods and intensity histogram methods. The algorithm uses repeated clustering to increase the number of foreground pixels. As long as the minimum amount of foreground pixels is not reached, the remaining background pixels are clustered into two groups and the group with the higher intensity pixels are assigned to the foreground. In addition, the number of outlier pixels in the segmentation result is reduced with mask matching. It estimates the expected spot shape and is used to filter the data, improving the results of the cluster algorithm. As a consequence of the filtering step, pixels are divided into three groups, namely foreground, background and deletions. This allows a separate treatment of artefacts and their elimination from further analysis.

### 5.3.9 Inference-Based Approaches

Markov random field (MRF) modeling has been adapted for microarray spot segmentation by Demirkaya et al. (2005). In this approach, the spot foreground and background intensities are modeled as exponential distributions. In addition to the intensity information, spatial information is also taken into account by modeling the neighborhood pixel labelings. A disadvantage of this approach is that an initial classification into spot foreground and spot background is needed to provide the basis for the MRF segmentation, therefore the final spot intensity is affected by the initial assignment. The initial classification can be done manually or via an automated thresholding algorithm. A related approach is available with the model-based segmentation algorithm (Li et al., 2005). This is a two-step method for spot segmentation involving model-based clustering of pixel values and spatial extraction of connected components. The first step applies model-based clustering to the distribution of pixel intensities, using the Bayesian Information Criterion (BIC). This is followed by the identification of large spatially connected components in each cluster of pixels. These inference-based approaches can deal with common artefacts such as doughnuts and irregular spot morphologies very well.

### 5.3.10 Matarray

The Matarray package provides an integrated approach to the spotfinding and segmentation approach. As with the inference methods describe above, both intensity and spatial information are combined to improve spot detection. Of particular interest, the authors introduce and iteratively apply their algorithm to improve performance. In addition they introduce a quality scoring system that includes signal-to-noise ratio, local background level and variation, spot size, and saturation status to permit selection of only the most reliable spots. The developers suggest that the use of quantitative spot assessment criteria substantial improves final data quality. The method has been adapted to utilize a third channel, in the published example, cDNA amplicons were generated using fluorescein-labeled primers but in principal any third channel could be used (see below).

## 5.4 METHOD COMPARISON

Inaccurate segmentation was originally thought to be a minor contributor to data extraction error compared to errors associated with background estimation methods (Yang et al., 2002). However, more recently it has become clear that the segmentation method can have a large influence on the accuracy of microarray data (Ahmed et al., 2004). Figure 5.4 demonstrates that different methods for spot segmentation lead to different pixels being included as foreground signal. There is no universally applicable segmentation method that works for all images and no segmentation technique is perfect, so how can one decide which algorithm to implement? Unfortunately, despite the variety of available segmentation methods, very few studies have extensively compared the performance of different algorithms (Korn et al., 2004). Lehmussola et al. (2006) report an evaluation of the performance of nine different microarray segmentation algorithms. They used simulated microarray images, thus allowing segmentation error to be evaluated at single pixel level. Overall, the results of their study show considerable differences between the performance of the different algorithms in terms of pixel identification and the derived 'gene expression' level. Algorithm performance depends upon image quality, with some algorithms performing extremely well on high quality images but very poorly when image quality is reduced. The most efficient algorithms were the $k$-means and hybrid $k$-means, both of which accurately segmented images with high correlations between replicates. The good performance is primarily due to effective detection of low intensity spots and spots with abnormal shape. However, the algorithms score less highly with images containing high intensity artefacts such as dust speckles. The simple fixed circle algorithm, although incapable of dealing with any spot to spot shape variations, performs robustly and is particularly beneficial

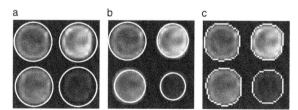

**FIGURE 5.4**   Illustration of the different results obtained from a selection of spot segmentation methods. All pixels within the white line are included in the foreground measurement. (a) Fixed circle method, (b) adaptive circle, and (c) watershed segmentation.
*Source*: Hovatta et al. (2005).

in cases where noise levels are high. The $k$-means and fixed circle algorithms demonstrate an interesting trade-off apparent in segmentation: the $k$-means algorithm, relying solely on intensity information, can identify spots of any shape but is extremely sensitive to high intensity artefacts, whereas the fixed circle, although constrained by the assumption that the spots are circular, is insensitive to all other errors except shape variation.

Interestingly, the adaptive circle and Mann–Whitney methods gave rather unstable results in this evaluation. The weak performance of the adaptive circle algorithm can be attributed to the failure of matched filtering to detect weak spot edges. The poor performance of the Mann–Whitney algorithm is probably due to the fact it involves an element of randomness and therefore results for the same image differ with repeat segmentations. This is undesirable, since one of the main motivations behind automated image analysis is output consistency when the same image is analyzed multiple times.

The performance differences between different segmentation algorithms should make it clear that the choice of software is an important factor that needs to be carefully considered to obtain the very best quality microarray data. However, one size does not fit all and a method that performs very well with one type of array may be less effective for another. For example, if doughnuts are a common feature with a particular array then one of the $k$-means or inference methods may be best but for arrays where spot morphology is even and highly reproducible (for example with some in situ synthesized arrays) then a fixed or adaptive circle may perform better – evaluation with the arrays being used is strongly advised. Since the chosen segmentation algorithm influences final data interpretation, it is important to report the algorithm used when submitting data to a public repository, as suggested by the MIAME standard (Brazma et al., 2001). It is difficult to overcome the errors originating from faulty segmentation, so it is perhaps even better to archive the raw images for alternative future analysis.

## 5.5 DATA EXTRACTION

Once the pixels which constitute foreground and background have been distinguished for each spot, the next step in the image analysis pipeline is to calculate the foreground and background intensities, and calculate spot quality measures. Histogram-based methods measure spot foreground and background values directly, whereas other methods only segment the image and require a subsequent intensity calculation step. In most cases the microarray analysis packages define foreground intensity as the mean or median value of pixels within the segmented spot mask. The mean value is generally considered to be more reliable since it is less prone to distortion from small numbers of pixels with outlying intensities in the signal distribution.

The methods available for background calculation, a measure of the nonspecific hybridization of labeled sample to the substrate, are much more varied with a range of techniques employed by different image analysis packages. Some software allows the user to choose between different methods, other packages have a fixed method, while some commercial packages use proprietary algorithms that do not disclose the exact methods employed. A typical method for background calculation takes the median of the values in selected regions around the spot mask. These regions can differ in their location (Figure 5.5).

Some packages consider background to be all of the pixels that are not within the spot mask but are within a square centered at the spot centre (ScanAlyze). The median value of these pixels is used as the local background measurement. Rather than a square, the area between two concentric circles can be used to define the pixels for background calculation (QuantArray). Since some segmentation methods are less accurate at effectively defining spot boundaries, ignoring the pixels immediately surrounding the spot makes the background estimation less sensitive to the performance of the segmentation procedure. To ensure that the background calculation is as independent as possible from the segmentation result, a set of pixels that are equidistant from four surrounding spots can be selected (Spot). These regions are referred to as the array 'valleys' and the local background can be estimated by taking the median of all pixels within these four valleys. The exact positioning of the valley region differs depending on the software being used. Since methods that include pixels close to the spot foreground are more likely to be corrupted by pixels belonging to the spot, they tend to have a bias for higher background estimates. While using pixels more remote from the spot reduces this bias, there is a trade-off because they tend to use fewer pixels and therefore have a higher variance in background estimation. The background adjustment method can substantially affect the variability in background-corrected spot values, especially with low intensity spots (Yang et al., 2001). In addition, it is not unknown for the pixel values within the spot to be lower than the background pixels, an effect visualized as dark spots. This can occur if the printed spot material excludes the substance generating the background pixel values. In such cases, background subtraction using 'off spot'

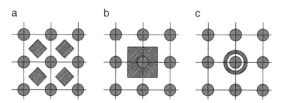

**FIGURE 5.5**  Alternative methods of local background definition. The spots are shown as dark gray circles and the area defined as background by the hatched areas. (a) A set of pixels that are equidistant from four surrounding spots: the array 'valleys'. (b) All of the pixels that are not within the spot mask but are within a square centred at the spot centre. (c) The area between two concentric circles, a fixed number of pixels away from the foreground circle.

pixels is clearly unsuitable and if background correction is needed then blank, spotting buffer only or nonhybridizing probe sequences are needed to assess the 'on spot' background.

The foreground and background measurements typically reduce the total data from an image file of megabytes to kilobytes of summary data. However, the data at this stage do not give meaningful information about gene expression. In order to obtain estimates of gene expression, the data need to be further processed and we shall describe these procedures in Chapter 6.

### 5.5.1 Current Tools and Methods Available

An ever increasing number of spotfinding and quantitation software sources are available either to purchase, or freely available to academic institutions. It would be futile to try and provide a comprehensive list of all available packages, but Table 5.1 lists features of some of those that are commonly used. The software mostly differs in cost, ease of use, flexibility, the level of automation or potential for manual intervention, the number of channels they can deal with, and the possible level of throughput. The exact methods for grid placement, addressing, spot segmentation and background subtraction algorithms (often not disclosed) differ greatly and, as discussed above, the method of choice depends on the array type. The majority of the spotfinding tools can be run on a Windows graphical user interface, whereas some require an UNIX-like operating system. Those for which the source code is freely supplied are particularly useful if you wish to tailor certain features to your own specific needs.

### 5.5.2 Automation/Manual Intervention

Microarray experiments generate a large amount of data, typically a single array has several thousand spots. It is therefore clear that manually identifying or segmenting every spot is unfeasible and automation of the image analysis process is essential. Ideally, a fully automated system should only need input from the microarray layout (number of rows and columns of spots and subgrids) and a list of image files to process. The software should be able to search the image for the grid position, identify the array configuration, localize the spots and perform measurements without the need for user intervention. The main value of automation is obviously speed, enabling large numbers of images to be analyzed in a short space of time. The second advantage is to ensure that the analysis is consistent and reproducible: manual analysis has the potential to introduce biases. Unfortunately, since microarray images are never perfect, some degree of user intervention is inevitable. For example, it is not unusual for the grids to be misaligned, perhaps shifted by one column or row, meaning that subsequent data will be associated with the wrong probe. It is therefore essential to be able to manually adjust the grids at the spot addressing stage.

**TABLE 5.1** Summary of Spotfinding Tools and Their Main Features

| Tool name | Source | Main features |
|---|---|---|
| ArrayPro Analyzer | Media Cybernetic | Image processing<br>Highly automated spot identification<br>Simple statistical analysis<br>Numeric and graphic feedback |
| BlueFuse | BlueGnome | Large numbers of images easily managed in batch, and<br>Fully automatic single click processing<br>Advanced statistical modelling technology used to separate signal from noise |
| Dapple 0.88 | Buhler et al. (2000) | Automated spot location, quality evaluation and quantification<br>Quality of putative spots judged automatically using a classifier trained to match the investigator's judgment<br>Source code available |
| F-SCAN | Munson et al., Center for Information Technology, NIH, Bethesda | Fluorescent array quantification and analysis<br><br>Scatterplots generated<br>Multiple image comparison<br>Data merging and web links<br>Free for academic use |
| GenePix Pro 6 | Molecular Devices | Spot identification, scatter plot, histogram, normalization<br>Designed for Axon scanner but can also analyze images acquired with other scanners |
| GridGrinder | Corning Incorporated | Automated origin location, skew detection, and grid placement<br>Multiple channel capability<br>Can process files in batch mode.<br>Source code available |
| ImaGene 8.0 | BioDiscovery | Multiple channel capability<br>Flexible grid design and alignment<br>Various segmentation, quantification and visualization tools<br>Customizable automated quality flagging<br>Batch automation module |

**TABLE 5.1** (*Continued*)

| Tool name | Source | Main features |
|-----------|--------|---------------|
| Matarray | Wang et al. (2001) | Spot detection, signal/background segmentation, signal intensity determination and quality determination<br>A simple algorithm uses both spatial and intensity information for spot detection and signal segmentation<br>Procedure can be iterated to improve performance<br>Available free of charge |
| P-SCAN | Munson et al., Center for Information Technology, NIH, Bethesda | Quantification and analysis of membrane microarrays<br>Scatterplots<br>Multiple image comparison<br>Data merging and web links<br>Free for academic use |
| QuantArray | Packard Biosciences | Choice of quantitation methods, spot quality measures and normalization<br>Multiple channel capability |
| ScanAlyze 2.5 | Eisen lab, Lawrence Berkeley National Laboratory | Semi-automatic definition of grids and complex pixel and spot analyses<br>Source code available |
| Spot | CSIRO Mathematical and Information Sciences | Automatic grid location<br>Flexible spot segmentation<br>Morphological background estimation<br>Free for academic use |
| TIGR Spotfinder 3.1 | Saeed et al. (2003) | Semi-automatic grid construction<br>Automatic and manual grid adjustments<br>Choice of segmentation methods<br>Optional background subtraction<br>Manual intervention possible<br>Source code available |
| UCSF Spot | Jain et al. (2002) | Automated spot identification and quantitation<br>Can use a counterstained DAPI image to aid segmentation<br>Free for academic use |

Similarly, after segmentation, deviations from perfect spot size, shape, location, homogeneity or intensity mean that spotfinding may fail for some spots. It is therefore useful to have an added layer of manual intervention to accurately address noisy arrays when automated processing is unreliable. For example, if it is clear that the foreground pixels in a spot are in fact due to a speck of dust or other such background it is desirable to flag this spot for rejection.

### 5.5.3 Third Channel as Method for Locating all Spots

Spot segmentation can be inaccurate with low intensity spots since the segmentation algorithm can have difficulty distinguishing spot foreground pixels from background pixels. One potential way to deal with this issue is to use a third array channel in which all spots are clearly visible. There are a number of ways this can be achieved, depending upon the array platform and probe type. As we describe above, Wang et al. (2003) used fluorescein labeled primers to prepare amplicon probes from cDNA libraries. Fluorescein is compatible with Cy3 and Cy5, reducing crosstalk concerns, and, by scanning the arrays prior to hybridization, offers an excellent printing quality control check. In principle, the method provides pixel level resolution of both probe location and abundance. A more general implementation, for example when using oligonucleotide arrays, is to mix every probe with a control probe that does not cross-hybridize with any sequence in the genome of interest or synthesize each oligonucleotide probe with a tag sequence (the latter is clearly more expensive). This latter approach is similar to the method employed by Illumina in their bead array technology where, since the beads containing gene-specific probes are distributed at random on microwell substrates, every array needs to be hybridized to determine the location of each probe. It is absolutely critical that any sequence used in this approach will not interact with the target genome: both sense and antisense strands need to be checked. In these cases, the fluorescently labeled complement of the tag is added to the labeled samples during hybridization: all spots should give a similar signal, proportional to the amount of deposited probe. Rather than add a tag and hybridize with a control probe, an alternative is to counterstain the array after hybridization with a DNA stain, such as DAPI (4′, -di-amidino-2-phenylindole) (Jain et al., 2002).

### 5.6 QUALITY CONTROL BASICS

### 5.6.1 Factors Affecting Quality

There are a large number of processes involved in microarray fabrication, hybridization and image analysis that have the potential to impact upon data quality. We describe some of the printing artefacts that can arise in Chapter 3 and as we suggest there, these problems may be detected by postprinting quality control. Any major problems identified at this stage may lead to the entire print run being discarded or, at the very least, flag potential problem areas on the array. Similarly the hybridization processes may create problems such as uneven or low signal intensity, high or uneven background, incomplete hybridization, uneven distribution of the two dyes across an array or within individual spots, dust particles, sample precipitation or desiccation around the array edges. While many of these problems can be minimized by careful handling of the arrays in a dust-free environment and by optimizing the sample preparation and hybridization process it is nevertheless rare for the arrays to have absolutely uniform background signal. All the above factors, which reduce the quality of the image, will have a knock-on effect for image analysis.

## 5.6.2 Slide Rejection Criteria

Ideally, badly manufactured arrays will be discarded before hybridization of precious samples in real experiments. However, it is important to have some criteria for rejecting poor arrays after hybridization. A variety of quality control and diagnostic plots of the raw data prior to normalization can help identify problems. For example, spatial bias heat maps can be used to observe the variation in foreground and background intensity values across each array. Any arrays displaying localized regions of high background can be dealt with either by rejecting all the spots in that region, or if the problem is more severe, then it may be necessary to reject the whole array from further analysis. Box plots can be used to visualize the distribution of intensity log ratios ($M$) for each print-tip group separately, thus enabling any problematic pins to identified. Pairwise scatter plots of raw intensity values with correlation coefficient calculations are particularly useful for comparing biological or technical replicates and identifying any channels that differ significantly from the other replicates, since channels containing the same sample type should have a correlation coefficient value close to 1. Scatter plots of log ratios versus the average of log intensities ($MA$ plots) are useful for detecting any intensity-dependent biases. Comparison of $MA$ plots for each array within a replicate group can help identify any arrays, which are behaving differently and are candidates for rejection. Chapter 6 describes these plots and their use in more detail.

## 5.6.3 Spot Rejection Criteria

Despite many new approaches emerging that focus on quality control for each spot, there is currently no standardized approach for minimizing variability in microarray data. It is important that spots of poor quality (e.g. Figure 5.6) are removed from subsequent analysis, otherwise they will give rise to inaccurate and misleading gene expression measurements. Automated image analysis software usually incorporates a method for assigning a spot 'accept' or 'reject' status and some also use a 'suspect' flag. This is typically based on some kind of quality measure such as variability in spot foreground and background intensity

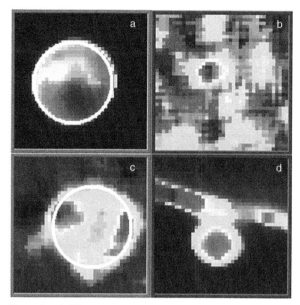

**FIGURE 5.6**   Examples of different spot morphologies. (a) Round spot with good signal. (b) Very low signal spot indistinguishable from background. (c) Irregular-shaped spot. The pixels comprising the foreground signal will be assigned differently depending on the algorithm used. (d) Spot bordered by dust, scratch or other artefact causing ectopic fluorescence.

distributions, relative foreground to background signal intensity, spot circularity measures or spot size variability. Some software allows user-defined thresholds whereas other, less flexible, packages use predefined values. The method of choice will depend on the quality of the image and the level of background and other artefacts. Most often an intensity threshold is determined empirically and any spots below it are eliminated. There is therefore always a trade-off between quality and quantity, that is by keeping all the data, including those from spots of low quality, one accepts that some gene expression measurements will be unreliable. Conversely, by keeping only high quality data, information for many genes may be discarded. The ultimate aim of the hybridization and data acquisition processes is to output signal intensity measurements for each spot, with some sort of reliability measure, ready for downstream analysis.

## 5.7 SUMMARY

The hybridization and scanning of microarrays can present a number of pitfalls for the novice user. The key to success is a rigorous series of pilot studies that optimize the wet protocols to ensure that signal is maximized while background is minimized. Even with commercial arrays, it is advisable to evaluate performance from time to time and certainly when first using a particular platform.

While it may seem a waste of money to sacrifice a few arrays on pilot optimization studies, it can pay dividends in terms of the quality and reliability of the experimental data, increasing confidence in the acquired data and saving money in the long run. If arrays are being generated in-house then there may be more opportunity for dedicated optimization studies. With good wet lab protocols in hand, the scanning and primary data extraction is another area that needs attention if data quality is to be maximized. Whatever the array type and the scanner, regular calibration and delimiting optimal settings during the image acquisition stage can help avoid systematic errors while maximizing the signal and dynamic range. These wet lab and data acquisition steps can be experimentally tested and optimized. What is perhaps more difficult is determining the best image processing approach. It is difficult to offer definite recommendations since so much depends on the quality and type of arrays being used. With our in house spotted long oligonucleotide arrays we acquire images with an Axon laser scanner and use the Dapple package for spotfinding and segmentation. Other colleagues using similar arrays and scanners swear by the Axon GenePix software. In some cases commercial software is expensive and an investment may lock you into an unsuitable solution. We recommend that any spotfinding or segmentation package is evaluated with a set of array images, preferably as close as possible to those that will be routinely generated. This can generally be done relatively easily with open source or publicly available software, commercial software providers should provide demonstration packages or allow evaluation in your lab. Whatever software is chosen, we recommend that the package has some flexibility and has an interface that makes different options readily available to the user. Finally, record the settings that are used during image acquisition, the information may be useful to others who wish to use your raw data.

## REFERENCES

Adams R, Bischof L. (1994) Seeded region growing. *IEEE Trans Pattern Anal Mach Intell* **10:** 6.

Ahmed AA, Vias M, Iyer NG, Caldas C, Brenton JD. (2004) Microarray segmentation methods significantly influence data precision. *Nucleic Acids Res* **32:** e50.

Bengtsson H, Jonsson G, Vallon-Christersson J. (2004) Calibration and assessment of channel-specific biases in microarray data with extended dynamical range. *BMC Bioinform* **5:** 177.

Bozinov D, Rahnenführer J. (2002) Unsupervised technique for robust target separation and analysis of DNA microarray spots through adaptive pixel clustering. *Bioinformatics* **18:** 747–756.

Brazma A, Hingamp P, Quackenbush J, Sherlock G, Spellman P, Stoeckert C, Aach J, Ansorge W, Ball CA, Causton HC, Gaasterland T, Glenisson P, Holstege FC, Kim IF, Markowitz V, Matese JC, Parkinson H, Robinson A, Sarkans U, Schulze-Kremer S, Stewart J, Taylor R, Vilo J, Vingron M. (2001) Minimum information about a microarray experiment (MIAME)-toward standards for microarray data. *Nat Genet* **29:** 365–371.

Buhler J, Ideker T, Haynor D. (2000) Dapple: improved techniques for finding spots on DNA microarrays. *University of Washington Technical Report*, UWTR 2000-08-05.

Chen Y, Dougherty ER, Bittner ML. (1997) Ratio-based decisions and the quantitative analysis of cDNA microarray images. *J Biomed Opt* **2:** 363–374.

Cheung VG, Morley M, Aguilar F, Massimi A, Kucherlapati R, Childs G. (1999) Making and reading microarrays. *Nat Genet* **21:** 15–19.

Dai H, Meyer M, Stepaniants S, Ziman M, Stoughton R. (2002) Use of hybridization kinetics for differentiating specific from non-specific binding to oligonucleotide microarrays. *Nucleic Acids Res* **30:** e86.

Demirkaya O, Asyali MH, Shoukri MM. (2005) Segmentation of cDNA microarray spots using markov random field modelling. *Bioinformatics* **21:** 2994–3000.

Dodd LE, Korn EL, McShane LM, Chandramouli GV, Chuang EY. (2004) Correcting log ratios for signal saturation in cDNA microarrays. *Bioinformatics* **20:** 2685–2693.

Dorris DR, Nguyen A, Gieser L, Lockner R, Lublinsky A, Patterson M, Touma E, Sendera TJ, Elghanian R, Mazumder A. (2003) Oligodeoxyribonucleotide probe accessibility on a three-dimensional DNA microarray surface and the effect of hybridization time on the accuracy of expression ratios. *BMC Biotechnol* **3:** 6.

Dudley AM, Aach J, Steffen MA, Church GM. (2002) Measuring absolute expression with microarrays with a calibrated reference sample and an extended signal intensity range. *Proc Natl Acad Sci U S A* **99:** 7554–7559.

Forster T, Costa Y, Roy D, Cooke HJ, Maratou K. (2005) Triple-target microarray experiments: a novel experimental strategy. *BMC Genomics* **5:** 13.

Hessner MJ, Wang X, Hulse K, Meyer L, Wu Y, Nye S, Guo SW, Ghosh S. (2003) Three color cDNA microarrays: quantitative assessment through the use of fluorescein-labeled probes. *Nucleic Acids Res* **31:** e14.

Hessner MJ, Wang X, Khan S, Meyer L, Schlicht M, Tackes J, Datta MW, Jacob HJ, Ghosh S. (2003) Use of a three-color cDNA microarray platform to measure and control support-bound probe for improved data quality and reproducibility. *Nucleic Acids Res* **31:** e60.

Hua J, Liu Z, Xiong Z, Wu Q, Castleman KR. (2004) Microarray BASICA: background adjustment, segmentation image compression and analysis of microarray images. *EURASIP J Appl Sig Proc* **1:** 92–107.

Hovatta I, Kimppa K, Lehmussola A, Pasanene T, Saarela J, Saarikko I, Sahrinen J, Tiikkainen P, Toivanen T, Tolvanen M, Vihinen M, Wong G, Editors Tuimala J, Laine M. (2005) *DNA Microarray Data Analysis*, 2nd edition. CSC – Scientific Computing Ltd., ISBN 952-5520-11-0.

Jain AN, Tokuyasu TA, Snijders AM, Segraves R, Albertson DG, Pinkel D. (2002) Fully automatic quantification of microarray image data. *Genome Res* **12:** 325.

Korkola JE, Estep AL, Pejavar S, DeVries S, Jensen R, Waldman FM. (2003) Optimizing stringency for expression microarrays. *Biotechniques* **35:** 828–835.

Korn EL, Habermann JK, Upender MB, Ried T, McShane LM. (2004) Objective method of comparing DNA microarray image analysis systems. *Biotechniques* **36:** 960–967.

Ku WC, Lau WK, Tseng YT, Tzeng CM, Chiu SK. (2004) Dextran sulfate provides a quantitative and quick microarray hybridization reaction. *Biochem Biophys Res Commun* **315:** 30–37.

Lehmussola A, Ruusuvuori P, Yli-Harja O. (2006) Evaluating the performance of microarray segmentation algorithms. *Bioinformatics* **22:** 2910–2917.

Li Q, Fraley C, Bumgarner RE, Yeung KY, Raftery AE. (2005) Donuts, scratches and blanks: robust model-based segmentation of microarray images. *Bioinformatics* **21:** 2875–2882.

Lyng H, Badiee A, Svendsrud DH, Hovig E, Myklebost O, Stokke T. (2004) Profound influence of microarray scanner characteristics on gene expression ratios: analysis and procedure for correction. *BMC Genomics* **5:** 10.

Maratou K, Forster T, Costa Y, Taggart M, Speed RM, Ireland J, Teague P, Roy D, Cooke HJ. (2004) Expression profiling of the developing testis in wild-type and dazl knockout mice. *Mol Reprod Dev* **67:** 26–54.

McQuain MK, Seale K, Peek J, Fisher TS, Levy S, Stremler MA, Haselton FR. (2004) Chaotic mixer improves microarray hybridization. *Anal Biochem* **325:** 215–226.

Peytavi R, Raymond FR, Gagne D, Picard FJ, Jia G, Zoval J, Madou M, Boissinot K, Boissinot M, Bissonnette L, Ouellette M, Bergeron MG. (2005) Microfluidic device for rapid (<15 min) automated microarray hybridisation. *Clin Chem* **51:** 1836–2844.

Qin L, Rueda L, Ali A, Ngom A. (2005) Spot detection and image segmentation in DNA microarray data. *Appl Biol Inform* **4:** 1–11.

Rahnenführer J, Bozinov D. (2004) Hybrid clustering for microarray image analysis combining intensity and shape features. *BMC Bioinform* **5:** 47.

Relogio A, Schwager C, Richter A, Ansorge W, Valcarcel J. (2002) Optimization of oligonucleotide-based DNA microarrays. *Nucleic Acids Res* **30:** e51.

Romualdi C, Trevisan S, Celegato B, Costa G, Lanfranchi G. (2003) Improved detection of differentially expressed genes in microarray experiments through multiple scanning and image integration. *Nucleic Acids Res* **31:** e149.

Saeed AI, Sharov V, White J, Li J, Liang W, Bhagabati N, Braisted J, Klapa M, Currier T, Thiagarajan M, Sturn A, Snuffin M, Rezantsev A, Popov D, Ryltsov A, Kostukovich E, Borisovsky I, Liu Z, Vinsavich A, Trush V, Quackenbush J. (2003) TM4: a free, open-source system for microarray data management and analysis. *Biotechniques* **34:** 374–378.

Shi L, Tong W, Su Z, Han T, Han J, Puri RK, Fang H, Frueh FW, Goodsaid FM, Guo L, Branham WS, Chen JJ, Xu ZA, Harris SC, Hong H, Xie Q, Perkins RG, Fuscoe JC. (2005) Microarray scanner calibration curves: characteristics and implications. *BMC Bioinform* **6:** S11.

Staal YC, van Herwijnen MH, van Schooten FJ, van Delft JHBMC. (2005) Application of four dyes in gene expression analyses by microarrays. *Genomics* **6:** 101.

Ting AC, Lee SF, Wang K. (2003) An easy setup for double slide microarray hybridisation. *Biotechniques* **35:** 808–810.

Vanderhoeven J, Pappaert K, Dutta B, Van Hummelen P, Desmet G. (2005) DNA microarray enhancement using a continuously and discontinuously rotating microchamber. *Anal Chem* **77:** 4474–4480.

Vanderhoeven J, Pappaert K, Dutta B, Van Hummelen P, Desmet G. (2005) Comparison of a pump-around, a diffusion-driven, and a shear-driven system for the hybridization of mouse lung and testis total RNA on microarrays. *Electrophoresis* **26:** 3773–3779.

Vernier P, Mastrippolito R, Helin C, Bendali M, Mallet J, Tricoire H. (1996) Radioimager quantification of oligonucleotide hybridization with DNA immobilized on transfer membrane: application to the identification of related sequences. *Anal Biochem* **235:** 11–19.

Wang X, Ghosh S, Guo SW. (2001) Quantitative quality control in microarray image processing and data acquisition. *Nucleic Acids Res* **29:** E75–E75.

Wang X, Jiang N, Feng X, Xie Y, Tonellato PJ, Ghosh S, Hessner MJ. (2003) A novel approach for high-quality microarray processing using third-dye array visualization technology. *IEEE Trans Nanobiosci* **2:** 193–201.

Waukau J, Jailwala P, Wang Y, Khoo HJ, Ghosh S, Wang X, Hessner MJ. (2003) The design of a gene chip for functional immunological studies on a high-quality control platform. *Ann N Y Acad Sci* **1005:** 284–287.

Wixforth A. (2005) Controlled agitation during hybridization: surface acoustic waves are shaking up microarray technology. *Methods Mol Med* **114:** 121–145.

Woo Y, Krueger W, Kaur A, Churchill G. (2005) Experimental design for three-color and four-color gene expression microarrays. *Bioinformatics* **21:** i459–i467.

Yang YH, Buckley MJ, Speed TP. (2001) Analysis of cDNA microarray images. *Brief Bioinform* **2:** 341–349.

Yang YH, Buckley MJ, Dudoit S, Speed TP. (2002) Comparison of methods for image analysis on cDNA microarray data. *J Comp Graph Stats* **11:** 108–136.

Yauk C, Berndt L, Williams A, Douglas GR. (2005) Automation of cDNA microarray hybridization and washing yields improved data quality. *J Biochem Biophys Methods* **64:** 69–75.

Yuen PK, Li G, Bao Y, Muller UR. (2002) Microfluidic devices for fluidic circulation and mixing improve hybridization signal intensity on DNA arrays. *Lab Chip* **3:** 46–50.

Zhang Y, Hammer DA, Graves DJ. (2005) Competitive hybridization kinetics reveals unexpected behavior patterns. *Biophys J* **89:** 2950–2959.

Chapter 6

# Data Preprocessing

Preprocessing includes data evaluation, cleaning and the analytical methods that are applied to obtain reliable estimates of the relative abundance of each gene in every sample. It is an essential requirement before any downstream statistical analysis or data clustering. With the appropriate data transformation, assumptions about the data structure can be met before statistical tests are applied and thus the identification of 'true answers' is more likely. Much of the preprocessing stage involves a process called data normalization, which can be broken down into three main steps: background correction, data transformation ('within-array' normalization) and 'between-array' normalization. The rationale for these steps is to adjust for systematic effects that arise from variation in the experimental technology rather than from biological differences between the RNA samples. The difference between the latter two steps is that the data transformation step is a normalization procedure that mainly deals with 'within-array' differences to account for dye, intensity and spatial dependent bias, whereas the 'between-array normalization' step mainly deals with multiple array differences and addresses the comparability of intensity distributions between arrays. In this chapter we discuss the various filtering and normalization tools currently in common use, highlighting where certain tools are appropriate. Some of the statistical concepts introduced in this chapter are discussed more fully in Chapter 7.

## 6.1 PREPROCESSING RATIONALE

We may consider the analysis pipeline as a series of stages, from experimental design to gene-set analysis, as outlined in Figure 6.1. While most microarray users generally consider only the three main steps in data analysis described above, in practice there are often more steps involved in an analysis pipeline. Broadly these steps may be classified as:

    **i.** Image analysis.
    **ii.** Background correction.
   **iii.** Probe correction.
    **iv.** Probe summarization.
    **v.** Treating missing values.
   **vi.** Data filtering and flagging.
  **vii.** Data transformation.
 **viii.** Between-array normalization.
   **ix.** Quality control.

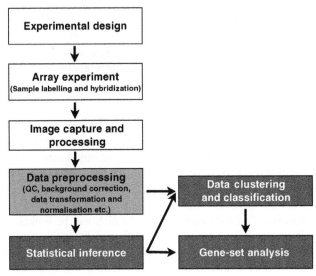

**FIGURE 6.1**   The data analysis pipeline: data processing is essential before any further downstream analysis since both statistical inference and data clustering approaches require appropriate transformation and normalization of the data.

Although all of these procedures are interdependent and some are array platform-dependent, they need not always be carried out in this order.

This chapter introduces these preprocessing steps, with the exception of the image analysis and processing steps covered in Chapter 5, presenting several important considerations before embarking on further data analysis. Many of the methods we introduce are freely available as tools or packages from the Bioconductor project (Box 6.1) and examples of these will be described where appropriate. Bioconductor packages that focus on preprocessing for two-channel spotted arrays include: `arrayMagic`, `arrayQuality`, `limma`, `maanova`, `marray`, and `vsn`. Packages exclusively for the analysis of Affymetrix arrays include: `affy`, `affycomp` and `affPLM`. For preprocessing Illumina bead arrays, the following packages can be used: `beadarray`, `BeadExplorer` and `lumi`. The main differences in data preprocessing between the Affymetrix and dual-channel array platforms are that the former requires probe correction and summarization steps since there are multiple probes per transcript. This area is discussed in more detail in Sections 6.18–6.20. In addition, background correction is also treated very differently for the two platforms and this area is also discussed. There are other sections in this chapter that are also relevant for the analysis of Affymetrix arrays, including sources of systematic error, treatment of missing values, data flagging and filtering, between-array normalization and more general quality control considerations.

---

**Box 6.1  The Bioconductor project**

---

- Website: www.bioconductor.org
- An open-source and open-development software project.
- Provides bioinformatics and computational biology software that aims to be both durable and flexible.
- Initial focus on Microarray data analysis but now also includes tools for proteomics and cell-based assays.
- Software is mainly written in an R environment, which is a high level interpreted programming language that lies at the heart of the Comprehensive R Archive Network (CRAN; http://cran.r-project.org).
- CRAN provides statistical analysis software for the wider community and also provide some useful tools for microarray data analysis.
- 'Vignettes' (documentation or user guides) are provided for all tools and training courses on using and writing various tools for Bioconductor are also widely available.
- More details on the Bioconductor project and various packages are available from Thomas Girke's 'R and Bioconductor manual' url: http://faculty.ucr.edu/%7Etgirke/Workshops.htm and from Dudoit et al. 2003; Parmigiani et al., 2003; Gentleman et al., 2005; Scacheri et al., 2006; Huber et al., 2007; Hahne et al., 2008.
- For further reading on the R statistical environment, refer to Venables and Smith (2002).

---

## 6.2 SOURCES OF SYSTEMATIC ERROR

As we have described in previous Chapters, the experimental process leading to the signal measurements that represent microarray data is complex and there are a large variety of sources of systematic global differences between measurements. These differences can reflect the following significant sources of technical variation:

- Sample preparation: for example, a time delay between surgical removal and cryo-preservation.
- Differential RNA extraction efficiency between samples.
- Overall amplification yield (if employed).
- Overall labeling yield.
- Overall hybridization efficiency and washing stringency.
- Incorporation efficiency for different dyes during direct labeling.
- Fluorescence gain of the excitation or detection system for different dyes.

Further sources of variation include specific effects of the manufacturing process, such as differences in:

- Characteristics of individual printing pins.
- Properties of specific probe source plates.

- Batch effects (with regard to slides, buffers, etc.).
- Sequence specific label incorporation bias.
- Spatial variation in hybridization efficiency or washing stringency.
- Nonlinear effects: for example, nonlinear dye fluorescence response or interactions between dyes (i.e. cross-talk and quenching).
- Differences in scanner settings.

Moreover, the operator and the date of the experiment can be significant sources of variation. The general difficulty encountered when attempting to combine data from different laboratories is well-known and technical effects may often dominate biologically relevant signals in multicenter studies. Unusually, ozone is known to have a destructive effect on the Cy5 dye and the levels of ozone can vary throughout the day and between seasons, summer being the worst time of year for high ozone levels (Fare et al., 2003; Lynch et al., 2007a,b). Branham et al. (2007) showed that simple installation of carbon filters in the laboratory air handling system resulted in low and consistent ozone levels. Also, the array features themselves can contribute towards unwanted effects. Lynch et al. (2007a,b) observed probe-design effects in Agilent dual-channel arrays due to GC content bias. Typically, in order to identify real biological differences, it is important to compensate, where possible, for the systematic differences in measurement. Although the aims of data transformation and normalization for all arrays are similar, as we indicate, the issues and techniques used in normalization of two-color arrays differ from those useful for normalization for Affymetrix arrays.

## 6.3 BACKGROUND CORRECTION

### 6.3.1 Additive Background Correction

Background correction is needed to adjust the foreground spot intensities for the background noise within the spotted region. Background noise can arise from many sources, such as nonspecific binding of the labeled sample to the array surface, process effects such as salt deposits or debris left after the washing stage, or optical noise from the scanner. In general, image-processing software will produce an absolute expression value ($A$) and a background measurement ($B$) for each spot (see Chapter 5 for further details). However, if $A$ is the result of signal and additional background noise then it is a biased estimate of the true hybridization measurement and is likely to be higher than expected. To obtain an unbiased measurement, some microarray technologies use an additive background approach. As the assumption that background and signal intensities are additive, this may involve the subtraction of a local background intensity from each spot intensity, ($A - B$), or the use of a small number of background measurements to calculate a mean background intensity, which is subtracted from many or all spot intensities.

## 6.3.2 The No Background Approach

There is some dispute as to whether background intensities should be subtracted from the spot intensities. For example, Qin and Kerr (2004) found a dramatic and detrimental effect of background adjustment. Since both $A$ and $B$ are both estimates, the variability of $A - B$ is larger than that of $A$ alone, and therefore, subtracting background adds variance. This is especially problematic with spots in the low intensity range, where $B$ can be of the same order of magnitude as $A$. In addition, background subtraction can generate negative values and zeroes with the resulting estimated ratio being both unreliable and likely to generate extreme ratios. On scatter plots or $MA$ plots (see Section 6.7.3 for general details on these plots), major 'fish tail' effects or fans can be seen (Figure 6.2), reflecting identical intensities in one channel but varying intensities in the other channel.

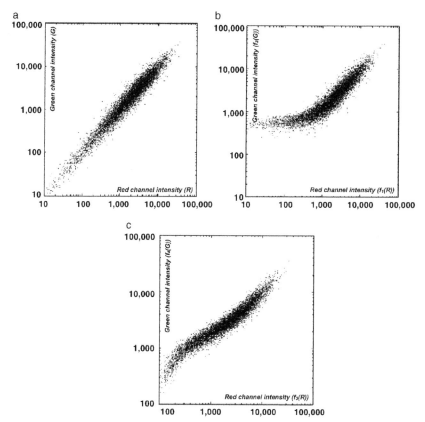

**FIGURE 6.2   Data anomalies**: example scatter-plots reflecting various response conversions for different fluorescent channels. A) Intensities for both channels are equivalent and are said to be 'linear'. B) 'Banana' or 'fishtail' effect. C) Sinusoid or 'S'-shape. Taken from Balagurunathan et al. (2002).

Negative values can be removed but this will result in loss of information (Bakewell and Wit, 2005). To add to the problem, Yang et al. (Yang et al., 2001, 2002a; Yang and Speed, 2002) have found that the background estimates produced by some popular image-processing software are not reliable since they tend to overestimate the background. An alternative is therefore to avoid background subtraction altogether (equivalent to setting both background channels to zero) and only use $A$ to estimate the expression level.

### 6.3.3 Avoiding Negative Values

Different approaches have been proposed for avoiding negative background-corrected values, for example, Smyth (2004) suggests replacing negative values with smaller positive values, while Edwards (2003) proposes a log-linear interpolation approach. One alternative uses a Bayesian method (Kooperberg et al., 2002), however, this is computationally intensive and does not exclude all spots with dominant background noise. The `NormExp` method (available as a function `normexp` in the `limma` package in Bioconductor) is based on the same model used to background correct Affymetrix data as part of the popular RMA algorithm (Irizarry et al., 2003b). A slight variation on this method (`normexp+offset`) is to add a small positive offset to move the corrected intensities away from zero. This variance-stabilizing method is analogous to the approach described by Rocke and Durbin (2003). Another variance stabilizing approach, the *arsinh* transform can handle negative values and is available from both the `vsn` package and as a function in `limma` (Huber et al., 2002). More details on this method are discussed later in Section 6.6.5. Recently, Ritchie et al. (2007) compared eight different approaches and the top performers were those that best stabilize the variance as a function of intensity, such as `normexp + offset` and `vsn`.

### 6.4 DATA FILTERING, FLAGGING AND WEIGHTING

Data flagging is the marking of data that may be untrustworthy or unreliable. There are two ways of dealing with flagged data: it is either filtered and removed or marked but not discarded. Filtering and removing unwanted data can provide more biologically meaningful results than those based on noisy data. For example, when given the option of removing low intensity values, there is a certain scanning limit, below which the intensity values are considered unreliable. These cutoffs depend on scanner precision but typically the cutoff is about 100–200 for Affymetrix data and 100–1000 for spotted dual-channel arrays. However, these cutoffs are arbitrary and may lead to loss of potentially valuable information so should be applied with caution. On the other hand, rather than removing the data entirely, information may be gained rather than lost. Data may appear unreliable at first glance but may still provide valuable information with regard to any further analysis. For example, for spots that have been flagged as 'saturated' in one channel, there may be a considerably lower intensity level for the same spot in another channel. In this situation, rather than filtering out saturated spots, the data is retained and flagged since the resulting fold-change may be large and genuine although the actual numerical value of the ratio may be unreliable.

Another approach is to assign weights to the data. Smyth et al. (2003) suggest a method of weighting spots for quality and this is available in the Bioconductor package, limma. In this approach, spots are weighted according to their area, defined as the number of pixels in the segmented foreground region of the spot. For example, the authors use the weight function to downweight spots smaller or larger than 165 pixels, the area of an ideal circular spot in their arrays. Each spot is assigned a weight between 0, for spots with zero size, and 1, for ideal circular spots (Smyth et al., 2003). The values from the weight function are used as relative weights in all the loess regressions used in subsequent data normalization. Obviously other spot quality metrics computed during the image analysis stage could be used in the same way to provide weights. A prerequisite for using weights is that they should be numerical and inversely proportional to the variances of the $M$ values.

The Bioconductor packge, limma can also assign weights to arrays to improve the ability to detect differentially expressed genes (Ritchie et al., 2006). This approach allows poorer quality arrays, which would otherwise be discarded, to be included in the downstream analysis. During the preprocessing stage, poor quality arrays are identified but often these arrays are not entirely bad. Once identified, the low quality arrays are normalized as usual with the other arrays and then downweighted for the subsequent analysis. The approach involves assigning weights to each microarray by fitting a heteroscedastic linear model with shared array variance terms.

## 6.5 TREATING MISSING VALUES

Missing values result from the spot quantification process because the spot is empty (intensity $= 0$) or because the background intensity is higher than the spot intensity (background corrected intensity $<0$). Commonly, values are marked missing in a quality filtering preprocessing step. Missing values can lead to problems in downstream data analysis as they can easily interfere with statistical tests or affect the stability of gene clusters (de Brevern et al., 2004). Rather than removing entire rows of the gene expression matrix[1] or filtering and removing rows based on a percentage of missing data, there are alternative options to choose from:

i. Replace the missing values with zero.
ii. Replace missing values with the average for a particular gene (reporter) over all sample conditions.
iii. Replace missing values with estimated values, a process called imputation.

---

1 Large matrices of gene expression are comprised of rows and columns, where each row represents a gene expression across multiple samples (the columns).

The problem with data deletion methods is that they run the risk of removing valuable information and may consequently bias the results if the remaining data are under-representative of an entire sample group. However, the commonly used mean and zero substitution methods have been demonstrated to distort relationships and correlations among variables and artificially reduce the variance of the variable in question (Troyanskaya et al., 2001). By far the best solution is the third option, imputation.

As summarized in Table 6.1, there are several imputation methods available: both Brás and Menezes (2006) and Brock et al. (2008) provide a recent review and evaluation of many of the commonly used methods. Troyanskaya et al. (2001) suggest a method called weighted $k$-nearest neighbors imputation (KNN or KNNimpute), which is available via various Bioconductor packages. The approach reconstructs missing values and, for each reporter, the weighted average of the most similar reporters is used as the imputation value. Two reporters are considered to be similar when the Euclidean distance between their expression patterns is small. Other similar approaches have been suggested (Kim et al., 2004, 2005; Ouyang et al., 2004; Scheel et al., 2005; Sehgal et al., 2005) and are based on the same principle of using some form of average over the neighboring reporters as the imputation value. A more recent alternative is the weight nearest neighbor imputation (WeNNI), which uses a continuous spot quality weight integrated into the imputation method (Johansson and Hakkinen, 2006).

The recent evaluation of eight different imputation algorithms (Brock et al., 2008) showed the best methods to be:

- Least squares adaptive (LSA), which combines gene-based and array-based imputation estimates (with the latter involving multiple regression) using an adaptive procedure to determine the weighting of the two estimates (Bo et al., 2004).
- Local least square (LLS or LLSimpute), which selects neighbors based on the Pearson correlation and performs multiple regression using all $k$ nearest neighbors (Kim et al., 2005).
- Bayesian Principal Component Analysis (BPCA), which involves Bayesian estimation to fit a probabilistic PCA model (Oba et al., 2003).

The evaluation also demonstrated that the success of each method can depend on the underlying 'complexity' of the expression data, where complexity indicates the difficulty in mapping the gene expression matrix to a lower-dimensional subspace. Global-based imputation methods (PLS, SVD, BPCA) performed better on microarray data with lower complexity, while neighbor-based methods (KNN, OLS, LSA, LLS) performed better on higher complexity data. The lowest sample size examined in the study was 14. If there are less than 10 samples and a large percentage of missing values ($>10\%$), then performance is likely to deteriorate and in such cases, a global imputation method may be preferable to a local one. However, the effectiveness of these methods with

**TABLE 6.1** Summary of Various Imputation Methods

| Name | Imputation method | Bioconductor package or R code | References |
|------|-------------------|-------------------------------|------------|
| KNN* | Weighted average k-nearest neighbors imputation | Impute; sam; pam; maanova | Troyanskaya et al., 2001 |
| SVD* | Singular value decomposition | svdImpute function in pcaMethods | Troyanskaya et al., 2001 |
| SKNN | Sequential k-nearest neighbors imputation | SeqKnn | Kim et al. 2004 |
| WeNNI | Spot quality weight nearest neighbor imputation | Only as a BASE plug-in[1] | Johansson and Hakkinen, 2006 |
| GMC | Gaussian mixture clustering with model averaging imputation | No only in Matlab | Ouyang et al., 2004 |
| LinImp | Linear model impuation | LinImp | Scheel et al., 2005 |
| CMVE | Collateral missing value estimation | Stand-alone Windows tool[2] | Sehgal et al., 2005 |
| BPCA* | Bayesian principal component analysis | arrayImpute | Oba et al., 2003 |
| LLS* | Local least squares imputation | llsImpute function in pcaMethods | Kim et al., 2005 |
| OLS* | Ordinary least squares | Not known | Brock et al., 2008 |
| PLS* | Partial least squares | Not known | Nguyen et al., 2004 |

Summary of various imputation methods. Those marked with an asterix were evaluated by Brock et al. (2008) and they produced their own R code for KNN, OLS, PLS, SVD, and BPCA. Future plans include implementing some of these methods into C for faster performance (Guy Brock, personal correspondence).

[1] WeNNI BASE plug-in available here: http://baseplugins.thep.lu.se/wiki/WeNNI.
[2] CMVE tool available here: http://personal.gscit.monash.edu.au/~shoaib/cmve.html.

smaller sample sizes has not been evaluated and some caution must be taken when applying any methods to small sample size experiments or data sets with large missing value levels. If there are a large percentage of missing values, many of the imputation methods will start by using some simple method (e.g. replacing with a gene average), and then proceed with one of the more sophisticated algorithms (Guy Brock, personal correspondence).

Other forms of missing data include missing whole channel data from two-color arrays. This can arise, for example, when due to the destructive effects of ozone on the Cy5 dye, the Cy-5 channel dual-channel arrays that are unprotected from a high ozone environment may be partially or completely corrupted. Options for handling this kind of data include: discarding the arrays that have a problem in the Cy-5 channel; discarding the Cy-5 channels and performing a single-channel quantile normalization; ignoring the problem with some precaution; or abandon the array entirely. In a simulation experiment, Lynch et al. (2007a,b) examined these various options to deal with this situation and found to a certain extent that the choice of method depends on the level of damage. The single-channel approach performs well when a majority of arrays are damaged but is outperformed when only a small number of arrays are damaged, by an approach based on simply discarding those arrays. The approach that generally performed the best was the 'combined' approach to linear modeling in estimating $k$, defined as 'the ratio of the variance of the residuals arising from the arrays providing only one channel'. First $k$ is estimated and then the linear model is applied using `limma`. By far the worst approach is to include all the damaged arrays.

## 6.6 DATA TRANSFORMATION

It is common practice to transform DNA microarray data before proceeding with any analysis in order to improve comparability and signal to noise ratio. Values calculated after image analysis, with or without background correction, are usually skewed or have an abnormal distribution (asymmetric with a long tail at the high expression end), which will result in spurious statistical interpretations. To avoid this, a change in measurement scale is usually applied and the most common approach is to transform data to the logarithmic (log) scale. While, as discussed below, a log intensity ratio transformation is not the most ideal method for removing various technical effects, graphical representations of the log intensity ratio can be used to reveal the presence of any technical variation.

There are several data transformations available that can be applied to remove technical effects, which are generally either intensity dependent or spatial. It is important to note that most of the approaches described here rely on the assumption that most genes are not differentially expressed or that the proportion of overexpressed and underexpressed genes between a pair of samples is approximately symmetric. In situations where this assumption is not true, it may be necessary to use an approach that employs control probes and targets for data normalization. However, one array platform, the Illumina bead array, has the unique advantage of having a sufficient number of randomly distributed probe replicates to apply a transformation that does not make this assumption (Lin et al., 2008). In a nutshell, transformation methods are essential to avoid misleading results caused by common technical artefacts in microarray data. The choice of normalization approach depends entirely on whether there is a problem with the data and what exactly the problem is. This is normally assessed through quality control procedures involving interpretation of various diagnostic plots. In addition, no matter which approach is chosen, it is necessary to apply the same transformation to all of the arrays in an experiment to ensure data consistency.

## 6.6.1 Common and Basic Transformations

The *intensity ratio* is calculated from the background corrected or uncorrected data:

$$\text{Intensity ratio} = \frac{R}{G}$$

where $R$ (red) is the intensity value for the Cy5 channel and $G$ (green) is the intensity value for the Cy3 channel. For Affymetrix data, $R$ and $G$ can be substituted with intensities from the sample and control microarrays. The intensity ratio is one for unchanged expression, less than one for downregulated genes, and greater than one for upregulated genes. However, the intensity distribution ratio is highly asymmetric or skewed and such distributions are not useful for statistical testing. The skew is due to upregulated genes having values between one an infinity and downregulated genes having values squeezed between zero and one.

An alternative approach, which produces a symmetric distribution of intensity ratios is to calculate the *fold change*:

$$\text{For expression values} > 1, \text{ fold change} = \frac{R}{G}$$

$$\text{For expression values} < 1, \text{ fold change} = \frac{1}{(R/G)}$$

where $R$ is the intensity value for the Cy5 channel and $G$ is the intensity value for the Cy3 channel, and both upregulated and downregulated genes can take values between zero and infinity.

## 6.6.2 Logarithmic Ratio Transformations

A more symmetric (normal or Gaussian-like) distribution can be achieved with a *log-transformation* and the most commonly used log transformation is the base 2 ($\log_2$):

$$\text{log ratio} = \log_2(\text{intensity ratio}) = \log_2 \frac{R}{G}$$

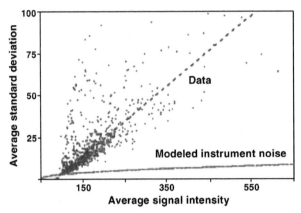

**FIGURE 6.3    Noise**: increasing measurement variance with higher hybridisation signal. Note that the noise associated with the scanner is negligible in comparison. Taken from Brown et al. (2001).

where $R$ is the intensity value for the Cy5 channel and $G$ is the intensity value for the Cy3 channel. After any log transformation ($\log_e$, $\log_2$ or $\log_{10}$ etc) is applied, the unchanged expression is zero, and both upregulated and downregulated genes have values from zero to infinity.

Logarithm to base 2 can be converted from the natural logarithm[2] using the following equation:

$$\log_2 x = \frac{\log_e x}{\log_e 2}$$

For the purposes of statistical analysis, it is usual to decouple random error and signal intensity (Cui et al., 2003; Kreil and Russell, 2005). A logartithmic (log) scale converts multiplicative effects[3] (ratios) into additive effects (differences), which are easier to model. It collapses the original range of the signal (including all extreme outliers), which can span over five orders of magnitude, and stabilizes the variance of high intensity spots. A log-transform decouples a random multiplicative error $e^{\eta}$ from the true signal intensity $\mu$ in the fluorescent intensity measurement $y$ (Rocke and Durbin, 2001; Durbin and Rocke, 2003):

$$y = \mu e^{\eta} \log y = \log \mu + \eta$$

In line with such a model, it is generally accepted that the variance of microarray data increases with signal intensity. Larger intensity values have larger variations when repeatedly measured and this phenomenon is known as heteroskedasticity. The source of this variance is likely to be caused by the hybridization process rather than by instrument error (see Figure 6.3).

---

2 The antilog of natural log $x$, $\log_{10}x$ or $\log_2 x$ are exp $(x)$, $10^x$ and $2^x$ respectively.

3 Effects on intensity of microarray signals tend be multiplicative, for example, doubling the amount of RNA should double the signal over a wide range of absolute intensities.

However, a log transform can introduce systematic errors at the lower end
of the expression value distribution. It can artificially truncate negative values
from background correction and inflates variance for low signal intensities. A
purely multiplicative error unrealistically predicts vanishing measurement
error for very small signals and this can be seen in log–log plots of 'same
versus same' (or 'self–self') comparisons (Figure 6.4). On a linear scale (top
panel), via a scatter plot,[3] the variance increases with signal intensity, whereas

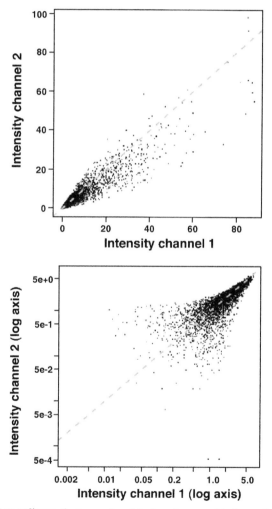

FIGURE 6.4   **Log or linear**: the two scatter plots show the same data from a 'same-same' comparison. The top plot is on a linear scale and the bottom plot is on a log scale. Taken from Kreil and
Russell (2005).

on a log scale (bottom panel), the variance is more or less constant for medium and high signal intensities. This is in line with the purely multiplicative noise model, however, the variance for low intensity signals is larger than expected (Figure 6.5).

Systematic dependence of the ratios on fluorescent intensity can be revealed by 'Ratio by Intensity' (RI) plots, although, there is a precedent for calling them *MA* plots (Yang et al., 2002a; Yang and Speed, 2002). These plots are essentially scatter plots of log ratios versus the average of log intensities and are used to detect intensity-dependent biases in two channel spotted microarrays. The characteristic curvature seen in these plots can result from background differences in the two dye channels. Some of the features commonly observed in these plots are shown in Figure 6.6. Scanner saturation results in truncation at the high end of intensity range (A and B). Variation at the low and high ends of the intensity range are due to channel specific additive errors (C and D) or multiplicative errors (E and F). Curvature results from mean background differences between channels (G and H) with slopes differences reflecting channel-specific differences in the linear relationship (I and J). Spatial heterogeneity shows dramatic bifurcations in the plot (K and L).

The log differential expression ratio for each spot is calculated as follows:

$$M = \log 2 \frac{R}{G} \quad \text{or} \quad M = \log R - \log G$$

Finally, the log intensity of the spot (a measure of the overall brightness of the spot) is:

$$A = \frac{1}{2} \log 2 \, RG \quad \text{or} \quad A = \frac{(\log R + \log G)}{2}$$

Note that the letter $M$ is a mnemonic for minus, while $A$ is a mnemonic for add. It is also convenient to use base 2 logarithms for $M$ and $A$ so that $M$ is in units of twofold change and $A$ is in units of twofold increase in brightness. On this scale, $M = 0$ represents equal expression, $M = 1$ represents a twofold change among the RNA samples, $M = 2$ represents a fourfold change, and so on.

---

3 A scatter plot visualizes data on a linear scale. However, it is more informative to produce a scatter plot of the log-transformed intensities in order to have better visual representation of the low intensity range. If the data points fit a straight line, then the data or the relationship between the two channels, can be considered to be linear. On a scatter plot of a 'self–self' comparison, if the two channels are behaving similarly, then one would expect the cloud of data points to approximately form a straight line, and the linear regression line through the data should have a gradient (the correlation coefficient, $R^2$) of 1 and an intercept of 0. Deviations from these values indicate nonlinearities in the intensity responses of the different Cy dye channels. See Figure 6.5 for pair wise scatter-plots and correlation coefficients for four arrays in a biological replicate group.

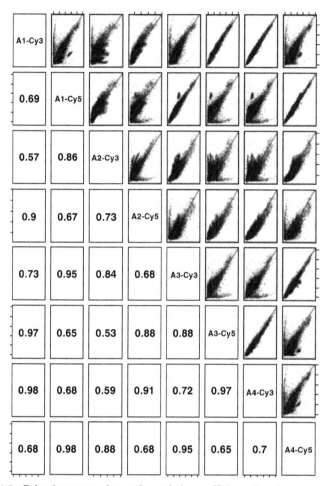

**FIGURE 6.5   Pair-wise scatter plots and correlation coefficients.** The plots are generated for four arrays in a single replicate group. The upper panels show pair-wise scatter plots of raw intensity values between the different slide channels within the replicate group. The lower panels show the correlation coefficients. The R-code for generating these plots is available from the FlyChip website (http://www.flychip.org.uk/code).

Scatter plots can be converted into *MA* plots by rotating them through 45° and scaling the two axes appropriately. *MA* plots are generally more powerful tools than scatter plots for visualizing and quantifying both linear and nonlinear differential responses of both array channels (see Figure 6.7 for a comparison). If the two channels behave similarly, then the data appears symmetrical around the horizontal line through zero and any deviations from this line represent different responses of the two channels. These linear and nonlinear regressions are more

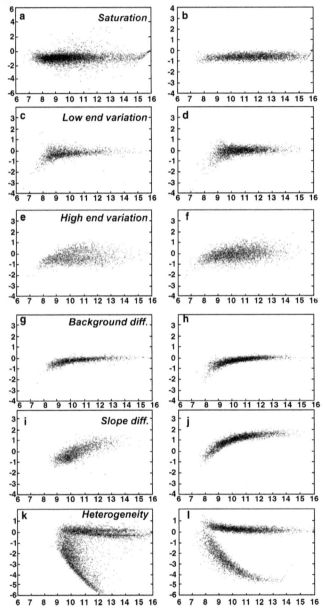

**FIGURE 6.6**   Common artefacts. Comparison of RI plots from real (left) and simulated (right) data. See text for details, taken from Cui et al. (2003).

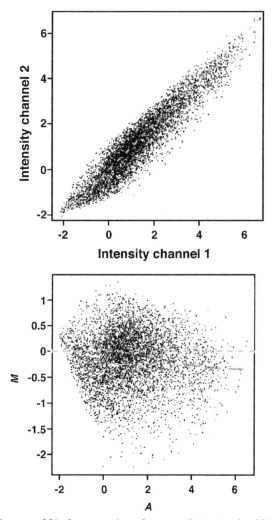

**FIGURE 6.7   Scatter or MA plot**: comparison of a scatter-plot (top) and an MA-plot (bottom) of the same data set after '*asinh-scale*' (via `vsn` in Bioconductor), show that the inflated fluctuations at low signal intensity end have disappeared. However, closer inspection of the MA-plot, where $M$ is the difference and $A$ is the mean between the two channels can reveal artefacts that are more obvious. The grey dashed lines show a loess smoother, giving an indication of local trends in the plot with an 'S'-shaped distortion observed.

robust and reproducible than performing regressions of one channel against the other as they treat the two channels equally. Additionally, the human eye and brain are more effective at perceiving differences from horizontal and vertical lines than from diagonal lines.

As discussed in more detail in Cui et al. (2003), if both the multiplicative and additive errors[4] for the two channels are large and approximately equal, the low intensity end of the *MA* plot will show large variance and will have a symmetrical distribution around a horizontal line. Additionally, if the additive errors in the two channels have different variances, the low intensity end of the *MA* plot will tilt the channel with smaller variances to form a curved shape. The following alternative approaches address either one or both these issues of intensity-dependent effects on the log ratio.

### 6.6.3 Shift Transformation

Kerr et al (Kerr et al., 2002) proposed a shift-log transform to correct the curvature seen in *MA* plots and an implementation is available with the Bioconductor maanova package. The approach adjusts log ratios by adding a constant to the signal intensities of one channel and subtracting the same constant from signals in the other channel before the log transformation:

$$\text{Shift} - \log R = \log_2(R + C)$$

$$\text{Shift} - \log G = \log_2(G - C)$$

The constant *C*, is estimated by minimizing the absolute deviation of each shift-log ratio from the median log ratio of the array. The shift-log transformation minimizes any curvature seen in RI plots. Alternatively, there are other methods available such as lowess smoothing, should this approach fail to correct the curvature.

### 6.6.4 Curve Fitting Transformation

Another curve fitting transformation method used to overcome these data trends is lowess smoothing[5] and is available from the loess function in the statistics environment *R* or from limma and maanova Bioconductor packages. In terms of robustness,[6] the lowess approach can tolerate some asymmetry in differential expression between the samples being hybridized provided that the majority of genes are not differentially expressed (Yang et al., 2002b). The approach detects systematic deviations in the *MA* plots and corrects them by fitting a local weighted linear regression. The fitting requires a choice of smoothness ('span' or 'window length') that determines which data

---

4 The multiplicative error is related to RNA labeling, scanning and spot features while the additive error is related to local background.

5 Lowess stands for *locally weighted polynomial regression*. 'Lowess' with a 'w' involves using a linear function whereas 'loess' without the 'w' involves using a quadratic function.

6 Robustness refers to the ability of a statistical method to follow a major data set trend and to ignore outlier values.

are local. It works by performing a large number of local regressions in over-lapping windows along the length of the data. The choice of span is subjective and 20% is usually chosen (Yang et al., 2002a). However, if the span is too big, there will be no effective removal of curvature, but if the span is too small, there is a major risk of over-fitting the data and introducing errors larger than those being removed.

A possible cause for curvature ('smile'- or 'S'-shaped trends) in *MA* plots may be differences in the nonlinear fluorescence response of dyes. The effects are clearly seen in 'self–self' experiments, where two identical mRNA samples are labeled with different dyes and hybridized to the same slide (Dudoit et al., 2002). Here, no differential expression is expected and the green and red channels are expected to be equal: in practice, the red signal intensities tend to be lower than the green intensities. In addition, the imbalance in the red and green channels is usually not constant across the spots within and between arrays and can vary depending on, for example, the overall spot intensity or the array location. However, simulations by Stivers et al. (2002) have shown that such observed nonlinear trends in *MA* plots can also be explained if 20% of genes are specifically expressed in only one of the samples. A situation that can easily arise in comparisons of different tissue types. Dye swap experiments are therefore important in that they can help detect such situations (Figure 6.8) thus avoiding indiscriminate application, for example, of a loess smoother to detrend $M(A)$, since inappropriate detrending would remove biological signal and create further technical artefacts.

## 6.6.5 Variance Stabilizing Transformation

The generalized log (glog) transformation, including the *arsinh* transform, is an alternative to the commonly used log transformation. At low intensity, the signals are small and the additive component of error dominates, especially since background subtraction often exaggerates this effect. Since a log transform can inflate the variance and create extreme differences between small signal intensity values, an alternative transformation method, a variance stabilizing transformation (VST) can be applied to stabilize the variance of these low intensity values (Huber et al., 2002). The transformation might prove superior to the log transformation, as most genes typically have low signal values in microarray experiments.

For a data matrix $x_{ki}$, the variance stabilizing approach fits the following transformation:

$$x_{ki} h_i(x_{ki}) = g \, log \frac{x_{ki} - a_i}{b_i}$$

where $k$ indexes the probes; $b_i$ is the scale parameter for array $i$ (or array and colour channels for two-channel arrays); $a_i$ is a background offset, and glog is

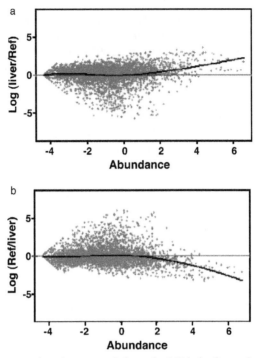

**FIGURE 6.8  Dye-swapping**: the top panel shows the *M(A)* plot for a microarray experiment comparing liver tissue with a reference sample comprising a pool of liver, kidney and testis tissue (Stivers et al., 2002). As seen in the bottom panel, the trend shown by the solid line (a loess smoother) reverses under dye swap.

the generalized logarithm (Rocke and Durbin, 2003). The Bioconductor packages vsn and limma can apply the variance stabilizing approach. They involve the *arsinh* transform as the generalized logarithm to stabilize the asymptotic variance of microarray data across the full range of expression intensities and makes the data more symmetric. The variance stabilizing approach is based on the general error model (Tibshirani, 1988; Rocke and Durbin, 2001; Huber et al., 2002), which allows for both multiplicative and additive error terms. Raw intensity measurements always demonstrate an intensity-dependent (nonlinear) measurement variation and the relationship between variance and intensity differs among different equipment and arrays. The model can be defined as follows:

$$y = a + b\mu e^{\eta} + v$$

where $\mu$ is the noise-free intensity value; $e^{\eta}$ and $v$ are the additive and multiplicative error terms respectively; $a$ and $b$ are constants that model a global background and gain, respectively. Unlike the log transform, the *arsinh* transform will decouple $\mu$ from the from both, $e^{\eta}$ and $v$. As illustrated in Figure 6.9, it

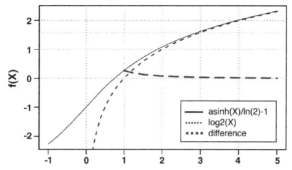

**FIGURE 6.9**   **asinh**: comparison of an asinh and a log2 transform. Note the deviation for values close to zero that are more pronounced with the log2 transform. (Kreil and Russell, 2005).

is approximately linear for low values and for larger values, it is approximated well by the log transform. For each sample, the parameters *a* and *b* are obtained by fitting the model to the data and all the genes share the same parameters determined by the fitting of the model.

After obtaining values for *a* and *b*, differences in global background and gain are addressed. The approach is based on the assumption of a quadratic equation between variance and signal intensities at the original scale. The iterative trimmed least-squares fit implemented in the vsn package will identify a subset of nondifferentially expressed nonoutlier genes across the arrays with a high breakdown point of 50%. This means that up to 50% of genes can be differentially expressed and, in contrast to many other methods, the approach can work reliably if there are unequal numbers of up and down regulated genes. Using the strata argument, the vsn package or function in limma, can also accommodate a more parameterized model to account for spatial effects. This involves adjusting for different grids or blocks that require different scale and offset parameters.

The variance stabilizing approach uses raw untransformed data (no ratios) and combines both background correction and normalization into one single procedure. This is in contrast to many other methods that consider both these procedures as separate tasks. One advantage of the combined approach is that information is shared across multiple arrays to estimate the background correction parameters and can be used for both within-array and between-array normalization. When a dye swap experimental design is incorporated, it works quite well at reducing variance between arrays, while also compensating for the intensity-dependent dye bias for two-channel arrays. The variance stabilization can be verified using *MA* plots and mean versus standard deviation plots (Figure 6.10).

Using the same general error model but with a different approach for estimating the parameters, the VST has been applied to Illumina's bead array data and is available from the lumi package in Bioconductor (Lin et al., 2008). This

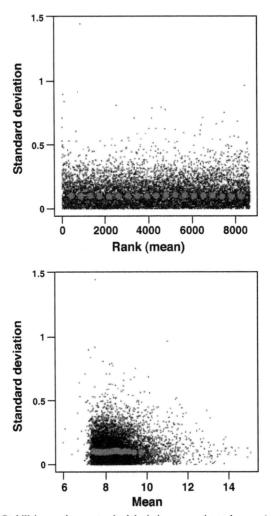

**FIGURE 6.10   Stabilising variance**: standard deviation versus the rank mean (top) and the mean (bottom). The dots connecting a line show the running median of the standard deviation. The rank ordering distributes the data evenly along the x-axis. The plot, which should show a straight horizontal line indicating that the variance is independent from the mean, is used to verify whether vsn has been effective. Taken from the vsn vignette.

array platform is unique in that it contains on average, 30 replicate probes (feature types) per array and these are randomly distributed so that every array is different in terms of probe location. The design and layout helps to reduce random noise and systematic bias such as spatial differences. Lin et al. have made use of these design features by modeling the functional relationship between the mean and the variance for each array, which was previously impossible with most other array platforms (Lin et al., 2008). Based on this, VST can be applied to each array separately to model the mean–variance relationship directly before applying a normalization method, such as quantile normalization, to account for between array differences. Unlike the previous approach used for other array platforms, which requires inter-array measurements, the VST approach applied to the Illumina platform does not rely on the assumption that most genes are not differentially expressed. In addition to being more robust and faster, this particular VST approach can also work with other array normalization methods in tandem, but of course, due to the unique array design and layout, this approach is only suitable for normalizing Illumina bead array data. However, this method relies on Illumina's proprietary software BeadStudio, which has a limited set of pre-processing and quality assessment methods. For further information on Ilumina arrays and preprocessing considerations refer to two recent papers (Dunning et al., 2007, 2008), the `beadarray` package and the corresponding vignette in Bioconductor. A combination of using both `beadarray` and `lumi` for data preprocessing and normalization is almost certainly a better alternative than BeadStudio.

### 6.6.6 The Linlog Transformation

The `linlog` transformation, proposed by Cui et al. (2003), is available from `maanova` in Bioconductor. It also behaves like a logarithm for high intensity values and is linear for low intensity values to stabilize the variance. The authors suggest setting a transition point that includes 30% (default setting) of data points in the linear range to stabilize the variance. Similar to *arsinh*, `linlog` is a transformation intended for stabilizing the variance of observations at an achievable scale, therefore, it does not provide any parameters for changing signal values, nor does it correct for curvature in *MA* plots. The authors suggest combining `linlog` with *shift-log* (linlogshift) or *lowess* transformation, to both stabilize the variance and minimize curvature simultaneously. They recommend applying the gentlest transformation approach that corrects the observed problem since, in some cases, attempted corrections may introduce biases greater than the ones they remove. In general, it is best to correct for biases at the technical level and through sensible experimental design (such as the use of dye swaps) rather than rely on data adjustments at the analysis stage.

### 6.7 DEALING WITH SPATIAL EFFECTS

Most of the data transformation methods described so far deal with intensity-dependent effects on log ratios, however, it is important to consider how spatial effects on microarrays can be dealt with. Spatial variance is typically 2–3 times stronger than residual noise in microarray experiments and is clearly a more significant issue. Spatial effects can be described as regions of the array that appear considerably brighter than others. Well-designed microarrays have replicate probes distributed at random across the slide surface to allow separation of spatial artefacts from biological signal. However, many early microarrays from both the academic and commercial sector have not followed this design criterion for reasons of convenience in manufacturing or limitations in spotting density. For such arrays, computational approaches are prone to fail when trying to separate spatial artefacts from biological signal. Hybridizing every other replicate slide rotated by 180° in the hybridization station (if one is used) may alleviate the problem but this is not ideal.

Recently, most fabricated arrays do tend to follow a randomized design, and the spatial trends in signal intensity are usually a sign of technical artefacts caused, for example, by array printing, unequal coverage of hybridization solution, washing, imaging of microarrays. For instance, with regard to hybridization, more of the sample may be present at one end of the array, or along one side, or some areas are thinner than others, especially around the corners of arrays as they tend to dry up more easily than other areas, leading to certain regions of the array being brighter than others. In addition, spatial bias can arise from the array not being completely flat in the scanner.

Here, the approaches that deal with spatial bias are mainly loess-based. However, it is important to note that loess approaches that deal with intensity-dependent effects do not remove spatial biases while the spatial-dependent loess approaches, such as the print-tip group loess normalization, do not remove intensity-dependent bias. These two steps can be used in sequence to remove both biases, as suggested by Cui et al. (2003). Alternatively, as described above, the `strata` argument used with `vsn` can adjust for different array grids or blocks that require different scale and offset parameters. However, here, the focus is mainly on loess-based approaches and on the use of diagnostic plots to visualize spatial bias effects.

Spatial heterogeneity can be detected by representing the log ratios using various diagnostic plots, many of which can be created using the following Bioconductor tools: `marray`, `arrayQuality`, `arrayQualityMetrics` and `limma`.

### 6.7.1 Spatial Images

Two-dimensional (2D) spatial images comprising different shades of grey or colour can be used to represent values of a spot statistic for each spot on an array (Figure 6.11). Each box within a grid corresponds to a particular array spot and its coordinates represent the array location. The spot statistics can be the intensity log ratio $M$, a measure of spot quality, or a test statistic. Pseudo-images of the Cy3 and Cy5 background signal intensities may reveal hybridization artefacts such as scratches on the microarray slide, dust or cover-slip effects.

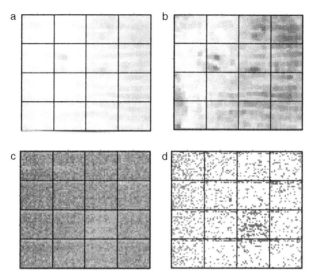

**FIGURE 6.11    Spatial trends**: images produced by `marray` showing background intensity for the Cy3 (**A**) and Cy5 (**B**) channels of an array. Images of pre-normalisation log-ratios *M* for all spots (**C**) and only spots with the highest and lowest 10% log-ratios are highlighted (**D**). Taken from the `marray` vignette in Bioconductor.

## 6.7.2 Spatial Bias Heat Maps

Spatial bias heat maps can be used to observe spatial differences and trends for either the foreground or background (Figure 6.12). The coordinates represent the location on the array, while the shading and contours indicate differences

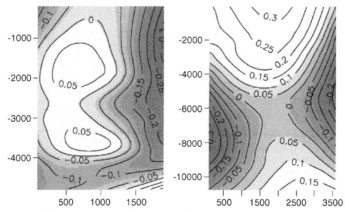

**FIGURE 6.12    Heat maps**: spatial bias can be plotted in form of heat maps. The x- and y-coordinates represent location on the microarray, whilst the shading and contours indicate the difference between the Cy5 and Cy3 channels. The R-code for creating these plots is available from the Flychip website: http://www.flychip.org.uk/mtip.

between red and green channels. These plots can also be used to identify problems associated with hybridization.

### 6.7.3 Boxplots

Boxplots visualizing print-tip group of spot statistics such as the *intensity* log *ratios M* can be used to identify hybridization artefacts (Figure 6.13). A box plot shows a distribution as a central box, the inter-quantile range (IQR), which is defined as the difference between the upper quantile and lower quantile (the difference between the 75th and 25th percentiles). The size of the box is the median absolute deviation from the median of the distribution. The line through the center represents the median or the 50th percentile of the distribution, a measure of central location of the data. The two horizontal lines bracketing the box represent where the extreme values of the distribution lie: 1.5IQR of the 75th percentile and 1.5IQR of the 25th percentile. Extreme outliers above and below these respective IQRs are usually plotted individually.

### 6.7.4 Scatterplots and MA Plots

As discussed earlier in Section 6.6, scatter plots and *MA* plots can be used to visualize the concordance between the red and the green channels on a single slide, with the latter plots being more effective at identifying interesting features of the data. The print-tip group can be highlighted in a particular colour on an *MA* plot to identify any problems with the array printing (Figure 6.14) or an *MA* plot for each print group can be visualized simultaneously (Figure 6.15).

### 6.8 PRINT-TIP GROUP LOESS NORMALIZATION

Print-tip group (or block-by-block) loess normalization is a one-dimensional loess regression approach that can be used to correct for various printing effects. Systematic differences between print-tips may include, for example, slight differences in the length or opening of the tips and deformation of the tips after many hours of printing. In limma (See Figure 6.15 for pre and post normalization plots), the approach involves subtracting the corresponding value of the tip group loess curve from each *M*-value and corrects the *M*-values both for within-array spatial variation and for intensity based trends:

$$N = M - \text{loess } i(A)$$

where, loess $i(A)$ is the loess curve function of $A$ for the $i$th tip group (see Section 6.6.4 for more details on loess). However, the disadvantages of using this method are twofold. First, each grid may contain a relatively small number of data points and it is possible that most of the features within a particular intensity range could be differentially expressed. This does not comply with the assumption

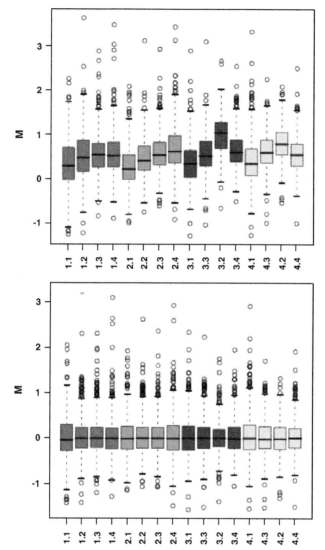

**FIGURE 6.13  Boxplots**: plots of pre- and post-normalisation intensity log-ratios $M$ (top and bottom panels respectively) by print-tip group using `marray`. Taken from the `marray` vignette in Bioconductor.

that the majority of genes are not differentially expressed and thus fitting the loess curve may result in the elimination of genuine differential expression measures, losing important information. Second, spatial bias may arise from the array not being flat in the scanner, which obviously has no relation to print-tip variability.

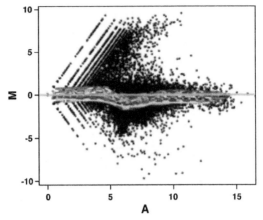

**FIGURE 6.14**   MA-plot with shaded curves for each pin (all spots are shown, including rejected spots), generated with `marray` and usually in colour.

## 6.9 TWO-DIMENSIONAL (2D) LOESS

An alternative, and generally much better method, for modeling spatial variation is to perform a 2D loess regression on the microarray data. In `limma`, the normalized log ratio $N$ is calculated by fitting a smooth 2D polynomial surface to the data instead of a curve to deal with changes in print tip group:

$$N = M - \text{loess}(r, c) - \text{loess}(A)$$

where $\text{loess}(r, c)$ is a two-dimensional loess curve, which is a function of the overall row position $r$, and the column position $c$, of the spot on the array.

## 6.10 COMPOSITE LOESS NORMALIZATION

Boutique arrays are custom-made arrays that have a small number of genes, often selected because they either have similar biological functions or were previously found to be differentially expressed. Since there are fewer irrelevant probes, the specificity of the array is increased and this results in a lower false discovery rate when identifying differentially expressed genes. Therefore, such arrays can be used for high-throughput assays to interrogate genes at minimal cost. However, they do pose problems for normalization since it is possible that more than half the probes on the array may represent differentially expressed. This violates the assumption of most transformation approaches such as those involving lowess (or loess) and variance stabilizing methods for technologies where there is no significance within array probe replication.

In such situations, the best strategy is to include a large number of nondifferentially expressed control spots, randomly distributed across the array. If the probes of interest have been selected from an initial large-scale microarray

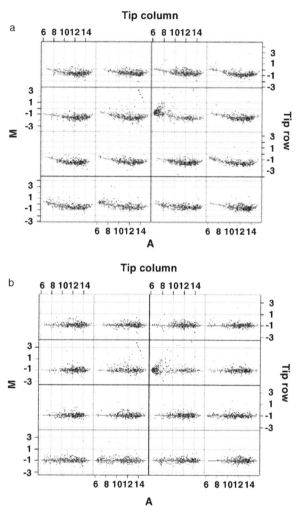

**FIGURE 6.15** *MA*-plots generated using `limma` (a) for each print-tip group on an array together with the loess curves (usually in red, here in grey), which will be used for normalisation. (b) for each print-tip after print-tip group loess normalisation of *M*-values. Here the curvature seen in the raw data at the top has been corrrected. Taken from the `limma` vigenette in Bioconductor.

experiment, a set of probes known not to vary under the experimental conditions being used may be selected. If this is not the case, control probes designed against 'housekeeping genes', spike-in controls designed against a distinct sequence (for example, the ERCC set) or a titration series of pooled whole-library spots may be considered (Yang et al., 2002b; Oshlack et al., 2007). Since true housekeeping genes are difficult to identify, the use of probes for exogeneous spikes is becoming more widespread. However, spike-in approaches involve spike-in RNA that is not extracted with the main RNA sample and has to be added separately, this may result in the spike-in probes not following the same intensity dependent normalization curve as the regular probes (in such cases the spike-in probes are usually seen as being off-set from gene probes on diagnostic plots). If this is the case, the control probes have to be normalized independently. A more effective and reliable approach seems to be the use of a whole microarray sample pool (MSP). The MSP method does not require foreign RNA to be added to the samples since the MSP probes interact with the samples under study and consequently any sample-specific biases should be minimal. In addition, the MSP approach has the effect of simulating the average expression that would be observed on a microarray constructed from an entire cDNA library and therefore it has the properties required for lowess normalization.

Typically, any control RNAs should span a wide range of intensities to make the most of the loess normalization, which requires a titration series of the corresponding probes. Whether MSP or spike-in probes are used, the probes are titrated at a series of different sample concentrations. For the spike-in approach, where a distinct or unique sequence is the probe, a known and equal amount of exogenous RNA spike is added when labeling each sample. With this experimental setup, a composite loess normalization approach (Yang et al., 2002b) can be applied to the data set using the following calculation available from limma:

$$N = M - p(A)\text{loess } C(A) - \{1 - p(A)\}\text{loess } i(A)$$

where, loess $C(A)$ is the loess curve through the control spots and $p(A)$ is the proportion of spots on the array with $A$-values less than $A$. However, this approach requires a sufficient number of unbiased probes at low intensities, where the loess curve generated from the gene probes has the largest influence on the normalization adjustment. Additionally, problems can also occur if gene probes reach higher or lower intensities than the control probes.

## 6.11 WEIGHTED LOESS NORMALIZATION

Alternatively, with the use of MSPs, Oshlack et al. (2007) proposed a weighted normalization approach called '*wlowess*', which is available as a function within the limma package. It uses probe-specific quantitative weights in intensity dependent lowess normalization as a more flexible, reliable and accurate method for normalizing boutique arrays (See Figure 6.16 for a comparison). The approach aims to be robust against probe selection bias at all intensities. As an extension from the lowess smoothing procedure, it defines a set of quantitative weights that are applied to each probe and an estimated curve on an *MA* plot is then used for normalization. The approach ensures that optimal use of whatever mix of control and regular probes are available on an intensity-dependent basis. The procedure can be applied to account for spatial bias, provided that there are a sufficient number and range of MSP probes printed with each print-tip. For very

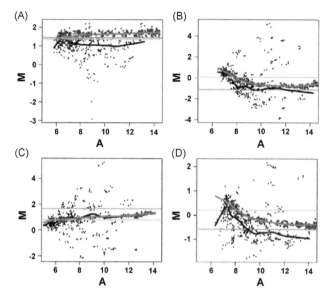

**FIGURE 6.16** *MA*-plots to visualize a *lowess* versus *wloess* normalization comparison for a boutique array designed to profile 109 genes during the late stages of B-lymphocyte differentiation. The cDNA clone, shown in black, is spotted on the array four times. The MSP titration series are shown as the grey (originally in blue) points. The black line is the *lowess* fit through the gene probes only. The grey line (originally in red) is the weighted lowess (*wlowess*) fit with MSP probes up-weighted. Differential expression levels of two house keeping genes, *HPRT* and *GAPDH*, are shown as the straight, horizontal dotted and dashed lines respectively. Plots **A**, **B** and **D** show examples where the gene probe *lowess* and the *wlowess* curves are considerably different from each other. Plots **B**, **C** and **D** show examples where house keeping genes give very different intensities from each other and from the *wlowess* curve. Taken from Oshlack et al. (2007). See color plate section.

small boutique arrays, however, there may be an insufficient number of probes available to perform a print-tip lowess normalization.

## 6.12 PROBE REPLICATES USED FOR EFFECTIVE NORMALIZATION

It seems that whether the approach is gene, spike or MSP based, all methods are affected by spatial effects as well as overall hybridization and labeling yield. A critical factor is the need for a sufficient number of replicated probes on the array to reduce noise. In addition, multiple replicated probe types are required to avoid sequence specific artefacts. Recently, Tuchler and co-workers (CAMDA 2008; personal correspondence, University of Natural Resources and Applied Life Sciences, Vienna, Austria) have shown that if at least 10% of probes on the array are control probes for exogenous spikes (and are also randomly distributed and designed to multiple sequences), a 10-fold improved spatial normalization can be achieved. Their spatial detrending approach is based on a Gaussian process framework to determine quantitative characteristics of spatial effects in microarrays. It is reported to outperform traditional spatial normalization procedures relying on nonreplicated gene probes and *MA* detrending. At a cost of the number of genes or feature types on the array, custom arrays have the flexibility of designing randomized feature arrays to account for spatial effects by having more probe replicates. However, other than the Illumina bead arrays, most commercial arrays do not have this kind of design feature. This is particularly short-sighted now that many technologies allow much higher feature densities on individual arrays.

## 6.13 'BETWEEN-ARRAY' NORMALIZATION

The previous section on data transformation discussed general transformation or normalization approaches that mainly deal with 'within-array' differences to account for independent- and spatial-based effects from two-channel spotted arrays. This section discusses common 'between-array normalization' approaches that address the comparability of intensity distributions between arrays. In this case these may be either two-channel spotted arrays or single-channel Affymetrix arrays. Here, each array hybridization reaction may be slightly variable, therefore there may be a difference in the overall signal intensities between different arrays. For the purpose of comparing samples hybridized to different arrays on an equal level or scale, it is important to correct for technical variability introduced by using multiple arrays.

Common approaches for visualizing several array distributions simultaneously include box plots (for a further explanation on box plots see Section 6.7.3) and histograms. Boxplots are important quality control tools for viewing the data before and after normalization. They provide a check for identifying any unusual array distributions and to verify whether a normalization procedure has been successful (Figure 6.17 for examples). Additionally, the skewness and normality distribution of the data should be checked before and after normalization, partly in order to check for normalization success and partly to check for statistical viability of the normalized data. This can be achieved by plotting *histograms* or *density plots* of the distribution of the data (log transformed intensity ratios) before and after data transformation and normalization (Figure 6.18). Regarding histograms, there should be a sufficient number of columns in the histogram for reliable results. In order to check for skewness, a comparison of the mean and median should also be made in conjunction with the plots.

Normality is an indication of how well the data fits to the normal distribution and this is important to assess since many statistical procedures assume that the data is normally distributed. Even the most basic of statistical methods applied to the data can be misleading if the distribution is highly skewed: the standard deviation provides information on the spread of the data and is meaningless if the distribution significantly deviates from normality.

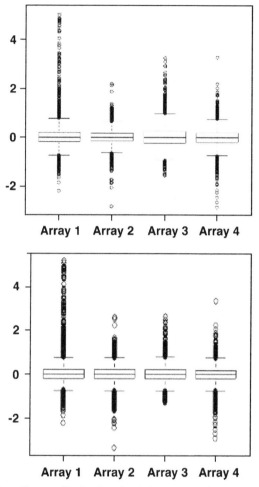

**FIGURE 6.17   Quantile normalisation**: boxplots (using `limma`) of pre- and post quantile normalisation (top and bottom panels respectively). Taken from the `limma` vignette in Bioconductor.

The commonly used methods for between-array normalization are:

  **i.** Scale normalization.
 **ii.** Centering normalization.
 **iii.** Quantile (or distribution) normalization.
 **iv.** Variance stabilizing normalization (vsn).

These techniques can be applied by using the following Bioconductor packages: `limma` and `vsn` (for two channel arrays) and `affy` (for Affymetrix arrays). For between-array normalization to be effective it is important to avoid any missing values in the log-ratio data that may arise from negative, or zero corrected

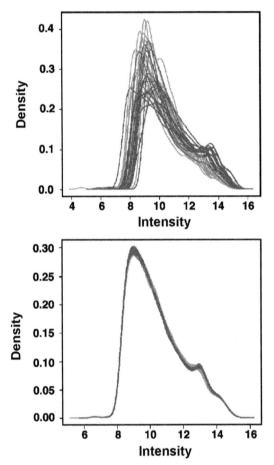

**FIGURE 6.18  Density plots**: pre- and post quantile normalised data (top and bottom panels respectively). The normalisation approach produces similar distributions across all the arrays, however, there appears to be some noise still present. Taken from the `limma` vignette in Bioconductor. See color plate section.

intensities. See the earlier section on treating missing values (Section 6.5) for more details.

### 6.13.1 Scale and Centering Normalization

Scale normalization can involve either the mean or the median of all the intensity distributions and ensures that all the means or medians of all the distributions are equal. The mean (or median) log ratio (or log intensity) of all the data on the array is subtracted from each log ratio (or log intensity) measurement on the array. The median provides a more robust measure of the average intensity on an array in instances where there are outliers or intensities that do not follow a normal distribution. Scaling is used by Affymetrix in version 5.0 of their Microarray Suite software (MAS 5.0) and is available from both `affy` and `affycomp` in Bioconductor. In `limma`, the scale normalization procedure involves the scaling of the $M$-values from a series of arrays so that each array has the same median absolute deviation.

A slight variation on the scaling approach, a method called centering: an initial scaling and then for each measurement on the array, the mean (or median) measurement of the array is subtracted and divided by the standard deviation (or median absolute deviation if the median is used). This normalization approach results in the mean (or median) measurements on the array being zero, and the standard deviation being one.

## 6.13.2 Quantile Normalization

As discussed by Bolstad et al. (2003), the aim of quantile normalization is to ensure that all the intensity distributions on each array are identical. It involves initially centering the data (see above), and then for each array, the centered measurements are ordered from lowest to highest. Next, a new distribution is calculated whereby the lowest value is the average of the values of the lowest expressed gene on each of the arrays and this is repeated for each subsequent order of intensity values up to the average value of the highest values from each of the arrays. Each measurement on each array is then replaced with the corresponding average value in the distribution. This normalization approach results in data where the mean measurement of each array is zero, the standard deviation is 1 and all the arrays have identical distributions. The procedure assumes that the distribution of gene abundances is nearly the same in all samples and that a majority of the expressed genes are not differentially expressed. It works quite well at reducing variance between arrays while also compensating for the intensity-dependent dye bias observed with two-channel arrays. In `limma`, a quantile normalization can either be applied to the $A$-values or directly to the individual red and green intensity channels. This type of quantile normalization is applied by both the RMA and gcRMA methods used for normalizing Affymetrix arrays (Irizarry et al., 2003b): these are explained further in Section 6.19 and are available from one or more of the following Bioconductor packages: `affy`, `affycomp`, `rma` and `gcrma`.

## 6.13.3 Variance Stabilizing Normalization (VSN)

As described in more detail in the previous data transformation section (Section 6.6.5), variance stabilizing normalization (VSN) is an *arsinh* transformation approach that estimates an additive and multiplicative offset and stabilizes the asymptotic variance of microarray data across the full range of expression intensities, resulting in more symmetric data (Huber et al., 2002). The approach is available from the following Bioconductor packages: limma, vsn and affy. In limma, VSN can either be applied to the $A$-values or directly to the individual red and green intensity channels. The VSN approach combines both background correction and normalization into one single procedure, effectively dealing with both within-array and between-array normalization issues. It is therefore important that no other data transformation method should be applied prior to using VSN (i.e. the input data for vsn should be untransformed raw data with no background correction). In addition, it is important to be aware that VSN assumes that most or at least half of the genes expressed show no change in expression. Therefore, like, all the other methods in this section, it is an inappropriate method for experiments where most expressed genes on the array are expected to show differential gene expression or for small arrays such as 'boutique' arrays where there is a focus on a particular biological area. See Section 6.10 for alternative approaches, such as composite normalization and further discussion on the issue of normalizing boutique arrays.

## 6.14 OTHER QUALITY CONTROL CONSIDERATIONS

Quality control is an important step that evaluates microarray data and assessing whether the transformation and normalization procedures applied are appropriate. The quality of microarray data is dependent on the quality of the array design and fabrication, on sample integrity, on the wet lab processes from sample extraction to hybridization and on appropriate image processing. Therefore, when troubleshooting microarrays, it is important to understand what the source of the problem may be. The scale of such problems can range from catastrophic failure (no signal) to data of apparently high quality but exhibiting unacceptably high levels of technical variation. At a glance, a good array may look to have good spot morphology and intensity range, low homogeneous background and a strong spot signal. However, these metrics are insufficient to properly assess the quality of an array. It is consequently important to implement a consistent range of methods for monitoring microarray data. As described above, various forms of graphical representations and diagnostic plots can be used to visually inspect the data and quality metrics can be used to obtain a measurement of variability or provide some weight to the data before any further analysis.

With the establishment of various microarray technologies in many areas of research in the life sciences, both academic and commercial scientists raised their expectations concerning the validity of the measurements. As a result, several consortia have been set up to address such issues and it is worth keeping up to date with them as they provide valuable recommended methods and guidelines. Important initiatives include the MicroArray Quality Control Consortium (MAQC), the External RNA Controls Consortium (ERCC) and

EMERALD (Empowering the Microarray-Based European Research Area to Take a Lead in Development and Exploitation) Consortium. Here we focus on the technical details associated with each of these efforts.

## 6.15 MICROARRAY QUALITY CONTROL CONSORTIUM (MAQC)

MicroArray Qualitry Control Consortium (MAQC) published a series of papers in *Nature Biotechnology* (2008) addressing important ongoing issues with regard to microarray data reliability (Canales et al., 2006; Guo et al., 2006; Patterson et al., 2006; Shi et al., 2006; Shippy et al., 2006; Tong et al., 2006). Important MAQC goals include:

i. The generation of reference data sets using multiple platforms produced across multiple arrays.
ii. The establishment of reference RNA samples.
iii. Quality metrics for microarray data reproducibility.
iv. The evaluation of various data analysis tools.

For a complete list of MAQC goals see Shi et al (Shi et al., 2006). MAQC's main conclusions confirm that with well-conceived experimental design coupled with appropriate data transformation, normalization and analysis, microarray data can be reproducible and comparable across different formats and laboratories irrespective of platform. The levels of variance among the different microarray experiments reported were low and mainly attributable to cross-platform differences in probe binding affinities to alternatively spliced transcripts or to probes showing high levels of cross-hybridsation with multiple targets. Although array performance was shown to be influenced by a diverse range of factors, from fluctuations in weather conditions to microarray fabrication batch quality and nuclease levels in sample tissues, experimental variability was shown to be sufficiently manageable. Importantly, differential expression measures derived from microarray studies were found to correlate closely with results from quantitative RT-PCR experiments.

## 6.16 THE EXTERNAL RNA CONTROLS CONSORTIUM (ERCC)

As we have previously discussed, the External RNA Controls Consortium (ERCC, 2005), has the objective of creating 'well-characterized and tested RNA spike-in controls useful for evaluating sample and system performance, to facilitate standardized data comparisons among commercial and custom microarray platforms'. The External RNA Controls (ERCs) are designed to be added after RNA isolation, but prior to cDNA synthesis, and evaluate whether the results for a given experiment are consistent with defined performance criteria. ERCs are extremely valuable for quality control because their true concentrations are known at the outset of the experiment. Since the actual microarray measurement is known, the performance of the microarray can be examined by the simple criteria of whether or not it accurately measures ERC levels. Recently, Tong et al. (2006) addressed the MACQ goal of evaluating data analysis methods for microarray data and examined datasets from hybridizations that examined ERCs for five different array platforms.

## 6.17 THE EMERALD CONSORTIUM

The EMERALD (Empowering the Microarray-Based European Research Area to Take a Lead in Development and Exploitation; http://www.microarray-quality.org) consortium aims to establish and disseminate quality metrics (QC), microarray standards and optimal laboratory practices (QA) throughout the European microarray community. The consortium is involved in two main projects:

i. *The Normalization and Transformation Ontology (NTO)* (as part of the microarray gene expression database group, MGED) for the controlled vocabulary of quality metrics, quality control, transformation and normalization methods.
ii. *ArrayQualityMetrics*, a Bioconductor package, which provides a HTML report with diagnostic plots for both single-channel and dual-channel arrays. The quality report contains an evaluation of individual array quality, reports the existence of any spatial effects, indicates the reproducibility of particular experiments, the homogeneity between experiments, any GC content effects, the mapping of the reporters and the evaluation of the biological signal to noise ratio. The report can be used as a quality control step for microarray data analysis and to compare the efficiency of different normalization approaches.

## 6.18 PREPROCESSING AFFYMETRIX GENECHIP DATA

Affymetrix provides a complete microarray platform, which is for most purposes also a closed platform, i.e. only Affymetrix chips can be read on Affymetrix scanners and their unique chip design requires specific analysis tools to interpret them. Affymetrix provides its own software to perform the data analysis, including the additional steps of interpreting perfect match (PM) and mismatch (MM) probe data, and condensing the 'Probe Set' intensities into a single expression index. Since a complete package is provided, it is often tempting to accept the results generated by the Affymetrix software at face value. However, early versions of the software had some peculiarities, particularly in relation to the interpretation of the mismatch probe data. The use of mismatch probe data is a contentious issue but also provides fertile ground for the development of tools for assessing array specificity. As we shall see in the discussion that follows, various approaches have been developed to deal with Affymetrix arrays. Moreover, despite the differences in the manufacture and design of Affymetrix chips, the arrays suffer from essentially the same problems as robotically spotted arrays with regard to manufacturing variability, hybridization and scanning effects. Some examples of rogue arrays are shown in Bolstad's *Chip Gallery* website: http://www.plmimagegallery.bmbolstad.com/.

As a consequence of the particulars of Affymetrix chip design, combined with the usual problems of microarray technologies in general, development of preprocessing and normalization approaches for Affymetrix GeneChip[7] data is an active area of research and there are many alternative procedures available for analyzing Affymetrix data. While these new procedures give users more flexibility in the treatment of their results, the proliferation of tools also begs the obvious question: what method should I use? To assist users in identifying the most suitable method for their data analysis, a benchmark was developed by Cope et al. (2004) involving scientifically meaningful summaries using the same assessment data and statistics. The Cope benchmark evaluates performance in terms of bias (lack of accuracy) and variance (precision). Tools implementing the assessment are included in the affycomp package from Bioconductor or as an independent webtool (http://affycomp.biostat.jhsph.edu/).

While the specifics of different tools for the analysis of Affymetrix data varies, the procedures all have to use the data generated by the Affymetrix Scanner Gene Chip Operating Software (GCOS)[8] and must attempt to make sense of Affymetrix' specific chip design approach. As a consequence, most approaches have fairly similar processing steps, although the order and specific methodologies vary quite considerably. Before discussing any of these tools, it is important to understand what quality control procedures are available to assess Affymetrix arrays and what data formats are generated. A more detailed account of quality control procedures is provided by Brettschneider et al. (2006). Standard reports are generated from GCOS, which can also be computed by the Bioconductor package simpleaffy; an outline of the procedure is described below.

---

7 For information on analyzing Affymetrix GeneChip *exon arrays* see Okoniewski and Miller (2008).

8 Various Affymetrix software can be downloaded from the following url: http://www.affymetrix. com/support/developer/tools/affytools.affx. The Affymetrix Power Tools (APT) software package contains a set of cross-platform command line programs that implement algorithms for analyzing Affymetrix arrays and can be downloaded here: http://www.affymetrix.com/support/developer/ powertools/index.affx. Annotations and query-based tools can be found at the NetAffx™ Analysis Center website: http://www.affymetrix.com/analysis/index.affx.

### 6.18.1 General Pipeline of Quality Control Procedures and Generation of Data Formats

**i.** Scanning of Affymetrix GeneChip.

**ii.** Visual inspection and assessment of image.

**iii.** Generation of a QC report for each array containing parameters for the following:

    **a.** *Scaling factor*: which provides a measure of the brightness of the array. It is a multiplicative factor applied to the signal values to make a 2% trimmed mean of signal values for selected probe sets equal to a constant. For HU133 chips the default is 500.

    **b.** *The average background*: a measure of the signal intensity caused by auto fluorescence of the array surface as well as nonspecific binding of target or the streptavidin–phycoerythrin stain. Affymetrix does not provide any official guidelines for this measure but mentions that values typically range from 20 to 100 for arrays scanned with the GeneChip scanner 3000.

    **c.** *The RawQ* (noise): which is a measure of the pixel-to-pixel variation in background intensities. There are no official guidelines on this because of the strong scanner dependence. It is recommended that data from the same scanner is checked for comparability.

    **d.** *Percent present*: the fraction genes called as 'present' is the number of probe sets called 'present' relative to the total number of probe sets on the array. A low Percent present rate may indicate poor chip quality but also depends on sample properties: e.g. cell or tissue type and biological or environmental stimuli. Typically, human and mouse arrays have 30–40 Percent present whereas yeast and *E. coli* have 70–90 Percent present.

    **e.** *The $3'–5'$ control probe ratio*: the $3'$-end and $5'$-end internal control probe sets of various housekeeping genes (i.e. actin, GAPDH, etc.) are used to give an idea of the integrity of the sample RNA and efficiency of first strand cDNA synthesis. Ideally, the $3'–5'$ ratio should not exceed 3 as ratios any higher are indicative of low RNA quality.

    **f.** *The presence of spiked control cRNAs*: *bioB, bioC, bioD* and *cre* are exogenously added RNAs that serve as hybridization controls.

**iv.** Generation of Affymetrix data using GCOS. The software produces the following file types:

    **a.** '.DAT': the original scanned image

    **b.** '.CEL': the intensity view of the GeneChip

    **c.** '.CHP': the data extracted from analyzing the CEL file

**v.** It is important to utilize other quality control checks, such as those available through certain Bioconductor packages e.g. `affy`, `affyPLM` and `arrayQualityMetrics`. Visualization plots, such as spatial plots, density plots, histograms, scatter plots *MA* plots and boxplots are all highly informative when assessing or comparing different normalization methods. A selection of these plots are shown in Figures 6.19–6.21.

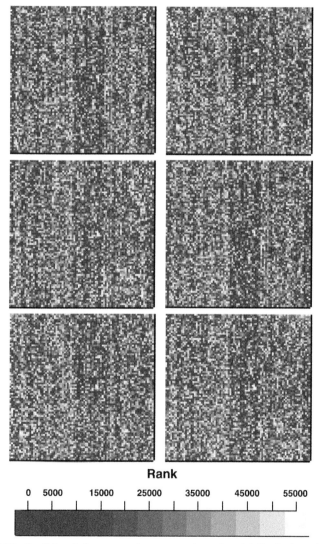

**Rank**

0    5000        15000        25000        35000        45000        55000

**FIGURE 6.19   Spatial variation**: greyscale representations of spatial distributions of raw probe (feature) intensities for six arrays using `arrayQualityMetrics`. The greyscale shown at the bottom is proportional to the ranks. These spatial plots may help in identifying for example, spatial gradients in the hybridization chamber or artefacts on the array caused by air bubbles or dust.

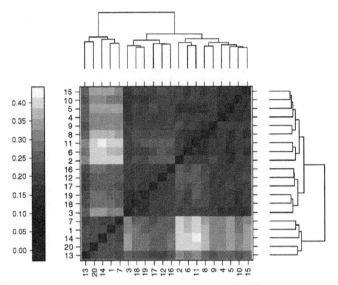

**FIGURE 6.20   Quality assessment**: heatmap representation of the distance between experiments using `arrayQualityMetrics`. The distances are computed (using the *mad* function in R) as the MAD of the M-value for each pair of arrays. Arrays with very different distance matrix entries give cause for suspicion. The dendrogram can be used to check whether the experiments cluster according to biological meaning or interest.

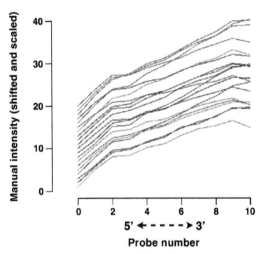

**FIGURE 6.21   RNA degradation plot**: the graph shows the mean expression from the 5' to the 3' end of the mRNA for control probesets. Each array is represented by a single line. In the ideal situation, lines are flat but this is usually not the case. When the lines are not flat, the slopes and profiles should be as similar as possible between arrays.

## 6.18.2 The Main Areas of Preprocessing Affymetrix Data

Typically preprocessing methods for Affymetrix data can be divided into four main areas:

i. *Background correction*: processing data to allow for background correction and within-array correction.
ii. *Probe correction*: the interpretation of perfect match (PM) and mismatch (MM) probe data.
iii. *Probe summarization*: the condensation of multiple probe intensities from a Probe Set to give a single index for each gene.
iv. *Normalization*: comparing multiple arrays to eliminate experiment-specific distortions of the expression measures on a given array.

A major difference between Affymetrix GeneChip and most other arrays types is the presence of multiple different oligonucleotide probes for each transcript on the Affymetrix chips. Each gene on the GeneChip is represented by one or more probe sets with each probe set consisting of between 11 and 20 probe pairs. Each probe pair consists of a perfect match (PM) probe and a corresponding mismatch (MM) probe, which has the same sequence as the PM probe except the central base is replaced with a mismatched nucleotide. The complete set of PM and MM probe pairs from each transcript is referred to as a 'Probe Set'.

Probe correction and summarization is required to obtain a single estimated measure of transcript expression from the multiple probes, once individual probes have been corrected for noise or cross-hybridization. Probe summarization is a particular source of disagreement in the community and, as we discuss later, there is even dispute about whether to use the mismatch probes. In addition, there are many background correction approaches for Affymetrix arrays and these can either be global and/or probe-specific. As with spotted arrays, there is a school of thought that does not agree with using any background correction and we also outline this approach. The normalization methods for Affymetrix arrays are generally similar to those applied to robotically spotted arrays and these will be discussed.

## 6.19 PROBE CORRECTION AND SUMMARIZATION

Probably the most important differences between the various approaches available to analyze Affymetrix data reside in the probe correction and summarization processes used. For the purpose of summarizing probe-level data after probe correction there are methods that employ the simple trimmed average (MAS4), the more robust average procedure based on Tukey's biweight (MAS5) and model based methods such as dChip (Li and Wong, 2001), PLIER (Probe Logarithmic Intensity Error: (Affymetrix, 2001a), RMA and GCRMA (Irizarry et al., 2003b).

### 6.19.1 Affymetrix Software

In early versions of MAS (versions 4 and below), the gene expression measure for each probe pair was calculated as a combination of the differences between PM probes and corrected MM probes using a very simple function, referred to as AvDiff. The AvDiff is calculated as the mean of the difference between PM and MM of each probe pair:

$$\text{AvDiff} = \frac{1}{N} \sum\nolimits_{j=1-N} (\text{PM}_j - \text{MM}_j)$$

where $N$ is the number of probe pairs in a probe set for a given gene and $j$ is an index to identify the $j$th probe pair. The actual software also applied various ad hoc corrections to allow for common 'problem' situations such as when the MM intensity is greater than the PM intensity or when background correction produced a negative signal. Other problems relate to how extreme intensities are dealt with. The signals from the MM probes were originally intended to represent cross-hybridization. Therefore, if the MM probes have a higher intensity than the corresponding PM probes, the signal is considered to be mostly cross-hybridization and is thus unreliable. In such situations the early versions of the MAS software generated negatives values, which are clearly not meaningful.

Affymetrix changed their algorithm in MAS version 5 (MAS5) so that it is no longer possible to obtain negative values. The latest software, entirely replacing MAS, is the GeneChip Operating System (GCOS), however, there is little change in the summarization algorithm between GCOS and MAS5. An expression index is calculated as follows:

$$\text{Signal} = \text{Tukey Biweight}\{log(\text{PM}_j - \text{CT}_j)\}$$

where $\text{CT}_j$ is a value derived from the MM signal intensity that is never bigger than its corresponding PM intensity.

As discussed below, several researchers have developed their own methods for correction and summarization of Affymetrix probe-level data. Some choose to ignore the MM probes entirely and use uncorrected PM probes alone, while others use detailed probe sequence information to develop more sophisticated PM correction methods.

### 6.19.2 dCHIP (MBEI)

Li and Wong (2001) proposed an alternative approach to the analysis of Affymetrix data. Their Model Based Expression Index (distributed in the dCHIP software package) assumes that the signal obtained from each probe pair from a series of $i$ different arrays is a function of the product of a true expression value ($\theta$) and a probe-specific affinity ($\phi$). The model is as follows:

$$\text{PM}_{ij} - \text{MM}_{ij} = \theta_i \phi_j + \epsilon_{ij}$$

Thus, the difference between the probe pair affinities for the $j$th probe pair on array $i$ is the product of the true expression level $\theta_i$ on the $i$th array multiplied by the probe-specific affinity of the $j$th probe pair $\phi_j$ plus an error factor $\varepsilon_{ij}$. The actual values of $\theta$ are determined using a maximum likelihood estimation.

In both MAS 5.0, GCOS and MBEI, the MM intensities are supposed to be a measure of nonspecific binding but recent studies indicate that the MM probes have a strong signal specific component and are thus not a good measure of nonspecific binding (Naef and Magnasco, 2003). As a consequence, it has been proposed that only the PM values should be analyzed.

### 6.19.3 RMA and GCRMA

The robust multi-array average model of Irizarry et al. (2003a) is a PM-only tool, i.e. mismatch probe data is discarded. RMA uses a similar approach to dCHIP, in that it assumes that the observed signal is the product of the true expression measure and a probe-specific affinity factor. RMA differs from MBEI in that the model uses log transformed intensities with model defined as follows:

$$Y_{ijn} = \mu_{in} + \alpha_{jn} + \epsilon_{ijn}$$

Where $Y$ is the log transformed, background corrected and normalized intensity measured, $\mu$ is the log scale expression measurement, $\alpha$ is the probe pair specific affinity factor and $\varepsilon$ is an error term. The data is fitted to the model using Tukey's median polish procedure (Tukey, 1977).

### 6.20 BACKGROUND CORRECTION

The optical detection system, nonspecific binding of labeled material to the surface of the Affymetrix GeneChip, labeling artefacts and any surface imperfections all combine to create a background noise that ideally should be removed from GeneChip data before further evaluation. However, determination of what is background is not trivial and there is always a danger that real signal may be lost in the attempt to remove noise.

It was intended that the difference between PM and MM would act as an effective background correction. However, this approach runs into problems if the intensity of MM > PM, which occurs in up to 30% of probes on some arrays (Irizarry et al., 2003b). Thus as we note above, it appears that MM probes do show a significant amount of sequence specific binding.

Naef and Magnasco (2003) suggest a very plausible explanation for many of the peculiarities of Affymetrix data. The mismatch probes are designed to have the complement of the central base of the PM sequence. This means if a pyrimidine (C or T) is present in the PM sequence this is swapped for a purine (G or A) in the MM sequence. Similarly, purines in PM are swapped for pyrimidines in the MM probe. This means that a cRNA that would bind to a pyrimidine containing PM probe will obviously have a purine at the position corresponding to the central base in the PM probe and the corresponding MM will also have a purine. In this situation, binding of the cRNA to the MM probe will be energetically unfavorable as the steric effect of the larger rings of two purines at the central position will interfere with hybridization. This means that MM probes with purines will tend to bind with much lower affinity than the PM probes and act more or less as intended. Conversely, a cRNA that binds to a purine containing PM probe will obviously have a pyrimidine at the position corresponding to the central base in the PM probe and the corresponding MM will also have a pyrimidine. Since pyrimidines are relatively small, two mismatched pyrimidines will not interact in the mismatch sequence and will 'dangle', thus having a small effect on hybridization. This is further compounded by the fact that the Affymetrix labeling method introduces biotinylated pyrimidines into the cRNA. The biotinylation increases steric strain and reduces binding affinity. This manifests itself as reduced affinity for PM binding but in the MM probes where two pyrimidines are dangling, there is more space to accommodate the label. This implies that for many MM probes with pyrimidines at the central base, the reduction in affinity that results from the label will be less than for the corresponding PM probes resulting in a brighter signal from the MM probe. The effect is shown below in Figure 6.22 (Naef and Magnasco, 2003).

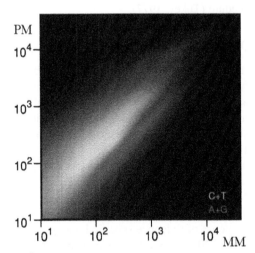

**FIGURE 6.22**   $17 \times 10^6$ (PM, MM) pairs from 86 human HG-U95A arrays were used to construct a two-dimensional histogram of PM versus MM probes. Probe pairs whose PM middle letter is a pyrimidine (C & T) are shown in the lightest shade, and purines (A or G) in the darker grey shade. The distribution for PM and MM shows strong sequence specificity: 33% of all the probe pairs are below the PM = MM diagonal; of these 95% have a purine as their middle letter. Reproduced from Naef and Magnasco. (Naef and Magnasco, 2003). See color plate section.

As a consequence of these observations, various more sophisticated approaches to dealing with background have been suggested. Each of the most popular tools implements a different approach (Irizarry et al., 2006), although it is possible with Bioconductor to mix and match. Approaches include: no or little background correction (e.g. VSN_scale), local background correction (MAS 5.0), global background correction (e.g. RMA; RSVD), probe-specific background correction (e.g. PLIER + 16; GCRMA). Irizarry et al (Irizarry et al., 2006) compared a number of preprocessing tools and concluded that the accuracy/precision (bias/variance) trade-off of a preprocessing methodology is driven mostly by background correction. Some methods are aimed at maximizing accuracy without taking precision into account, while others attempt high precision. In general the latter is preferred because the ability to reliably detect expression changes is superior when variance in the data is reduced. Methods that do not background correct have the worst bias for low expression values. However, some methods such as RSVD, ZL, PLIER + 16 and GCRMA appear to strike a balance between accuracy and precision, thus performing well across the entire expression level range. It is important to bear in mind that having replicate arrays in an experiment improves accuracy but not precision, and therefore, in this situation, a user may be willing to sacrifice accuracy for precision.

### 6.20.1 Affymetrix Software

MAS 5.0 and GCOS apply a local background correction (Statistical Algorithms Description Document; Affymetrix, 2002), in which the chip is sub-divided into a grid of $K$ predefined rectangular zones $Z_k$ ($k = 1, \ldots, K$; default $K = 16$). Control cells and masked cells are not used in the calculation. The lowest 2% of probe intensities in each region are considered to represent background ($bZ_k$) and the standard deviation these probes is calculated as an estimate of background variability. These background intensities for each grid region are used to calculate a correction for that grid region. The value $bZ_k$ for each zone is assumed to be the value for the center of that zone. For each cell $(x,y)$, the Euclidean distance ($d_k$) to the center of the $k$th zone is calculated. These distances are then used to calculated a weighting for the background value for the cell:

$$w_k(x, y) = \frac{1}{d_k^2(x, y) + \text{smooth}}$$

Where smooth is a factor added to ensure $d^2$ is never zero. The sum of all the weights to all zone centers is then used to calculate background for each cell $(x,y)$:

$$b(x, y) = \frac{1}{\sum_{k=1}^{K} w_k(x, y)} \sum_{k=1}^{K} wk(x, y)bZ_k$$

Affymetrix then applies various heuristics to this value to ensure that negative values are not obtained when this is subtracted from the actual cell intensities.

### 6.20.2 RMA

RMA applies a global background correction estimated using a model that considers that the perfect match signal ($S$) is comprised of a normally distributed background intensity ($Y$, with mean; $\mu$ and standard deviation; $\sigma$) and a signal ($X$) that is exponential ($\alpha$) with a multiplicative error. Background corrected signal, $E(X|S = s)$, is then:

$$\frac{a + b[\phi(a/b) - \phi((s - a)/b)]}{[\Phi(a/b) - \Phi((s - a)/b - 1]}$$

where $a = s - \mu - \sigma^2\alpha$, $b = \sigma$, $\phi$ and $\Phi$ are the normal and cumulative density, respectively.

### 6.20.3 GC-RMA

GC-RMA uses a different background correction model from RMA that is intended to improve the accuracy of expression values while retaining the precision of RMA (Wu et al., 2004). The modification was motivated by observations that the fold-changes predicted by RMA were less accurate but more precise than MAS 5.0 as a result of the way that RMA treats background. GC-RMA corrects background using a model of the sequence dependence of labeled cRNA binding to PM and MM probes developed from the work of Naef and Magnasco (2003). For any particular probe pair, GC-RMA assumes that the intensities PM and MM are as follows:

$$\text{PM} = O_{\text{PM}} + N_{\text{PM}} + S$$

$$\text{MM} = O_{\text{MM}} + N_{\text{MM}} + \phi_{\text{S}}$$

where $O$ represents optical noise, $N$ represent nonspecific binding noise and $S$ is a quantity proportional to RNA expression, i.e. the signal of interest. The parameter $0 < \phi < 1$ accounts for MM being quite specific in many cases. It is assumed that $O$ follows a log-normal distribution and that $\log(N_{\text{PM}})$ and $\log (N_{\text{MM}})$ follow a bivariate-normal distribution with means of $\mu_{PM}$ and $\mu_{MM}$ and the variance $\text{var}[\log(N_{PM})] = \text{var}[\log(N_{MM})] \equiv \sigma^2$ and that there is a correlation constant ($\rho$) across probes. GC-RMA also assumes $\mu_{\text{PM}} \equiv h(\alpha_{\text{PM}})$ and $\mu_{\text{MM}} \equiv h$ ($\alpha_{\text{MM}}$), where $h$ is a smooth and almost linear function and $\alpha$ is a model of probe affinity that accounts for position-specific effects and is defined by:

$$\alpha = \sum_{k=1}^{25} \sum_{j\in\{A,T,G,C\}} \mu_{j,k} 1_{b_k=j} \quad \text{with} \quad \mu_{j,k} = \sum_{l=0}^{3} \beta_{j,l} k^l$$

where $k = 1, \ldots, 25$ indicates the position along the probe, $j$ indicates the base letter, $b_k$ represents the base at position $k$, $1_{bk=j}$ is an indicator function that is 1 when the $k$th base is of type $j$ and 0 otherwise. The contribution to the overall probe affinity of the base $j$ at position $k$ is represented by $\mu_{j,k}$. This affinity model is fitted to a large number of arrays using a least squares fit to determine the affinity values used for the background correction model. The parameters $\mu_{PM}$, $\mu_{MM}$, $\rho$, $\sigma^2$ and $\phi$ are estimated from the data. The background adjustment procedure then becomes the problem of predicting $S$ given the observed values of PM and MM assuming $h$, $\rho$, $\sigma^2$ and $\phi$ are known. Optical noise is assumed to be an array-dependent constant. The actual calculation is discussed in detail elsewhere (Wu et al., 2004).

## 6.21 NORMALIZATION

MAS 5.0 uses a 'scaling' method (Affymetrix, 2001b), similar in principle to the total intensity normalizations discussed in Chapter 2 in relation to cDNA macroarrays. The scaling procedure in MAS 5.0 scales the intensities of each array in a data set to a baseline array. The precise details are not published but the software is assumed (Bolstad et al., 2003) to calculate the scaling factor for each array as follows:

$$\beta_i = \frac{\tilde{x}_{baseline}}{\tilde{x}_i}$$

where the value $x_{baseline}$ and $x_i$ are the trimmed means (i.e. the highest and lowest intensities are removed) of the baseline array and the $i$th array in the data set respectively. The baseline array is typically chosen as the array whose median intensity is also the median intensity of the arrays in the data set.

### 6.21.1 dCHIP

The MBEI (dCHIP) method uses a locally weighted regression (loess) correction (Li and Wong, 2001). The method attempts to identify a series of 'rank invariant' genes in the set of arrays that are being compared, i.e. genes that appear at the same position or rank, within predefined tolerances, in different arrays when the genes are sorted in order of intensity. The method relies on finding genes that are rank invariant over the whole intensity range. The mean intensities of each rank invariant set of genes are then calculated. The difference between the mean intensity of each rank invariant gene set and the intensity of the occurrence of the gene in each array is used to calculate a local error for each array. A normalization curve is fitted to these local errors and this is used to calculate the correction factor that is applied to each gene over each whole array.

### 6.21.2 RMA

RMA uses a process called 'Quantile' normalization (Bolstad et al., 2003). This method is based on the assumption, that if the same amount of labeled RNA is obtained from two samples and applied to two theoretically identical arrays, they should have the same overall intensity distributions. Thus, this method attempts to make the intensity distributions of two or more real arrays the same. This is done by projecting the data from $n$ arrays onto a line in $n$-dimensional space that is diagonal to all the $n$ expression vectors. It turns out that this is equivalent to sorting the $n$ arrays according to intensity, calculating the mean intensity for each intensity quantile and setting all the intensities in each quantile to the quantile mean. The normalized intensities are then resorted into the original gene ordering. This is an extremely simple method from a practical point of view and is computationally fast.

In a comparison of a number of normalization methods including quantile, scaling normalization and loess corrections using rank invariant genes, the quantile method performed better than the scaling method and comparably to the loess method but it reduced variance in the data and introduced fewer artefacts and biases than the loess method (Bolstad et al., 2003). GC-RMA continues to employ quantile normalization, differing only from RMA in the approach taken to effect background correction.

### 6.22 SUMMARY

Data preprocessing is a crucial stage in the microarray data analysis pipeline before any downstream analysis to identify differentially expressed genes or clustering-based methods. The three main steps (background correction, within-array' normalization and 'between-array' normalization) adjust for systematic effects that arise from variations in the technology. Additional preprocessing steps include imputation to treat missing values, data filtering, data weighting and quality control. Although not always done in this order, all these procedures are inter-dependent and some are platform-dependent, such as the probe correction and summarization step for preprocessing Affymetrix data. Affymetrix also uses a very different background correction approaches and it is generally agreed that both the RMA and GCRMA methods perform well. For dual-channel arrays, there are several methods for background correction, and those that perform well stabilize the variance as a function of intensity (e.g. VST or VSN). Regardless of array type, rather than filtering out missing values or data of poor quality, there are ways to dealing with such data to gain rather than lose information. For example, assigning weights to the data, may improve the ability to detect differentially expressed genes and depending on sample size, missing values can be estimated by various imputation procedures that are either global-based or neighbor-based. In addition to many downstream methods not being able to handle missing data, many also rely on appropriate data transformation and the assumption that the data is normally distributed. There are several methods available for making data more symmetrical, the most common of these being a log transformation. However, there are problems associated with this approach since it tends to introduce systematic errors at the lower end of the intensity value spectrum. More effective and reliable methods include those that are based on both an additive and multiplicative error model, such as VST or VSN. When dealing with spatial effects, several methods can be used, depending on the array platform, design and layout. For robotically spotted arrays, print-tip loess normalization methods can be used to correct for various printing effects. For whatever array type, provided there are sufficient numbers of randomly distributed control probes (whether based on replicate genes probes, spikes or whole microarray sample pool, MSP), either composite or weighted loess normalization are more effective at correcting for spatial effects. The latter method is especially useful for boutique arrays or experimental studies that expect more than 50% of genes to be differentially expressed. Uniquely, VST can be applied to Illumina bead array data to account for spatial effects without relying on this assumption since these arrays have sufficient numbers of replicated gene probes. To account for differences between both single- and dual-channel arrays, several methods can be used, and they include both variance stabilizing and quantile normalization methods. When applying any transformation and normalization method, quality control is a valuable and important step that involves evaluating microarray data and assessing whether the procedures applied are appropriate. Various graphical plots such as scatter and *MA* plots, histograms or density plots, boxplots, and spatial heat-maps can be used to visually inspect the data and quality metrics can be used to obtain a measurement of variability. Recently, several international consortiums (MAQC, ERCC and EMERALD) of both academic and commercial scientists have been set up to address issues related to the standardization of terminology, procedures, formats and measurements related to quality control for all platforms, and it is worth keeping up to date with their recommendations and practices.

## REFERENCES

Affymetrix. (2001a) *Affymetrix GeneChip Expression Analysis Technical Manual.* Affymetrix, Santa Clara, CA.

Affymetrix. (2001b) *Statistical Algorithms Reference Guide.* Technical Report, Affymetrix.

Bakewell DJ, Wit E. (2005) Weighted analysis of microarray gene expression using maximum-likelihood. *Bioinformatics* **21:** 723–729.

Balagurunathan Y, Dougherty ER, Chen Y, Bittner ML, Trent JM. (2002) Simulation of cDNA microarrays via a parameterized random signal model. *Biomed Optics* **7:** 507–523.

Bo TH, Dysvik B, Jonassen I. (2004) LSimpute: accurate estimation of missing values in microarray data with least squares methods. *Nucleic Acids Res* **32:** e34.

Bolstad BM, Irizarry RA, Astrand M, Speed TP. (2003) A comparison of normalization methods for high density oligonucleotide array data based on variance and bias. *Bioinformatics* **19**: 185–193.

Branham WS, Melvin CD, Han T, Desai VG, Moland CL, Scully AT, Fuscoe JC. (2007) Elimination of laboratory ozone leads to a dramatic improvement in the reproducibility of microarray gene expression measurements. *BMC Biotechnol* **7**: 8.

Brás LP, Menezes JC. (2006) Dealing with gene expression missing data. *Syst Biol* **153**: 105–119.

Brettschneider J, Collin F, Bolstad B, Speed TP. (2006) Quality Assessment for short oligonucleotide data. *Technometrics* arXiv:0710.0178v2.

Brock GN, Shaffer JR, Blakesley RE, Lotz MJ, Tseng GC. (2008) Which missing value imputation method to use in expression profiles: a comparative study and two selection schemes. *BMC Bioinform* **9**: 12.

Brown CS, Goodwin PC, Sorger PK. (2001) Image metrics in the statistical analysis of DNA microarray data. *Proc Natl Acad Sci U S A* **98**: 8944–8949.

Canales RD, Luo Y, Willey JC, Austermiller B, Barbacioru CC, Boysen C, Hunkapiller K, Jensen RV, Knight CR, Lee KY, Ma Y, Maqsodi B, Papallo A, Peters EH, Poulter K, Ruppel PL, Samaha RR, Shi L, Yang W, Zhang L, Goodsaid FM. (2006) Evaluation of DNA microarray results with quantitative gene expression platforms. *Nat Biotechnol* **24**: 1115–1122.

Cope LM, Irizarry RA, Jaffee HA, Wu Z, Speed TP. (2004) A benchmark for Affymetrix GeneChip expression measures. *Bioinformatics* **20**: 323–331.

Cui X, Kerr MK, Churchill GA. (2003) Transformations for cDNA microarray data. *Stat Appl Genet Mol Biol* **2**: A4.

de Brevern AG, Hazout S, Malpertuy A. (2004) Influence of microarrays experiments missing values on the stability of gene groups by hierarchical clustering. *BMC Bioinform* **5**: 114.

Dudoit S, Fridlyand J, Speed TP. (2002) Comparison of discrimination methods for the classification of tumours using gene expression data. *J Am Stat Assoc* **97**: 77–87.

Dudoit S, Popper Shaffer J, Boldrick JC. (2003) Multiple Hypothesis Testing in Microarray Experiments. *Statistical Science* **18**(1): 71–103.

Dunning MJ, Smith ML, Ritchie ME, Tavaré S. (2007) Beadarray: R classes and methods for Illumina bead-based data. *Bioinformatics* **23**: 2183–2184.

Dunning MJ, Barbosa-Morais NL, Lynch AG, Tavaré S, Ritchie ME. (2008) Statistical issues in the analysis of Illumina data. *BMC Bioinform* **9**: 85.

Durbin B, Rocke DM. (2003) Estimation of transformation parameters for microarray data. *Bioinformatics* **19**: 1360–1367.

Edwards D. (2003) Non-linear normalization and background correction in one-channel cDNA microarray studies. *Bioinformatics* **19**: 825–833.

ERCC. (2005) Proposed methods for testing and selecting the ERCC external RNA controls. *BMC Genomics* **6**: 150.

Fare TL, Coffey EM, Dai H, He YD, Kessler DA, Kilian KA, Koch JE, LeProust E, Marton MJ, Meyer MR, Stoughton RB, Tokiwa GY, Wang Y. (2003) Effects of atmospheric ozone on microarray data quality. *Anal Chem* **75**: 4672–4675.

Gentleman R, Carey V, Huber W, Irizarry R, Dudoit S. (2005) *Bioinformatics and Computational Biology Solutions Using R and Bioconductor (Statistics for Biology and Health)*. Springer.

Guo L, Lobenhofer EK, Wang C, Shippy R, Harris SC, Zhang L, Mei N, Chen T, Herman D, Goodsaid FM, Hurban P, Phillips KL, Xu J, Deng X, Sun YA, Tong W, Dragan YP, Shi L. (2006) Rat toxicogenomic study reveals analytical consistency across microarray platforms. *Nat Biotechnol* **24**: 1162–1169.

Hahne F, Huber W, Gentleman R, Falcon S. (2008) *Bioconductor Case Studies*. Springer.

Huber W, von Heydebreck A, Sultmann H, Poustka A, Vingron M. (2002) Variance stabilization applied to microarray data calibration and to the quantification of differential expression. *Bioinformatics* **18**: S96–104.

Huber W, Carey VJ, Long L, Falcon S, Gentleman R. (2007) Graphs in molecular biology. *BMC Bioinformatics* **8** Suppl 6: S8.

Irizarry RA, Wu Z, Jaffee HA. (2006) Comparison of Affymetrix GeneChip expression measures. *Bioinformatics* **22**: 789–794.

Irizarry RA, Bolstad BM, Collin F, Cope LM, Hobbs B, Speed TP. (2003) Summaries of Affymetrix GeneChip probe level data. *Nucleic Acids Res* **31**: e15.

Irizarry RA, Hobbs B, Collin F, Beazer-Barclay YD, Antonellis KJ, Scherf U, Speed TP. (2003) Exploration, normalization, and summaries of high density oligonucleotide array probe level data. *Biostatistics* **4**: 249–264.

Johansson P, Hakkinen J. (2006) Improving missing value imputation of microarray data by using spot quality weights. *BMC Bioinform* **7**: 306.

Kerr MK, Afshari CA, Bennett L, Bushel B, Martinez J, Walker NJ, Churchill GA. (2002) Statistical analysis of a gene expression microarray experiment with replication. *Stat Sinica* **12**: 203–217.

Kim H, Golub GH, Park H. (2005) Missing value estimation for DNA microarray gene expression data: local least squares imputation. *Bioinformatics* **21**: 187–198.

Kim KY, Kim BJ, Yi GS. (2004) Reuse of imputed data in microarray analysis increases imputation efficiency. *BMC Bioinform* **5**: 160.

Kooperberg C, Fazzio TG, Delrow JJ, Tsukiyama T. (2002) Improved background correction for spotted DNA microarrays. *J Comput Biol* **9**: 55–66.

Kreil DP, Russell RR. (2005) There is no silver bullet – a guide to low-level data transforms and normalization methods for microarray data. *Brief Bioinform* **6**: 86–97.

Li C, Wong WH. (2001) Model-based analysis of oligonucleotide arrays: expression index computation and outlier detection. *Proc Natl Acad Sci U S A* **98**: 31–36.

Lin SM, Du P, Huber W, Kibbe WA. (2008) Model-based variance-stabilizing transformation for Illumina microarray data. *Nucleic Acids Res* **36**: e11.

Lynch AG, Neal DE, Kelly JD, Burtt GJ, Thorne NP. (2007) Missing channels in two-colour microarray experiments: combining single-channel and two-channel data. *BMC Bioinform* **8**: 26.

Lynch A, Curtis C, Taveré S. (2007) Correcting for probe-design in the analysis of gene-expression microarrays. In: Barber S, Baxter PD, Mardia KV, editors. *Systems Biology & Statistical Bioinformatics*. Leeds University Press, pp. 83–86.

Naef F, Magnasco MO. (2003) Solving the riddle of the bright mismatches: labeling and effective binding in oligonucleotide arrays. *Phys Rev E Stat Nonlin Soft Matter Phys* **68**: 011906.

Nguyen DV, Wang N, Carrol RJ. (2004) Evaluation of missing value estimation for microarray data. *J Data Sci* **2**: 347–370.

Oba S, Sato MA, Takemasa I, Monden M, Matsubara K, Ishii S. (2003) A Bayesian missing value estimation method for gene expression profile data. *Bioinformatics* **19**: 2088–2096.

Okoniewski MJ, Miller CJ. (2008) Comprehensive Analysis of Affymetrix Exon Arrays Using BioConductor. *PLoS Comput Biol* **4**(2): e6.

Oshlack A, Emslie D, Corcoran LM, Smyth GK. (2007) Normalization of boutique two-color microarrays with a high proportion of differentially expressed probes. *Genome Biol* **8**: R2.

Ouyang M, Welsh WJ, Georgopoulos P. (2004) Gaussian mixture clustering and imputation of microarray data. *Bioinformatics* **20:** 917–923.

Parmigiani G, Garett E, Irizarry RA, Zeger SL. (2003) *The Analysis of Gene Expression Data.* Springer.

Patterson TA, Lobenhofer EK, Fulmer-Smentek SB, Collins PJ, Chu TM, Bao W, Fang H, Kawasaki ES, Hager J, Tikhonova IR, Walker SJ, Zhang L, Hurban P, de Longueville F, Fuscoe JC, Tong W, Shi L, Wolfinger RD. (2006) Performance comparison of one-color and two-color platforms within the MicroArray Quality Control (MAQC) project. *Nat Biotechnol* **24:** 1140–1150.

Qin LX, Kerr KF. (2004) Empirical evaluation of data transformations and ranking statistics for microarray analysis. *Nucleic Acids Res* **32:** 5471–5479.

Ritchie ME, Diyagama D, Neilson J, van Laar R, Dobrovic A, Holloway A, Smyth GK. (2006) Empirical array quality weights in the analysis of microarray data. *BMC Bioinform* **7:** 261.

Ritchie ME, Silver J, Oshlack A, Holmes M, Diyagama D, Holloway A, Smyth GK. (2007) A comparison of background correction methods for two-colour microarrays. *Bioinformatics* **23:** 2700–2707.

Rocke DM, Durbin B. (2001) A model for measurement error for gene expression arrays. *J Comput Biol* **8:** 557–569.

Rocke DM, Durbin B. (2003) Approximate variance-stabilizing transformations for gene-expression microarray data. *Bioinformatics* **19:** 966–972.

Scacheri PC, Crawford GE, Davis S. (2006) Statistics for ChIP-chip and DNase hypersensitivity experiments on NimbleGen arrays. *Methods Enzymol* **411:** 270–282.

Scheel I, Aldrin M, Glad IK, Sorum R, Lyng H, Frigessi A. (2005) The influence of missing value imputation on detection of differentially expressed genes from microarray data. *Bioinformatics* **21:** 4272–4279.

Sehgal MS, Gondal I, Dooley LS. (2005) Collateral missing value imputation: a new robust missing value estimation algorithm for microarray data. *Bioinformatics* **21:** 2417–2423.

Shi L, Reid LH, Jones WD, Shippy R, Warrington JA, Baker SC, Collins PJ, de Longueville F, Kawasaki ES, Lee KY, Luo Y, Sun YA, Willey JC, Setterquist RA, Fischer GM, Tong W, Dragan YP, Dix DJ, Frueh FW, Goodsaid FM, Herman D, Jensen RV, Johnson CD, Lobenhofer EK, Puri RK, Schrf U, Thierry-Mieg J, Wang C, Wilson M, Wolber PK, Zhang L, Amur S, Bao W, Barbacioru CC, Lucas AB, Bertholet V, Boysen C, Bromley B, Brown D, Brunner A, Canales R, Cao XM, Cebula TA, Chen JJ, Cheng J, Chu TM, Chudin E, Corson J, Corton JC, Croner LJ, Davies C, Davison TS, Delenstarr G, Deng X, Dorris D, Eklund AC, Fan XH, Fang H, Fulmer-Smentek S, Fuscoe JC, Gallagher K, Ge W, Guo L, Guo X, Hager J, Haje PK, Han J, Han T, Harbottle HC, Harris SC, Hatchwell E, Hauser CA, Hester S, Hong H, Hurban P, Jackson SA, Ji H, Knight CR, Kuo WP, LeClerc JE, Levy S, Li QZ, Liu C, Liu Y, Lombardi MJ, Ma Y, Magnuson SR, Maqsodi B, McDaniel T, Mei N, Myklebost O, Ning B, Novoradovskaya N, Orr MS, Osborn TW, Papallo A, Patterson TA, Perkins RG, Peters EH, Peterson R, Philips KL, Pine PS, Pusztai L, Qian F, Ren H, Rosen M, Rosenzweig BA, Samaha RR, Schena M, Schroth GP, Shchegrova S, Smith DD, Staedtler F, Su Z, Sun H, Szallasi Z, Tezak Z, Thierry-Mieg D, Thompson KL, Tikhonova I, Turpaz Y, Vallanat B, Van C, Walker SJ, Wang SJ, Wang Y, Wolfinger R, Wong A, Wu J, Xiao C, Xie Q, Xu J, Yang W, Zhang L, Zhong S, Zong Y, Slikker W. (2006) The MicroArray Quality Control (MAQC) project shows inter- and intraplatform reproducibility of gene expression measurements. *Nat Biotechnol* **24:** 1151–1161.

Shippy R, Fulmer-Smentek S, Jensen RV, Jones WD, Wolber PK, Johnson CD, Pine PS, Boysen C, Guo X, Chudin E, Sun YA, Willey JC, Thierry-Mieg J, Thierry-Mieg D, Setterquist RA, Wilson

M, Lucas AB, Novoradovskaya N, Papallo A, Turpaz Y, Baker SC, Warrington JA, Shi L, Herman D. (2006) Using RNA sample titrations to assess microarray platform performance and normalization techniques. *Nat Biotechnol* **24:** 1123–1131.

Smyth GK. (2004) Linear models and empirical Bayes methods for assessing differential expression in microarray experiments. *Stat Appl Genet Mol Biol* **3:** A3.

Smyth GK, Yang YH, Speed T. (2003) Statistical issues in cDNA microarray data analysis. *Methods Mol Biol* **224:** 111–136.

Stivers D, Wang J, Rosner G, Coombes K. (2002) Organ-specific differences in gene expression and UniGene annotations describing source material. *Proceedings of the Second International Conference for the Critical Assessment of Microarray Data Analysis.*(CAMDA), Durham, NC.

Tibshirani R. (1988) Estimating transformations for regression via additivity and variance stabilization. *J Am Stat Assoc* **83:** 394–405.

Tong W, Lucas AB, Shippy R, Fan X, Fang H, Hong H, Orr MS, Chu TM, Guo X, Collins PJ, Sun YA, Wang SJ, Bao W, Wolfinger RD, Shchegrova S, Guo L, Warrington JA, Shi L. (2006) Evaluation of external RNA controls for the assessment of microarray performance. *Nat Biotechnol* **24:** 1132–1139.

Troyanskaya O, Cantor M, Sherlock G, Brown P, Hastie T, Tibshirani R, Botstein D, Altman RB. (2001) Missing value estimation methods for DNA microarrays. *Bioinformatics* **17:** 520–525.

Tukey JW. (1977) *Exploratory Data Analysis.* Addison Wesley.

Venables WN, Smith DM. (2002) An Introduction to R. *Network Theory* .

Wu Z, Irizarry R, Gentleman R, Matinez Murillo F, Spencer F. (2004) *A Model Based Background Adjustment for Oligonucleotide Expression Arrays.* Johns Hopkins University, Dept. of Biostatistics Working Papers Working Paper 1.

Yang YH, Speed T. (2002) Design issues for cDNA microarray experiments. *Nat Rev Genet* **3:** 579–588.

Yang YH, Buckley MJ, Speed TP. (2001) Analysis of cDNA microarray images. *Brief Bioinform* **2:** 341–349.

Yang YH, Buckley MJ, Speed TP. (2002) Comparison of methods for image analysis on cDNA microarray data. *J Comput Graph Stats* **11:** 108–136.

Yang YH, Dudoit S, Luu P, Lin DM, Peng V, Ngai J, Speed TP. (2002) Normalization for cDNA microarray data: a robust composite method addressing single and multiple slide systematic variation. *Nucleic Acids Res* **30**(4): e15.

# Differential Expression

## 7.1 INTRODUCTION

The goal of most microarray expression experiments is to identify genes that are regulated by modifying conditions of interest. This involves a comparison of conditions, which asks whether different classes of subjects differ in their gene expression. The idea is that these differences or changes in gene expression might be responsible for, or caused by, the change in condition or phenotype. The main objectives for such experiments are to:

i. Identify which genes behave differently under different conditions.
ii. Determine a measure of confidence for these genes that behave differently.

Before embarking on any microarray experiment or indeed even collecting the samples, it is important to note and understand that good experimental design is essential in extracting meaningful information from an experiment and we have outlined the relevant approaches in Chapter 2. The development of an experimental plan can maximize the quality and quantity of information obtained. To distinguish between technical and biological variance it is important to avoid any kind of bias caused, for example, by certain conditions of the experiment, mRNA extraction or processing and so on. A good experimental design estimates the 'main effects'[1] of interest and minimizes or eliminates any confounding factors.[2] Key experimental design features such as sample randomization, balanced designs and the use of local controls are all very important. The choice of design may be dictated by the type of array platform being used, for example, 'dye-swap' designs for two-channel arrays are generally used to account for dye bias. For arrays that have a multiple probe set layout and can accommodate multiple samples on an single slide, such as the Illumina Bead-Array platform, randomization of samples is employed to avoid any 'top to bottom' slide effects, and a replicate from each condition group is used on the same slide to account for 'between array' differences and for reasons of robustness (it would be disastrous if all or most of the replicates for a condition group were placed on the same failed or poor quality slide but for small studies this may be unavoidable). It may be prudent to involve a statistician at the experimental design stage to ensure that all the research objectives are met by avoiding experimental design issues mentioned above and to help identify any necessary assumptions required for suitable statistical method(s). The potential consequences are clearly stated in the wise words of the evolutionary biologist, geneticist and statistician, Sir Ronald Fisher (1890–1962): 'To consult a statistician after the experiment is finished is often merely to ask him to conduct a post mortem examination. He can perhaps say what the experiment died of'.

Before applying any statistical method to microarray data, it is assumed that the data to be analyzed are of good quality, by assessing various QC plots and measurements, and have been appropriately transformed and normalized to

---

1 A 'main effect' is the simple effect of a factor on a dependent variable e.g. the effect of a particular drug treatment on gene expression response.
2 A 'confounding factor' is a hidden variable in a statistical or research model that affects the variables in question but is not known or acknowledged, and thus potentially distorts the resulting data. This hidden third variable may cause the measured variables to falsely appear to be correlated, or to be a 'causal' relation.

remove any experimentally introduced biases (see Chapter 6 for further details on data preprocessing and normalization). Dendrograms of clustered normalized data (Chapter 8) can provide an overview of whether the arrays (or samples) cluster into the expected 'condition' groups of interest. This may be important since sample or array mislabeling mistakes can happen either in the wet-lab or in silico when assigning identifiers to raw data files. Often such clustering may identify real but unexpected sample properties; for example, a clinical sample may be expected to classify with a particular disease subgroup but the clustering analysis may indicate that the patient is misdiagnosed or can be diagnosed under a new disease category. Of course samples may fail to behave as expected due to, for example, a new presence of technical variation that was not successfully removed by the data preprocessing or normalization procedures, or indeed the experimental design may not be adequate for removing such effects. In such cases, a 'postmortem' may be required, where laboratory notebooks are examined for any unanticipated experimental variability or the experimental design scrutinized for uncontrolled variables.

One or more filtering steps may be applied to normalized data before statistical analysis. The aim of nonspecific filtering is to remove probes or genes (the array features) that, due to their low overall intensity or variability across the samples in an experiment, are unlikely to contribute towards the phenotypes or conditions of interest. As we shall see, the objective here is to keep the number of statistical tests as low as possible while retaining the genes of interest in the selected subset. Filtering examples include one or more of the following:

- Removing low expression measurements using a particular intensity cutoff (e.g. typically anything $<100$ for Affymetrix GeneChips) across a certain percentage or group of arrays.
- Filtering out probes/genes with 'absent' (A) MAS5 calls across a certain percentage or group of Affymetrix GeneChips.
- Retaining the interquantile range (IQR) of measurements across the samples on the log base 2 scale that are at least 0.5.

Filtering on the variability is generally better than filtering on intensity alone. However, filtering out data before any statistical analysis is a contentious issue since intensity value cutoffs are arbitrarily chosen by the investigator and these may be difficult to judge: consequently, information may be lost rather than gained. It may therefore be more appropriate to apply filtering steps alongside the statistical assessment rather than applying them before any statistical analysis. Another form of filtering may be to focus on a particular biology of interest before applying a statistical test. The investigator may, for example, be interested in genes that belong to a particular molecular function via a Gene Ontology (GO) category and therefore focus attention on such genes. As we shall see below, selecting a subset of genes results in a reduction in the number of statistical tests and this may lead to relatively more significant genes being identified.

Quantifying differences in gene expression depends on the experimental setting. The experimental setting is distinguished according to the number of 'conditions' (sometimes also known as 'classes') being compared and this can either involve the comparison of two conditions or multiple conditions. Ideally, each condition should be represented by multiple independent biological samples. The sample size is positively related to statistical power and is dependent on the number of biological or technical replication. If only technical replicates are available, statistical testing is still possible but technical replicates only allow for the effects of measurement variability to be estimated and reduced, whereas biological replicates allow this to be done for both measurement variability and biological variability between samples. However, there are some experimental situations where technical replicates are required, such as those involved in quality control or optimizing technical protocols. If the samples are pooled (because of limiting material) this can affect how biological variance can be accounted for. It is more favorable to have many pools of biological replicates rather than fewer numbers of pools with larger numbers of biological replicates in each pool (e.g. four pools of four samples is better than two pools of eight samples) and multiple pools are required to estimate the variance between them for inference testing.

Although increasing numbers of multiple condition microarray studies are performed, the vast majority tends to be studies with two conditions (binary), for example where a control and a treatment are examined (e.g. wildtype versus genetically modified or normal versus diseased state). The number of cases or samples is frequently limited typically by cost and/or the availability of sufficient biological material, and this together with the noise inherent in microarray data is a significant challenge to any gene selection approach. However, in the absence of replicates, possibilities for data analysis are very limited.

Often when there are no or very few replicates, genes are selected using a fold-change (FC) criterion since it is simple and intuitive but there are severe drawbacks to using this approach and it should be discontinued. Typically, a FC is considered to be significant if it is at least twofold or threefold in difference. However, the FC is chosen arbitrarily and might often be inappropriate, especially since the microarray technology tends to have a bad signal-to-noise ratio for genes with low expression levels. In addition, this approach does not involve any assessment of the significance of differential expression in the presence of biological and experimental variation, which may differ from gene to gene. Using the FC approach alone is not considered to be a valid inferential statistic because it does not provide known and controllable long-range error rates,[3]

---

3 The long-range error rate is defined as the expected error rate if experiments and analyses of the type or condition under consideration were repeated an infinite number of times.

which are essential for inference. The main reasons for using inferential statistics is to assess differential expression in terms of statistical power and measures of confidence.

When applying a statistical method, care must be taken to ensure that the statistic used is appropriate for the microarray data as not all methods are appropriate for every situation. Different statistical methods give different $p$-values, and therefore using an inappropriate statistic will result in inaccurate conclusions. In this chapter we introduce various statistical methods for finding upregulated or downregulated (differentially expressed) genes, and the methods range from classical statistics to more modern theories that address the issue of statistically analyzing small sample sizes. Furthermore, we make recommendations regarding approaches (mainly packages available from Bioconductor or methods implemented in $R$), which are most suited to different data structures. For example, many of the modern methods mentioned here are more suited for analyzing studies with small sample sizes. Another, increasingly popular, statistical approach for analyzing microarray data is encapsulated in various forms of gene-set or category analysis and these are also discussed in this chapter. Most of these methods are statistical tests that compare the number of genes in a class (or phenotypic setting such as a GO category) that are 'significant' with the number that are 'expected' to identify overrepresentations of genes or 'gene enrichments' associated with the class in question. They alleviate the problem of interpreting a long list of differentially expressed genes by giving some biological meaning to them.

Before any of these statistical methods are mentioned in any more detail, some of the fundamental concepts and issues that underlie all these methods must first be introduced: statistical inference; univariate and multivariate statistics, replicates and how these affect statistical power; hypothesis tests and $p$-values; type 1 and type 2 errors and the problem of multiple hypothesis testing in microarray data analysis. A good review on microarray data analysis can be found in Allison et al. (2006). Suggested textbooks on using various Bioconductor packages and R include Hahne et al. (2008), Gentleman et al. (2005) and Venables and Smith (2002), and for statistics these include Everitt and Palmer (2005), Witmer and Samuels (2003) and Kanji (2006). More advanced textbooks include Dobson, A. (2008) for generalized linear models, Hair et al. (2007) for multivariate data analysis, and Sivia & Skilling (2006) and Gelman et al. (2003) for Bayesian data analysis.

## 7.2 STATISTICAL INFERENCE

*Statistical inference* concerns the problem of inferring properties of an unknown distribution from data generated by that distribution. As it is impractical to measure every individual in a population, a representative sample of individuals is selected and measurements are made on these individuals (Figure 7.1). An

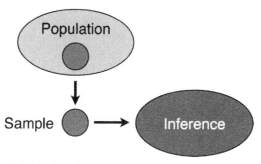

**FIGURE 7.1**   The basic idea behind statistical inference. It is impractical to measure every individual in a population, thus a representative sample of individuals from a population is typically chosen and measurements are made on these individuals. When making a statistical inference, the interest lies in the variations between individuals in the sample population to which an extrapolation is made.

extrapolation is then made on these measurements to make assertions about the population from which the representative sample was derived. To do this, it is important to capture as much *biological variability* in the experiment as possible, both at the experimental design stage and in the statistical analysis. As mentioned already, subject to any financial or practical constraints, it is important to maximize the number of biological replicates to capture this variability. Another important consideration may be to include population subtypes or additional variables such as sex, genotype or age.

According to the random sampling model, the data in an experiment is regarded as a *random sample* from a population and typically, for an experiment, only a single random sample is obtained from a very large population. The variability among multiple random samples from the same population is called *sampling variability* and a probability distribution that characterizes some aspect of sampling variability is termed a *sampling distribution*. In order to visualize sampling variability, a broader frame of reference is required that includes not merely one sample, but all possible samples that might be drawn from the population. This wider frame of reference is termed a *meta-experiment*. For example, in the case of an experiment where a random sample of $n = 10$ mice is drawn from a population and these are given a particular drug, the corresponding meta-experiment would consist of repeatedly choosing groups of $n = 10$ mice from the same population and administering the same drug under the same conditions indefinitely, with the members of each sample being replaced before the next sample is drawn. Essentially, a sampling distribution tells us how close the resemblance between the sample and population is like to be and the larger $n$ is, the more closely the sample resembles the population. A random sample is not necessarily a representative sample, but using sample distributions, the degree of representation to be expected in a random sample can be specified, for example a larger sample size is likely to be

more representative than a smaller sample from the same population. Therefore, sampling distributions provide what is known as 'certainty about
uncertainty'.

## 7.2.1 Univariate and Multivariate Statistics

The most commonly used statistical methods involve the use of a *univariate*
statistical test (e.g. Student's *t*-test). Univariate statistical methods are those
that utilize only a single *dependent variable* and one or more *independent
variables*. Independent variables are those whose values are controlled (or
selected by the experimenter) to determine their relationship to an observed
phenomenon (the dependent variable). In such an experiment, an attempt is
made to find evidence that the values of the independent variable determine
the values of the dependent variable (that which is being measured). For
example, in a microarray study to compare the following two condition
groups: 'the wild-type mouse' and 'the transgenic mouse', the statistical
analysis involves comparing the two groups of conditions as the two independent variables to evaluate differential gene expression, which is the
dependent variable. There are two types of variables: *numerical variables*
(also known as *measurement variables*, for example gene expression ratio,
blood pressure, height, weight, age, and probability of illness.), which are
treated as continuous since the relative magnitude of the values is significant
(e.g. a value of 2 indicates twice the magnitude of 1 but only if there is a
meaningful zero point, which is not the case for some parameters such as
temperature in °C). The second-type, *character variables* (e.g. sex: M for
male and F for female, which can also be denoted as 1 for male an 2 for female
and sometimes referred to as *nominal variables*), are treated as categorical
since their values function as labels rather than as number values for a
particular measurement. Some independent variables can have two levels or
more, for example gender can be either male or female, the time of drug
treatment can have two levels, pretreatment and posttreatment, whereas the
severity grade of breast cancer or drug dosage could have three levels such as
low, medium and high.

Characteristics of the independent and dependent variables determine the
statistical test and sampling distribution that will be used in order to assess
significance: many of these are summarized later in the chapter. If the study
has more than one dependent variable, then the procedures used are referred
to as *multivariate*. In other words, the labels univariate and multivariate are
only concerned with the number of dependent variables without reference to
the number of independent variables. Example univariate methods include:
regression (simple and multiple), *t*-test (dependent and independent)
and all forms of analysis of variance (ANOVA) or analysis of covariance (ANCOVA). Example multivariate methods include factor analysis,

multivariate analysis of variance (MANOVA) and multivariate analysis of covariance (MANCOVA).

## 7.2.2 Replicates and Power

Univariate statistical methods require a sufficient size (number) of sample replicates to achieve statistical power and sample size is positively related to power. Studies that do not find statistical significance with small sample sizes may have low power, however, had there been a larger sample size, the findings may have been different. Studies with larger sample sizes may find statistical significance even with small correlations or small differences between condition groups. When there are small differences in the population, larger sample sizes tend to have sufficient power to reliably detect such differences, although these small differences may or may not be meaningful. While the optimum number of replicates will vary between different experiments, for the most common design comparing two groups of conditions to evaluate differential gene expression, evidence indicates that at least a minimum (and not the optimum) of *five biological replicates per condition group* should be analyzed (Allison et al., 2002; Pavlidis et al., 2003; Tsai et al., 2003). Power[4] can loosely be defined as the ability of a test statistic to detect change given that one exists and depends on the number of replicates along with many other factors. Therefore, the most straightforward way of increasing statistical power is to increase the number of replicates examined. Jørstad et al. (2007) provides a good explanation of understanding sample size and how important it is in gaining sufficient power to detect differentially expressed genes. The main `take-home' message here is that designing a microarray experiment with the appropriate number of replicates is cost efficient. An undersized study will be wasteful in terms of resources since it will not have sufficient power to yield good results. In contrast, an oversized study will consume more resources than necessary since increasing the number of replicates beyond a certain point will have little impact on the power.

For medium to large-scale studies that can utilize classical statistical methods, power can be used to estimate the number of replicates required for the research objectives and there are several tools available to do this. These include sizepower, ssize and Ocplus, which are all available from Bioconductor. Some methods have been developed for estimating sample sizes for classification studies to develop, for example prognostic disease markers (Dobbin and Simon, 2005). Power analysis tools can also be used to return the power of the experiment when the number of replicates in the study is supplied. The formulae

---

4 Power is classically defined as the probability of rejecting a null hypothesis that is false, however, it has been defined in many ways for microarray studies.

for power analysis is complicated, however, when making power analysis calculations it is important to appreciate that power typically depends on the following factors:

- *The number of replicates* (but use biological replicates and not technical replicates since the analysis is making a statistical inference about the population from which the replicates derive).
- *The type of analysis* (e.g. paired tests are more powerful than unpaired tests).
- *Expected difference in mean*[5] (e.g. for paired data, this is the difference in mean and zero; for unpaired data, this is the difference in mean between two groups – the log ratio or the *M*-value in a microarray analysis).
- *Expected standard deviation*[6]*of the population variability* (this has the assumption that errors in gene expression are log normally distributed, otherwise approaches such as boot-strapping methods of power analysis must be used).
- *The significance threshold of the test* (e.g. 0.05, which is related to the confidence required, e.g. 95%).

Generally, power analysis tools are not appropriate for small sample sizes but many microarray studies are small mainly due to limitations in cost and sample material. For small sample sizes, the sample size formulae are problematic mainly because they assume that the variance is known. However, for small sample sizes an estimated variance is typically calculated and this may be unreliable.

### 7.2.3 Hypothesis Tests and *p*-Values

A hypothesis test is a method for making statistical decisions about experimental data: a hypothesis is a statement regarding the predicted outcome of a particular experiment. For microarray studies, hypothesis-testing approaches allow us to determine whether a gene is differentially expressed. There are two main schools of thought about hypothesis testing, the *classical* and the *Bayesian*. As explained by Wit and McClure (2004), classical statisticians (the 'frequentists') believe that 'parameters[7] such as the population mean are fixed but these are assumed to be unknown quantities and only the observations are truly random'. In contrast, Bayesians believe that 'data can be fixed after being observed but the knowledge about the parameters is assumed to be truly random'. This difference leads to alternative approaches for hypothesis testing.

---

5 The mean is the average value of all members of the group.
6 The standard deviation is a measure of how much the values of individual members vary in relation to the mean.
7 A parameter is a quantity, such as mean, that characterizes some aspect of a (usually theoretically infinite) population.

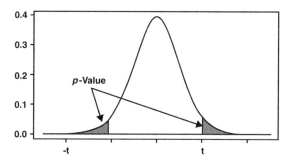

**FIGURE 7.2**   The Student's $t$-distribution. The $y$-axis is the density and the $x$-axis is the $t$-statistic value, $t$. The $p$-value of the test is the two-tailed area under the $t$-distribution at the two tail ends beyond the two values $-t$ and $+t$.

### 7.2.3.1 Classical Hypothesis Testing

The classical approach tests for each gene $g$ the hypothesis that that particular gene is not differentially expressed. The negative hypothesis is the *null hypothesis*, $H_0$ and based on this, a hypothesis test builds a probabilistic model for the observed data. Hypothesis testing requires a differential expression score, which is calculated for each gene and is also referred to as the *test statistic, $t_s$*. This score can take a variety of forms (of which the $t$-statistic is among the most common) and essentially summarizes the evidence in the data about whether a gene is 'active' or inactive' on the array and is used to form the basis of deciding whether a gene is differentially expressed.

For example, after the $t_s$ is calculated, the essence of the $t$-test procedure is to locate the observed $t_s$ in the Student's $t$-distribution, which is similar to the normal distribution except the tails of the bell curves are wider. If the $t_s$ is near the centre of the distribution then the data are regarded as compatible with $H_0$ because the observed differences between the means of the two samples being compared can be readily attributed to chance variation caused by sampling error. $H_0$ predicts that the sample means will be the same, since it says that the population means are equal. Alternatively, if $t_s$ falls within the far tails of the distribution, then the data are regarded as incompatible with $H_0$ because the observed data cannot be readily explained as being due to chance variation. A yardstick or cutoff for locating where $t_s$ is in relation to the two distribution tails is provided by the $p$-value. The $p$-value of the test is the area under the $t$-distribution (curve) in the double tails beyond two values in the $x$-axis: $-t_s$ and $+t_s$ (Figure 7.2). If the $p$-value is less than a particular critical value, alpha (provided by the investigator) then $H_0$ is rejected.

Studies that test directional predictions have more power than those that do not. If there is a directional hypothesis, that is if the investigator predicts that there is a positive or negative relationship between variables, a one-tailed test of significance is used since it looks for only one definite direction, either an increase or a decrease, in the parameter. In contrast, if there is no direction to the prediction then a two-tailed test is used since it looks for any change (an increase or a decrease) in the parameter (i.e. when looking at two condition groups, both groups are different but you do not care how). These two terms relate to how the critical value is determined for determining statistical significance.

For most practical purposes, the result of a hypothesis test is the $p$-value, a quantitative measure reporting the result of a hypothesis test. It is also known as an absurdity probability and is closely related to the *false positive rate* (FPR), the expected fraction of false positives. False positives represent type 1 errors, declaring genes to be differentially expressed when they are not (see Section 7.2.4). Depending on the data, there are several ways to calculate the $p$-value and they tend to be mathematically complicated. For example, the Student's $t$-distribution is used to calculate $p$-values when using $t$-statistics on large sample sizes (>30 in each sample) or when samples are from a normal distribution. When the data is neither large nor normally distributed, then $p$-values can be calculated using, for example, bootstrap methods (see Section 7.5.2 on boot-strapping methods).

The $p$-value translates the value of the test statistic into a probability that expresses the absurdity of the null hypothesis. The lower this probability, the stronger the evidence against the null hypothesis: if the $p$-values are close to zero and are less than a particular predefined cutoff, then the $H_0$ is absurd and is rejected in favor of the *alternative hypothesis*, $H_{g,1}$, where gene $g$ is differentially expressed. For example, a $p$-value of 0.01 would mean that there is a 1% chance of observing at least this level of differential gene expression by random chance. The $p$-value represents how likely the observed data would be, if in fact the null hypothesis were true. More formally, the frequentist interpretation of probability is a measure of limiting relative frequency of outcome in a set of identical experiments. Good comprehensive reviews on classical hypothesis testing in the context of microarray studies can be found in Ge et al. (2003) and Dudoit et al. (2003). Box 7.1 shows a basic structural overview of hypothesis testing.

### 7.2.3.2 Bayesian Hypothesis Testing

It is important to understand that classical $p$-values alone do not directly answer the main question that most investigators ask: 'whether a gene of interest is differentially expressed'. They provide the probability of observing a difference as large as or larger than those actually observed, if in fact there is no differential expression of the gene. If the probability ($p$-value) is relatively very low, it is concluded by deduction that the gene was probably differentially expressed. The Bayesian approach to statistical inference can provide a more direct approach to this primary question of whether a gene is differentially expressed. The frequen-tist approach to statistical inference is based on the distribution of data when given a fixed constant value for the parameter. In contrast to this, the Bayesian inference is predicated on defining a parameter as a random variable and the

---

**Box 7.1   The General Structure of Classical Hypothesis Testing**

Although there are some variations on the structure, typically, it is important to do the following:

   i. Define the research hypothesis and set the parameters for the study.
  ii. Set out the null and alternative hypothesis (or a number of hypotheses).
 iii. Operationally define what is being studied and set out the variables to be studied, in other words, explain how the concepts are being measured.
  iv. Set the significance level.
   v. Make a one- or two-tailed prediction.
  vi. Determine whether the distribution of the data is normal as this has implications for the types of statistical tests that can be used.
 vii. Select an appropriate statistical test based on the variables defined and whether or not the distribution is normal.
viii. Perform the statistical tests on the data and interpret the output.
  ix. Accept or reject the null hypothesis.

---

probability distribution of the parameter is obtained when given the data and the prior distribution of the parameter.

Bayesian data analysis involves making inferences from data using probability models for observed quantities and for uncertain quantities that we wish to learn about. The parameter of interest in the Bayesian approach is a binary one. In the context of studying microarrays and hypothesis testing, the Bayesian approach assumes that for each gene $g$ there is an unobservable variable, $v_g$, that defines the gene activity status, that is $v_g = 0$ if the gene is not differentially expressed and $v_g = 1$ if otherwise. The *posterior probability*[8] is the Bayesian probability that a hypothesis is correct, which is conditional on the observed data. It is used to express the *likelihood*[9] that $v_g = 0$, and this quantity is closely related to the *false discovery rate*[10] (FDR), the expected proportion of false positives. The Bayesian interpretation of probability is a measure of the degree of belief in a statement. Although the frequentist and Bayesian approaches may seem worlds apart, the main difference between them does not come from the difference in testing philosophy but from the choice of test statistic. Both Wit and McClure (2004) and Edwards and Ghosh (2006) discuss Bayesian approaches in relation to microarray studies in more detail.

---

8 The Bayesian approach of obtaining the posterior probability distribution involves multiplying the prior by the likelihood function. The prior, also known as the prior probability distribution, is generally referred to as prior knowledge, that is what was known before or some form of background knowledge available in addition to the data at hand.

9 In Bayesian terms, likelihood is the probability of obtaining observed data, given the model.

10 The false discovery rate (FDR), is defined as the expected fraction of false rejections among those hypotheses rejected. The FDR gives an idea of the expected number of false positive hypotheses that a practitioner can expect if the experiment is done an infinite number of times.

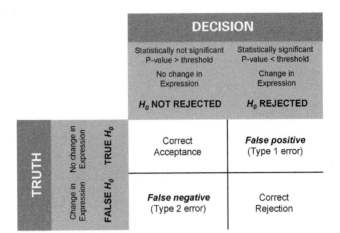

**FIGURE 7.3**   When testing a single pair of hypotheses, there are four possible outcomes in relation to the test result (the 'decision') and the true biological event, two of which are inferential errors: the type 1 error (a false positive) and the type 2 error (a false negative). A type 1 error is made when declaring a gene to be differentially expressed at the $p$-value threshold when it is not, and occurs when the null hypothesis is wrongly rejected. A type 2 error is made when not declaring a gene to be differentially expressed at the $p$-value threshold when it is, and occurs when the null hypothesis is wrongly accepted.

## 7.2.4 Error Rates and Power

As illustrated in Figure 7.3, there are two possible inferential errors when testing a single pair of hypotheses:

1. *Type 1 error* (also known as a *false positive*) occurs when the null hypothesis is wrongly rejected, for example declaring a gene to be differentially expressed at the $p$-value cutoff $p < \alpha$ when it is not.
2. *Type 2 error* (also known as a *false negative*) occurs when the null hypothesis is wrongly accepted, for example not declaring a gene to be differentially expressed at the $p$-value cutoff $p < \alpha$ when it is.

   The probability of making a type 2 error is indicated by the false negative rate (FNR), also referred to here as $\beta$. However, the probability of not making a type 2 error is indicated by the power of the test, that is one minus the FNR (i.e. $1 - \beta$). By reducing the significance level and therefore reducing the probability of a false positive, the power of the test tends to be reduced. The rule is that the lower the $\alpha$ value (i.e. the $p$-value cutoff), the greater the $\beta$ and thus the lower the power $(1 - \beta)$. In other words, as the probability of falsely accepting the null hypothesis increases the power of the test decreases. In practice, this means that the investigator should choose the largest permissible $\alpha$ level to increase the chance of finding a statistically significant differentially expressed gene, although this obviously increases the chance of making a type 1 error. However,

|  | NOT REJECTED | REJECTED | Total |
|---|:---:|:---:|:---:|
| Number of true $H_0$ (TRUE) | U | V | $m_0$ |
| Number of false $H_0$ (FALSE) | T | S | $m - m_0$ |
| Total | $m\text{-}R$ | R | $m$ |

**FIGURE 7.4**   Numbers of type 1 and type 2 errors of multiple hypotheses tests. $V$ is the number of falsely accepted null hypotheses (*Type 1 errors* or *false positives*); $U$ is the number of null hypotheses correctly rejected; $S$ is the number of correctly rejected null hypotheses; $T$ is the number of falsely rejected null hypotheses (*Type 2 errors* or *false negatives*); $m_0$, is the number of true null hypotheses; $m$, the total number of hypotheses; $m\text{-}R$, is the number of hypotheses not rejected; $R$, the number of rejected hypotheses.

it is important to note that $p = 0.05$, the most commonly used cutoff, is as high as one should really go for statistical significance.

## 7.2.5 Multiple Testing

'Multiple testing' (or multiple hypothesis testing) is a situation where more than one hypothesis is evaluated simultaneously and is one of the key statistical issues arising in microarray data analysis. When analyzing microarray data, statistical tests are performed on many genes to identify those that are likely to be differentially expressed based on standard $p$-values. However, since the statistical tests are applied to many genes in parallel this causes a problem known as multiplicity of $p$-values. Each condition or treatment comparison involves the testing of every gene on the array, often tens of thousands. Some microarray experiments may involve multiple conditions and so the total number of hypotheses tested in a particular experiment can easily exceed 100 000. Unfortunately, when many hypotheses are tested the probability that a type I error occurs increases sharply with the number of hypotheses. While this problem is not unique to microarray analysis, the magnitude in terms of the number of genes or features on typical arrays dramatically increases the problem. Dudoit et al. (2003) and Pollard et al. (2005) both provide good reviews on multiple testing.

When considering the problem of simultaneously testing $m$ null hypotheses, $H_j, j = 1, \ldots, m$, the number of rejected hypotheses is represented by $R$, as summarized in Figure 7.4 (Benjamini and Hochberg, 1995). For microarray studies, the null hypothesis is $H_j$ for each gene $j$, and rejection of $H_j$ corresponds to declaring the gene as being differentially expressed. It is assumed that the number of hypotheses $m$ is known, but the numbers $m_0$ and $m - m_0$ of the respective true and false null hypotheses are unknown. $R$ is an observable random variable, while $S$, $T$, $U$ and $V$ are unobservable random variables. The standard univariate approach involves pre-specifying an acceptable cutoff level, $\alpha$, for $V$ and finds tests that minimize $T$. This maximizes power within the class of tests with type 1 error rate at most $\alpha$. The confidence of a statistical test is commonly defined as the probability of not getting a false positive result, whereas the power of a statistical test is commonly defined as the probability of not getting a false negative result.

Neither Frequentists nor Bayesians can afford to ignore the issue of multiplicity. There are several methods that address multiple testing by adjusting the $p$-value from a statistical test and these are based on the number of tests performed. Such adjustments are said to control and reduce the number of false positives in an experiment at the level of $\alpha$,[11] if the error rate is less than or equal to $\alpha$ when the given procedure is applied to produce a list of $R$ rejected hypotheses. Many of these methods are reviewed in Dudoit et al. (2003) and Popper Shaffer (2006). Re-sampling based methods are reviewed in Pollard and van der Laan (2005) and van der Laan et al. (2004), while both Efron et al. (2001) and Storey (2001) discuss Bayesian interpretations of the FDR in the specific context of microarray data. There are two main categories of multiple testing procedures that are applied depending on which error rate is being corrected:

1. *Family Wise Error Rate (FWER) corrections* adjust the $p$-value so that it reflects the chance of at least 1 false positive being found in the list. Procedures include; Bonferroni (1935), Holm (1979), Westfall and Young (1993) and MaxT (Dudoit et al., 2004). The FWER is controlled at the level of $\alpha$ by a particular multiple testing procedure if FWER $\le \alpha$. The FWER is defined as the probability of at least one type 1 error:

$$\text{FWER} = P(V > 0)$$

2. *False Discovery Rate (FDR) corrections* adjust the $p$-value so that it reflects the frequency of false positives in the list. Examples of FDR corrections include: Benjamini and Hochberg (1995), Benjamini and Yekutieli (2001) and SAM (Statistical Analysis of Microarrays; Tusher et al., 2001; Storey and Tibshirani 2003a). The FDR is controlled at the level of $\alpha$ by a particular multiple testing procedure if FDR $\le \alpha$. The Benjamini and Hochberg (1995) definition of the FDR is the expected proportion of erroneously rejected null hypotheses (type 1 errors) among the rejected ones:

$$\text{FDR} = E\left(\frac{V}{R}\right) \text{ if } (R > 0), \text{ and } 0 \text{ if } (R = 0)$$

---

11 Alpha ($\alpha$) is the cutoff value used to decide if the probability of the results is high or low. In other words, alpha is the probability [$p$(outcome/$H_0$ is true)] which is low enough for you to stop believing that the null hypothesis is true. The most commonly used alpha is 0.05.

The FWER correction methods are more conservative than the FDR correction methods, but will give a higher number of false negatives (i.e. lower power). Probably the most well-known and traditional approach is the Bonferroni (1935), a single-step procedure that involves multiplying the $p$-value for each gene by $n$, the number of statistical tests performed, that is by the number of genes in the analysis. However, when analyzing microarray data this method is far too stringent since the resulting $p$-values are usually so large that no or very few genes are declared to be differentially expressed. A slightly less stringent approach is the Bonferroni step-down procedure, also known as the Holm procedure (Holm, 1979), which involves ranking the $p$-values from the smallest to largest. For the first $p$-value (the smallest), it is multiplied by $n$, and for the second $p$-value, it is multiplied by $(n - 1)$, and then for the third $p$-value, $(n - 2)$ and so on. Both the Bonferroni and Holm procedures are single-step procedures where each value is corrected independently.

A similar, but less stringent FWER correction procedure, is the Westfall and Young permutation method (Westfall and Young, 1993), which follows a step-down approach similar to the Holm procedure, but uses a resampling-based bootstrapping method to compute the $p$-value distribution (see Section 7.5 discussing nonparametric statistics for more information on boot-strapping). Unlike the Bonferroni and Holm procedures, the approach takes advantage of the dependence structure between genes[12] by permuting all the genes at the same time. The approach creates a pseudodata set by dividing the data into artificial condition groups and $p$-values for all the genes are computed on this data set. The new successive minima of $p$-values are retained and compared to the original ones. This whole process of creating pseudo data sets used to compute $p$-values, which are then retained etc, is repeated a large number of times (and can be computationally intensive). The adjusted $p$-value is then the proportion of resampled-data sets where the minimum pseudo $p$-value is less than the original $p$-value.

FDR control is a relatively new approach to the multiple comparisons problem and is usually acceptable for 'discovery' experiments, where a small number of false positives is acceptable. When some of the tested hypotheses are genuinely false, FDR control is less stringent than the FWER control, and thus FDR controlling procedures are potentially more powerful.

There are various FDR procedures, which have been well reviewed by Reiner et al. (2003). The linear step-up procedure (Benjamini and Hochberg, 1995) is probably the most well-known, indeed these authors first coined the terminology FDR. It involves ranking the $p$-values from each gene from smallest to largest and the largest $p$-value then remains as it is. The

---

12 Many genes are involved in multiple biological pathways and are therefore highly correlated.

second largest $p$-value is then multiplied by the total number of genes list, $n$, and then divided by its rank, $n - 1$: corrected $p$-value = $p$-value$(n/(n - 1))$. The procedure controls the FDR at the desired level $q$ for both independent and positively dependent test statistics, but it controls the FDR at a level too low by a factor of $m_0/m$. Therefore, adaptive methods are applied to try to estimate $m_0$ and $q^* = q(m/m_0)$ is used instead of $q$ to gain more power.

Adaptive (also known as FDR estimation) procedures include Benjamini and Hochberg (2000), which combines the estimation of $m_0$ with the linear step-up procedure, and SAM, which involves a two-stage $p$-value adjustment and FDR estimation using permutations of the data. Benjamini et al. (2001) suggest a similarly motivated two-stage procedure with proven FDR controlling properties. Resampling based FDR procedures such as SAM and Yekutieli and Benjamini (1999) allow the researcher to use quantile-quantile plots (q-q plots are discussed below in Section 7.3) of all $p$-value adjustments to decide on a meaningful rejection region while being warned of the overall type I error in terms of the FDR. Other FDR estimation procedures include mixture-model methods, MMMs (Pounds and Morris, 2003; Datta, 2005; Do et al., 2005). All these methods estimate a gene-specific FDR that is interpreted as the Bayesian probability that a gene declared to be differentially expressed is a false positive (Storey, 2001).

The `siggenes` Bioconductor package, which uses the SAM approach (Storey and Tibshirani, 2003c), provides a resampling-based multiple testing procedure involving permutations of the data (see Section 7.5 on nonparametric statistics for more details). `limma` provides several common options for multiple testing, including the Bonferroni and Holm FWER correction methods as well as FDR correction methods (Benjamini and Hochberg 1995; Benjamini and Yekutieli, 2001). The `multtest` package provides a wide range of resampling-based methods for both FWER and FDR correction (Pollard et al., 2005). The `qvalue`[13] package from CRAN uses a method that involves calculating $q$-values from $p$-values for controlling the FDR (Storey and Tibshirani, 2003b; Dabney and Storey, 2006). This method defines $q(t)$ as the minimal FDR that can occur when rejecting a statistic with value $t$. While the $p$-value is a measure of significance in terms of the false positive rate of the test, the $q$-value is a measure of strength of an observed statistic with respect to the FDR. The package can be used to create various graphical plots to assess quality and to identify an appropriate $q$-value cutoff (see Figures 7.5 and 7.6). However, for experiments where only a few genes are expected to be differentially expressed, the $q$-value method may not perform as well as other methods (Yang, 2004).

---

13 For availability and more information: http://genomics.princeton.edu/storeylab/qvalue/.

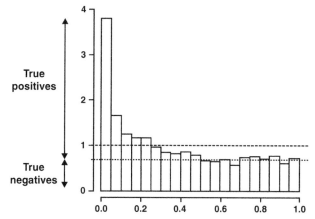

**FIGURE 7.5**   A density histogram of the 3170 $p$-values from the Hedenfalk and Duggan (2001) data created using the $q$-value package. Genes with no change in expression will contribute to a uniform frequency whilst genes that change with expression will tend to have a low $p$-value. The FDR process searches for a nonbackground effect. To assess this, this process needs to separate the background distribution from the changing distribution, which together, contributes towards the $p$-values distribution. The dashed line is the density expected if all genes were null (not differentially expressed), so it can be seen that many genes are differentially expressed. The dotted line is at the height of the estimate of the proportion of null $p$-values.
*Source*: Adapted and taken from Storey and Tibshirani (2003a,b,c).

## 7.3 PARAMETRIC STATISTICS

The most common methods of both classical and modern statistical approaches described in this chapter can be categorized by whether or not they assume normality in the data. While parametric statistics assume the data are normally distributed, nonparametric statistics do not and consequently tend to be used when there is a high amount of biological heterogeneity in the condition groups. Generally, classical statistical methods require large sample sizes and this is not usually the case with microarray experiments, however, there are some recent approaches mentioned here that help address this issue. The normality assumption does not matter so much for large sample sizes (>30 in each sample) since here the data tend to be normally distributed. However, it is always worth checking, since outliers or highly variable observations may skew the data and hence the results.

Parametric statistics use numerical data, such as the arithmetic mean, standard deviation, etc. to determine the significant differences between two groups or sets of data. To utilize these tests, assumptions are made on the normal distribution of the gene expression data and equality of variance among the groups. However, if these assumptions are not satisfied, then such methods become unreliable because the mean and variance will no longer completely describe the population. As discussed later, nonparametric tests are more

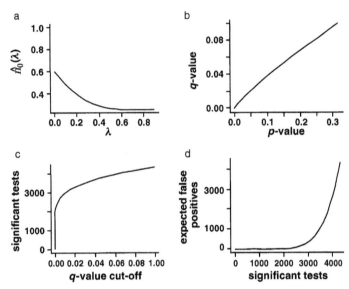

**FIGURE 7.6**   Examples of various plots from the `qvalue` package. (a) The estimate ($\pi_0 = m_0/m$) of the proportion of genes for which the null hypothesis is true versus the tuning parameter $\lambda$, is used to assess quality. As $\lambda$ gets larger the bias of the estimates decreases yet the variance increases. Here, the estimate = 0.251 can also be accessed and an estimate of the proportion of significant tests is 1 minus this number. (b) The $q$-values versus the $p$-values can reveal the expected proportion of false positives for different $p$-value cutoffs. (c) The number of significant tests versus each $q$-value cutoff. (d) The number of expected false positives versus the number of significant tests. The latter plot can be used to show how many tests are significant in relation to the expected false positives. A thorough discussion of these plots can be found in Storey and Tibshirani (2003a,b,c).

applicable under these circumstances. Figure 7.7 illustrates what a normal distribution is expected to look like. The most useful tool for assessing normality is through the use of quantile–quantile plots ($q$–$q$ plots), which are used to graphically determine if two data sets come from populations with a common distribution. Nonnormality of the population distribution in question can be detected by comparing it with the normal distribution. The points are formed from the quantiles of the two data sets as a scatter plot and if the two distributions are the same, this approximates a straight line, especially near the centre. Curvature of the points indicates departures of normality.

When assessing whether a population follows a normal distribution, the data should form a graph that looks like a bell curve with the following character-istics:

- There should be a single peak at the centre, which occurs at the mean ($\mu$).
- The curve should be symmetrical.
- The graph should continue towards infinity along the horizontal axis.
- The area under the curve should equal one.

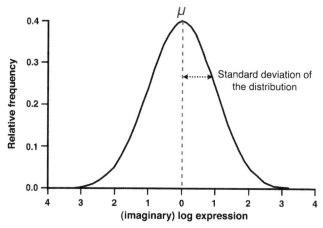

**FIGURE 7.7**   A standard Gaussian or normal distribution: the bell curve. The normal distribution is also called the Gaussian distribution, after the German Mathematician K.F. Gauss. A major use of the normal distribution is to describe sampling distributions that are used in the statistical analysis of the data, and many sampling distributions are approximately normal and described as having a bell curve shape.

- Approximately 68%, 95% and 99.7% of the data should lie within one, two and three standard deviations from the mean, respectively.

Parametric statistics test the hypothesis that one or more treatments have no effect on the mean and variance of a chosen variable and there are two versions of the classical parametric $t$-test, depending on whether the data is paired or unpaired. For example, a paired $t$-test is used when assessing the effects of a drug on an individual before and after treatment. For paired $t$-tests, it is the distribution of the differences that must be normal, whereas for unpaired $t$-tests, the distribution of both data sets must be normal.

All classical statistical tests require the measurements being analyzed to be independent. For example, for any given gene, two different patient measurements are independent if knowing the value of one measurement does not provide information about the value of the other. Example measurements that are not independent and therefore cannot be included as separate variables in a hypothesis test include: replicate measurements from the same patient; and replicate features or probes on the same array. To ensure that all data points in the analysis are independent, nonindependent variables can be combined into a single variable, for example, two measurements from the same patient are combined by subtracting one from the other.

The most commonly used parametric method for determining significantly expressed genes from replicated microarray data is the two-sample $t$-test and there are several variations (Dudoit et al., 2000). This test is generally used when there are no extreme outliers and where there are many observations ($>30$) in

**TABLE 7.1** A Summary of Parametric Microarray Data Analysis Methods and Their Availability

| Package | Availability | Parametric Method | Reference |
|---|---|---|---|
| EBarrays | R code | Empirical Bayes | (Newton, Kendziorski et al. 2001; Kendziorski, Newton et al. 2003) |
| limma | Bioconductor | Empirical Bayes | (Smyth 2004) |
| FEBarrays | R code | Empirical Bayes | (Lo and Gottardo, 2007) |
| timecourse | Bioconductor | Empirical Bayes | (Tai and Speed, 2004) |
| maSigPro | Bioconductor | Regression | (Conesa, Nueda et al. 2006) |
| maanova | Bioconductor | ANOVA | (Kerr, Martin et al. 2000; Cui, Hwang et al. 2005) |
| Envisage | R code[1] | ANOVA | Sam Robson |
| FSPMA | R code[2] | Mixed model ANOVA | Sykacek et al. 2005 |
| GlobalAncova | Bioconductor | ANCOVA | (Hummel, Meister et al. 2008) |
| ASCA-genes | R code | ANOVA & dimension reduction | Nueda, Conesa et al. 2007 |
| ANPCA | R code | ANOVA & PCA | (de Haan, Wehrens et al. 2007) |

Bioconductor: http://www.bioconductor.org.

[1] *The R-code for envisage can be downloaded from the following url: http://www2.warwick.ac.uk/fac/sci/moac/currentstudents/2003/sam_robson/linear_models/downloads/.*

[2] *The R-code for FSPMA (a Friendly Statistics Package for Microarray Analysis) can be downloaded from the following url: http://www.ccbi.cam.ac.uk/software/psyk/software.html#sykacek_furlong_2005.*

each sample. Other parametric methods include the analysis of variance approach (Kerr et al., 2000), a regression approach (Thomas et al., 2001), and several empirical Bayes methods (Newton et al., 2001; Kendziorski et al., 2003; Smyth, 2004). A semi-parametric hierarchical mixture model method for detecting differentially expressed genes has also been considered (Newton et al., 2004). The methods and the corresponding available packages or implementations are summarized in Table 7.1. In this section the focus is on describing classical parametric approaches that are standard methods implemented in $R$, the one sample and two-sample $t$-test, but we also introduce a modern parametric approach that involves empirical Bayes, limma (Smyth, 2004), which is available from Bioconductor. To accompany the sections on parametric and nonparametric statistics, a summary of characteristics for a test selection can be found in Table 7.2.

## 7.3.1 One-Sample Statistical Test

The *paired* t-test is applicable to data sets where there is one measurement variable and two nominal variables, for example, where there is a pair of gene expression measurements for each patient, before and after drug treatment. It is

**TABLE 7.2** Summary of Characteristics for a Selection of Classical Parametric and Nonparametric Statistics Used for Comparing Condition Groups

| Group comparison set number | Parametric test | Nonparametric test | Independent variable | Dependent variable | Essential feature example |
|---|---|---|---|---|---|
| Two groups | One sample t-test | Wilcoxon signed rank | One categorical independent variable (two levels) | One continuous dependent variable | Same people on two different occasions |
| Two groups | Two-sample t-test (Student's t-test) | Wilcoxon Rank Sum or Mann–Whitney U test | One categorical independent variable (two levels) | One continuous dependent variable | Different people in each group |
| Three or more groups, single factor | One-way ANOVA (between group) | Kruskal-Wallis | One categorical independent variable ($\geq 3$ levels) | One continuous dependent variable | Different people in each group |
| Three or more groups, single factor | One-way ANOVA (repeated measures) | Friedman test | One categorical independent variable ($\geq 3$ levels) | One continuous dependent variable | Same people on two different occasions |
| Four or more groups, multiple factor | Two-way ANOVA (between group) | Scheirer-Ray-Hare test | Two categorical independent variables ($\geq 2$ levels) | One continuous dependent variable | Two or more groups for each independent variable: different people in each group |
| Four or more groups, multiple factor | Mixed ANOVA (between-within) | None | One between-groups independent variable ($\geq 2$ levels); One within groups independent variable ($\geq 2$ levels) | One continuous dependent variable | Two or more groups with different people in each group, each measured on two or more occasions |
| | Multivariate ANOVA (MANOVA) | None | One or more categorical independent variables ($\geq 2$ levels) | Two or more related continuous dependent variables | |

Conditions (or factors) that are character variables are treated as categorical and those that are numerical variables are treated as continuous.

also known as the *one-sample* t-test since it is used, for example, when two biological samples are taken from each patient, that is where there is one statistical patient sample. These measurements are combined to form a single log ratio or *M*-value for each patient, and this single column of data, which is for one particular patient, is used to calculate the *t*-statistic, using the following formula:

$$t = \frac{\bar{x}}{s/\sqrt{n}}$$

where *x* is the average of the log ratios of each of the patient data, *s* is the standard deviation of the sample of patients, and *n* is the sample size (i.e. the number of patients in the experiment).

The *t*-test can be defined as the ratio of the difference of two group means and the standard error of this difference:

$$t = \frac{x_1 - x_2}{\sqrt{\sum(d_1 - d_2)^2/(n - 1)}}$$

where the standard error of the difference is calculated as the square root of the ratio of the variance of the difference scores for each individual:

$$\sum(d_1 - d_2)^2$$

and the sample number minus 1 ($n - 1$). The null hypothesis is that the mean difference between paired observations is zero. The *p*-value is calculated by comparing the *t*-statistic to a *t*-distribution with the appropriate number of degrees of freedom,[14] d.f. The d.f. is equal to the number of independent variables in the analysis, and for paired *t*-tests this is calculated as $n - 1$.

## 7.3.2 Two-Sample Statistical Tests

The Student's *t*-test, also known as the unpaired *t*-test or the two-sample *t*-test, is used to compare two condition groups that are both independent variables, for example, when one biological sample is taken from each of two patients. The Student's *t*-test is probably the most commonly used parametric statistical test. The *t*-test assesses whether the two means of two groups are statistically different from each other and the means are assessed relative to the spread or variance of the data. The Student's *t*-test can be defined as the ratio of the difference of two group means and the standard error of this difference:

$$t = \frac{\bar{x}_1 - \bar{x}_2}{\sqrt{\frac{S_1^2}{N_1} + \frac{S_2^2}{N_2}}}$$

---

14 Degrees of freedom (d.f.) are used to help create an unbiased estimate of the population parameters necessary for the sampling distribution.

where $x_1$ and $x_2$, $S_1^2$ and $S_2^2$, and $N_1$ and $N_2$ are the respective group means, standard deviations and sample sizes for group 1 and 2. The equation scales the differences between the means of the two sample groups by a factor relating to the amount of sample variation and sample size of the two groups. The more the noise in the sample, the less the systematic differences between the two means appear clear. The larger the sample, the smaller the noise impact of each individual observation, and this leads to a clearer systematic mean difference between the two groups.

There is also an alternative version of the equation that calculates a single standard deviation for all the data, but the version presented here is much better with regard to analyzing microarray data since it allows the standard deviation between the two groups to be different. The $p$-values are calculated using the Student's $t$-distribution with degree of freedom, d.f. Usually, d.f. is equal to $(n_1 + n_2) - 2$, but when the variances of two groups are not equal, a more complicated formula for d.f. can be used (Welch, 1947).

## 7.3.3 The One-Way Analysis of Variance (ANOVA)

In an experimental setting with more than two condition groups, the interest may be in identifying genes that are differentially expressed in one or more of the groups relative to the others. For example, given an experiment with three different patient groups, there are two ways to perform the analysis:

i. Apply an unpaired $t$-test three times to each pair of groups in turn and select genes that are significant in one or more of the $t$-tests.
ii. Use a statistical test that compares all three groups simultaneously and reports a single $p$-value.

The second method, a multiple condition statistics approach, is often taken by statisticians since the first method is problematic in two main ways:

i. By performing three tests, the problem of multiplicity increases the likelihood of seeing a significant difference between two of the groups as a result of measurement errors. This problem worsens as the number of groups increases.
ii. Since each of these comparisons is not independent of the other, it becomes hard to interpret the results in a meaningful way.

Analysis of variance is one of the most commonly used statistical tests for one or more factors. The one-way ANOVA can be used to test the hypothesis that one factor or variable of interest differs among groups and in situations where there are two or more factors, the two-way ANOVA can be used. Essentially, an ANOVA determines the overall variability within an analysis of multiple groups by employing multiple estimates of a population's variance. There are no restrictions on the number of groups, but where there are only two groups, ANOVA is equivalent to a $t$-test.

There are two estimates of variance for each group taken and they are based on either one of the following:

i. *The standard deviation*: this variance is not affected by any differences in the means of the groups being tested since the information is generated within each group. The variance should not differ as the test assumes equal variances among groups.

ii. *The variability between the means of each group*: if these estimates of each group's variability are the same, the overall variance among the groups is not expected to be different. However, if there are significant differences between the means, this may lead to changing the estimated population variance.

After determining these parameters, the one-way ANOVA statistic is calculated by dividing the population variance estimate of the means ('between group' variability) by the population variance estimate of the standard deviations ('within groups' variability). The larger the resulting $F$-value, the greater the difference among the group means relative to the sampling error variability (which is the within groups variability). In other words, the larger the $F$-value, the more likely it is that the differences among the group means reflect 'real' differences among the means of the populations from which they are drawn, rather than being due to random sampling error. The $F$-value is given by the equation:

$$F = \frac{\text{MSTR}}{\text{MSE}}$$

$$F = \frac{(n_1(x_1 - x) + \cdots n_k(x_k - x))/(k - 1)}{((n_1 - 1)s_1^2, \ldots, (n_k - 1)s_k^2 k)/(N - k)}$$

where MSTR is the treatment mean square; MSE is the error mean square; $n_1$, ..., $n_k$ is the number of individuals in each group; $(x_1 - x)$ is the mean of each group minus the average of all groups; $k$ is the number of classes; $s_1^2$, ... , $s_{k1}^2$ is the group variance for each class; and $N$ is the total number of individuals. The resulting $F$-value can be compared to the $F$-table (containing critical values for the theoretical $F$-distribution) to obtain the $p$-value using the following two degree of freedom calculations: number of classes minus one; and the total number of arrays minus the number of classes. The degree of freedom is a mathematical way of representing the number of replicates. If there are zero degrees of freedom, then this indicates no replicates and the one-way ANOVA cannot be performed.

The $p$-value associated with an $F$-value is the probability that a given $F$-value would be obtained if there were no differences among group means. This is a statement of the null hypothesis: therefore, the smaller the $p$-value, the less likely it is that the null hypothesis is valid. The differences among

**TABLE 7.3** Table of Treatment Conditions for Assessing Male and Female Patients (G) With Two Different Forms of Leukemia (L)

| Treatment conditions | Factor 1: disease type (L) | Factor 2: patient gender (G) |
|---|---|---|
| ALL male | ALL | Male |
| ALL female | ALL | Female |
| AML male | AML | Male |
| AML female | AML | Female |

group means are more likely to reflect real population differences as $p$-values decrease in size.

### 7.3.4 Multifactor ANOVAs

The real power of ANOVA is more apparent in multifactor (multiple factor) ANOVAs. Whereas the one-way ANOVA examines the effects of various levels of treatment conditions of one independent variable on a dependent variable, a two-way ANOVA is used to examine what effect two independent variables (multiple factors) may have on a dependent variable. In multifactor experiments, the effects of several variables are interrogated simultaneously and these experiments are usually more comprehensive than a set of single factor experiments. It is important to note that ANOVA requires well thought out experimental designs with enough replicates to ensure a sufficient number of degrees of freedom (Kerr and Churchill, 2001; Churchill, 2004). In addition they can be computationally intensive, while this tends to be less of a concern these days, large-data sets may still require considerable computational power.

The following is an example of a multifactor experiment with two independent variables. When assessing male and female patients with two different forms of leukemia, acute lymphoblastic leukemia (ALL) and acute myeloid leukemia (AML) there is a total of four different sample sets (Table 7.3). The two variables here are 'disease type' and 'patient gender'. A two-way ANOVA can be used to investigate whether gene expression depends on one or both of these independent variables. The two-way ANOVA statistic is calculated as follows:

$$F = \frac{\text{MSTR}}{\text{MSE}}$$

$$F = \frac{((1/m)(T_1^2) - (\sum x^2/n))/(k-1)}{\frac{(\sum x^2 - (\sum x^2/n)) - ((1/m)(T_1^2 \ldots T_k^2) - (\sum x)^2/n) - ((1/k)(B_1^2 \ldots B_k^2) - ((\sum x)^2/n))}{n-k-m+1}}$$

where $x$ is the mean for all groups; $m$ is the number of blocks; $k$ is the number of classes; $n = (k)(m)$ is the total number of pieces of data; $T_1^2 \ldots T_k^2$ is the squared sum of sample data for each treatment (class or column); $B_1^2 \ldots B_k^2$ is the squared sum of sample data for each block (row). When calculating the MSTR, this value is described by three components (instead of just two for the one-way ANOVA). When calculating the MSE, the sum of squares for the extra variable is also addressed. The resulting $F$-value can be compared to the F table to obtain the $p$-value using the following two different calculations of degree of freedom:

i.  d.f. $= k - 1$;
ii. d.f. $=$ (total number of arrays $- k$) $- (m + 1)$

With multifactor ANOVAs, two factors may behave together in an *additive* or *multiplicative* way. The response to the two factors is additive when the effect of one factor does not influence the effect of the other. However, a multiplicative response may be observed when, for example, a gene is differentially expressed in female AML patients relative to female ALL patients but not in male patients. Factors that behave multiplicatively can also be referred to as factors that have an *interaction* between one another. Some statistical packages can allow users to build interactions into the ANOVA models.

In the analysis of a multifactor experiment, the main effects of each of the factors and the interactions between the factors are investigated. A multifactor ANOVA is used to conduct formal tests for hypotheses of no interaction and hypotheses of no factor main effect (for each factor). As shown in Table 7.3, the assessment on two factors, patient gender (G) and leukemia disease type (L), is a $2 \times 2$ factorial design that has four different treatment conditions. There are four different 'main effects' to test for: main effect of G, main effect of L, the interaction effect between G and L; or the overall model effect (G × L). In a two-factor experiment, a mean plot can provide a visual indication of whether or not the two factors interact.

However, suppose two different drug treatments (D) were added to the design. This would then be a $2 \times 2 \times 2$, or $2^3$ factorial design with eight different treatment conditions and three 'main effects' needing to be estimated (differences between the two patient genders, the two disease types and the two drug treatments in each case averaging across the other factors). There would also be three two-way interactions: G × L; G × D; D × L; and one three-way interaction: G × L × D. If there are more than three factors, the model can be generalized in a similar way, but beware that such a multifactor ANOVA model quickly becomes very complicated to work with.

To test for the interaction effect or overall model effect, a regression model is used that contains the main effects for both factor 1 and factor 2 plus the interaction effect. To test for main effects, however, the interaction effect is not included in the model. Note that if the hypothesis of no interaction between two factors is rejected (i.e. interaction exists), then the two hypotheses of no factor main effects are irrelevant, since both factors clearly do affect the response variable through the interaction effect. In a multifactor ANOVA, if there is no interaction, and if only one of the factor main effects is significant, the model reduces to a one-way ANOVA model.

However, it is important to distinguish between terms that are either random or fixed in an experimental design since a failure to do so can lead to misleading results. To determine this, imagine repeating the experiment several times. If the effects stay the same in the repeated experiments then the term is fixed (such as mouse strain, diet or dye), otherwise it is random (such as mouse and array). This helps to ensure that any conclusions drawn from the analysis are repeatable in other contexts. If there are multiple sources of variation, the ANOVA model is referred to as a 'mixed model', as it contains a mixture of random and fixed effects. Construction of $F$ statistics to identify differential gene expression using mixed model ANOVA can properly account for the contributions of each of the sources of variations.

## 7.3.5 General Linear Models

An issue with multifactor ANOVA is that a single model is used to fit all the genes and this cannot account for the variability between genes. In addition, as more variables are included in the model, the problem of overfitting a model to all the genes becomes magnified. Many genes may show differential expression in response to only a subset of the experiment variables, so including more variables into the analysis may result in missing a lot of interesting effects for these genes. Therefore, a method must be used to fit a model to each gene individually.

Furthermore, ANOVA analysis is appropriate when the factors on which gene expression depends are all categorical variables (character variables). However, in some situations, some factors are continuous variables (numerical), for example a dose of a compound added to a sample or some phenotypic variables such as age, weight and height, and these covariates cannot be considered in this method. If it is thought that gene expression responds in a linear fashion to such a variable, then a *general linear model (GLM)*, can be used as an extension to ANOVA to allow for the inclusion of numeric variables.

The GLM can be written as:

$$Y = X\beta + \epsilon$$

where $Y$ is a vector of multivariate measurements; $X$ is the design matrix; $\beta$ is a matrix containing parameters that are usually to be estimated and $\epsilon$ is a vector of residuals (noise or errors).

A linear 'mixed' model can be written as follows:

$$Y = X\beta + Zu + \epsilon$$

where $X$ is the design matrix of fixed effects $\beta$; and $Z$ is the design matrix of random effects $u$ (Searle et al., 1992). A popular Bioconductor package that uses general linear models is limma (linear models for microarray data) and this will be discussed in more detail a little later in this chapter.

## 7.3.6 ANOVA Modeling for Both Nuisance and Biological Effects

One advantage of using ANOVA for microarray data analysis is that each source of variance is accounted for and therefore distinguishing between technical and biological variations is reasonably straightforward (Churchill, 2004). There may be factors in the microarray experiment that are of no biological interest but can influence the observed gene expression. For example, there may be a dye effect, more than one array batch or more than one scientist performing the hybridiza-tion. Some of these factors may be included into the statistical model as random effects.

Kerr and co-workers (Kerr et al., 2000; Kerr and Churchill, 2001) used ANOVA to model nuisance technical effects associated with microarrays along with the effects of interest in order to evaluate only the latter:

$$log(y_{ijkgr}) = \mu + A_i + D_j + AD_{ij} + G_g + VG_{kg} + DG_{jg} + AG_{ig} + S_{r(ig)} + \epsilon_{ijkgr}$$

where $log(y_{ijkgr})$ is the measured log ratio for gene, $g$ of variety $k$ (which refers to a condition such as healthy or diseased) measured on array $i$ using dye $j$; $\mu$ is the overall mean signal of the array; $A_i$ is the effect of the $i$th array; $D_j$ is the effect of the $j$th dye; $AD$ is the array by dye interaction; $G_g$ is the variation of the $g$th gene and captures the average intensity associated with a particular gene; $AG_{ig}$ denotes the interaction between the $i$th array and the $g$th gene and captures the effect of a particular spot on a given array; the spot term $S$ captures the differ-ences among the duplicated spots within an array and this drops out from the model if there are no duplicates; the variety by gene term, $VG_{kg}$ denotes the interaction between the $k$th variety and the $g$th gene and captures variations in the expression levels of a gene across varieties (the primary interest in the analysis); $\epsilon_{ijkgr}$ represents the error term for array $i$, dye $j$, variety $k$ and gene $g$. The error term is assumed to be independent and have a mean of zero. The index $r$ is linked with $i$ and $g$ to identify individual spots on each array and is only needed if there are multiple spots for the same gene on the array.

This one-step normalization and analysis procedure, which also involves 'bootstrapping' to deduce $p$-values (and retain the dependence structure between genes), has the advantage of carrying over the uncertainty of the normalization into the uncertainty of the effects of interest and avoids false confidence in the inaccurate results. However, the disadvantage of having a one-step procedure is that the data will have to be renormalized for any other analysis such as a new clustering or classification approach. As noted before, ANOVA can be compu-tationally expensive and if such analysis is being considered it should be noted that the computational performance of some *Bioconductor* packages can dramatically be improved using Linux clusters via the message passing interface (MPI).

A well-known Bioconductor package, maanova (MicroArray ANalysis Of VAriance; Kerr et al., 2000) can be used to analyze any microarray data, but is specifically suited for multifactor designs. Mixed effect models are implemented to estimate variance components and either *F*- or *t*-tests with multiple testing are performed for differential gene expression analysis. There are several methods available in maanova that form a workflow. These include: data quality checks, visualization plots and data transformation, ANOVA model fitting for both fixed and mixed effects, data summarization using tables and graphics such as volcano plots[15], a sampling and residual shuffling permutation approach to derive the *p*-value and bootstrap cluster analysis.

The maanova algorithm for computing ANOVA estimates involves fitting the model in two stages (using the same notations in the previous equation):

**i.** The normalization model fit to obtain residuals $r_{ijkgr}$ for the next stage:

$$Y_{ijkgr} = \mu + A_i + D_j + AD_{ij} + r_{ijkgr}$$

**ii.** The modeling of gene-specific effects, which involves iterative fitting of the model on a per gene basis:

$$r_{ijkr} = G + AG_i + DG_j + VG_k + \epsilon_{ijkr}$$

In addition, maanova offers four different F statistics and users have the option to turn them on or off. There are two different options also for data shuffling: residual shuffling and sample shuffling. The former works only for fixed effect models whereas the latter is used for mixed effect models. Missing values, zeroes and negative values cannot be tolerated by maanova and the recommendation is to filter these out of the data set before analysis. Alternatively, if the sample size is large enough, then imputation could be used (see Chapter 6 for a description of imputation methods). Further reading on all the algorithms behind maanova can be found in the *Bioconductor* vignette and in Wu et al. (2003) and Cui et al. (2003, 2005) with the basic concepts of

---

15 Volcano plots plot the log fold change in gene expression versus a non-negative statistic (e.g. negative $\log_{10}$-transformed *p*-values or the *B*-statistic), which tests the significance of the fold change (Cui & Churchill, 2003). Since the size of the statistic tends to increase with absolute log fold-change, such a plot has the characteristic shape of a volcano crater. Genes with statistically significant differential expression according to the gene-specific statistical test will lie above a horizontal threshold line. Genes with large fold change values will lie outside a pair of vertical threshold lines. Volcano plots can be used to emphasise that variability plays a role in significance as well as fold-change. They can be used for instance, to show that some large fold-changes are not necessarily statistically significant.

multifactor experimental design reviewed by Kerr and Churchill (Kerr and Churchill, 2001).

Other approaches that handle multifactor design include the well-known limma package, which will be explained in more detail in the next section, and the not so well-known tool, Envisage[16] (enable numerous variables in significance analysis of gene expression). With Envisage, the user can select from a wide range of predictor variables, including experimental parameters, phenotypic covariates and nuisance covariates to automatically create the ANOVA model. It is available as a graphical user interface or it can run in the *R* environment.

### 7.3.7 Other ANOVA Related Issues

Another ANOVA issue, which is also generally relevant to all classical test statistics, is that the simple 'gene by gene' testing approach often fails because there is not enough power to detect any population differences. This is because many microarray experiments have small sample sizes (replicate numbers) and this reduces the power to detect differential gene expression. One solution, which is more powerful and robust by allowing for gene-to-gene heterogeneity, involves 'shrinkage' of the variance component estimates by pulling each individual estimate closer to the mean values across genes (Cui et al., 2003, 2005) and this is implemented in both the maanova and limma packages.

Alternatively, principle components analysis (PCA, discussed further in the next chapter, Section 8.5.1) could be applied to combine data from multiple genes and then test for differential expression with this composite variable. de Haan et al. (2007) used PCA to summaries and visualize individual variance components of the ANOVA model. When identifying genes with significant group differences and then determining which groups are responsible, interaction effects can be visualized using biplots to display genes and variables together (Figure 7.8). This can provide insight into the effect of factors such as treatment or time on the correlation between different genes. Based on the interaction of interest, the authors demonstrated that the combination of ANOVA and PCA can provide a simple way to select genes. The R codes, anpca. synth.R and biplot.R are both available from: http://www.cac.science.ru. nl/research/software/.

### 7.4 LINEAR MODELS FOR MICROARRAY DATA (LIMMA)

Applying linear models to microarray data is a popular approach and is available from the Bioconductor packages, limma or limmaGUI. The latter version is a 'point and click' graphical user interface implementation of the approach

---

16 The *R* codes for Envisage are available from the following url: http://www2.warwick.ac.uk/ fac/sci/moac/currentstudents/2003/sam_robson/linear_models/downloads/.

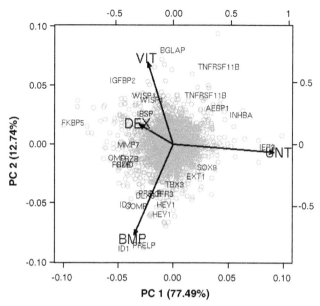

**FIGURE 7.8**  Biplot of two principle components (PC1 and PC2) from the PCA performed on the interaction between the factors gene and treatment. The points are the 22 283 probe sets of data that were used in the analysis, and those with gene symbols are a group of genes associated with skeletal development. PC1 (*x*-axis) indicates variation, which distinguishes between the three samples treated with the substances (DEX, VIT and BMP) and the untreated sample (UNT). This can be seen because the arrows, which indicate the loadings, point in the opposite directions in the plot for treated and untreated. Also shown is that BMP treatment is different from DEX and VIT.
*Source*: Taken from de Haan et al. (2007).

(Smyth, 2004; Wettenhall and Smyth, 2004). For more detailed discussions and command usage in relations to various case studies refer to Smyth (2005), Scholtens and von Heydebreck (2005) and the `limma` package vignette from the *Bioconductor* website.

## 7.4.1 Small Sample Size and 'Shrinkage' of Standard Deviations

The moderated statistics approach in `limma` has features specifically designed to stabilize the analysis even for experiments with small numbers of arrays, for both two-color arrays and single-channel arrays such as Affymetrix GeneChips or Illumina arrays. This is achieved through borrowing information across genes, and therefore 'strength' to improve precision. This 'information borrowing' statistic involves the empirical Bayes method[17] where there is a 'shrinkage' of the estimated sample variances towards a pooled estimate $s_0^2$

---

17 limma uses the empirical Bayes normal model for variance regularization.

based on all gene variances. The approach provides a more complex model of gene variance and involves moderating the standard errors of the estimated expression measurement values or $M$ values (the log differential expression values – see Chapter 6). The resulting test statistic is a moderated $t$-statistic, where instead of having single-gene (i.e. 'gene by gene') estimated variances $s_g^2$, a weighted average of $s_g^2$ and $s_0^2$ is used. This results in more stable inference and improved statistical power, especially for studies with small numbers of arrays. The approach is observed to be equally robust with a low number of samples and much more reliable than several other methods, including SAM (Jeffery et al., 2006), a nonparametric approach, which is discussed later. However, before applying the moderated $t$-statistics approach, linear modeling of the data is performed to aid the detection of differentially expressed genes especially over a range of related conditions or treatments. The procedure is explained below.

## 7.4.2 The Linear Model Fit

Expression ratios of conditions can be estimated directly from a single array and via a linear combination of ratios measured on other arrays. Additionally, it is possible to estimate expression ratios for comparisons not directly performed in the experiment. For example, when conducting a two-channel experiment to directly compare, $A$ versus $B$ and $A$ versus $C$, it is also possible to indirectly compare $B$ versus $C$, even though no direct hybridization of $B$ versus $C$ took place. In the context of simple loop designs, this concept allows for the estimation of a particular log expression ratio or $M$ value both directly and also through a combination of all the other interactions in the loop. For more complex designs, such as those involving interwoven loops, this linear modeling approach allows comparisons between conditions to be estimated simultaneously for each gene through many different combinations of intermediate results (see Chapter 2 for an introduction to loop designs). As there is the assumption that the relationships between $M$ values are linear, it is possible to estimate unknown $M$ values indirectly through a linear combination of more directly measured pair-wise $M$ values.

## 7.4.3 Parameterization

The linear model starts by identifying parameters on which to base the model. This is known as choosing a 'parameterization' and can be specified in terms of simple comparisons or 'contrast pairs' between different RNA sample types (condition groups) to define the treatment ratios or M values to estimate differential gene expression. This essentially requires the 'design matrix' and the 'contrast matrix' to be specified. The 'design matrix' indicates which RNA samples have been applied to each array and is a way to represent the 'main effects' (the effect of a factor on a dependent variable) in the study. The 'contrast matrix' specifies which comparisons are to be made by combining the coefficients defined by the design matrix into the contrasts of interest. Using the 'design matrix' and the 'contrast matrix', contrast pairs can be specified in `limma` from a formula via the function `model.matrix` or directly. The latter method allows investigators to have more flexibility and direct control over the construction of the matrices, especially when modeling more complex experimental designs. Although not everyone opts to use the advanced 'direct construction' approach, it is still worth understanding what this process entails.

The first step of the `limma` approach is to model the systematic part of the data by fitting a linear model for gene, $g$:

$$Y = X\beta + \epsilon$$

where $Y$ is a vector of multivariate expression measurements for $g$, where each row corresponds to the array; $X$ is the design matrix; $\beta$ is a *matrix of parameters* (to be estimated); and $\epsilon$ is a *vector of residuals* (noise or errors).

The main purpose of this step is to estimate the variability. Each row of the design matrix corresponds to the array and each column corresponds to a parameter or coefficient. Unlike two-channel arrays, single-channel arrays such as Affymetrix yield a single-expression value. For such arrays, design matrices can be formed from the biological factors underlying the experimental layout since competitive sample hybridization, dye bias and other two-color array considerations are absent. Therefore, for one-channel data, the number of parameters or coefficients will be equal to the number of distinct RNA sources. With two-channel data, there will be one fewer number of parameters than distinct RNA targets, or the same number if a dye effect is included in the design matrix or if the experiment is a reference design.

The next step, the contrast step, allows the fitted coefficients to be compared in as many ways as there are possible. The examples[18] below are by no means the most optimal in terms of experimental design but chosen as simple case studies that illustrate what is meant by parameterization and how the matrices are constructed for particular design types using two-channel arrays since these are generally more complicated to construct than single-channel arrays.

---

18 Some of the examples are taken from the Computational Biology Group (University of Cambridge, UK) introductory training course in microarray data analysis with the kind permission of Natalie Thorne: http://www.compbio.group.cam.ac.uk/teachning.html.

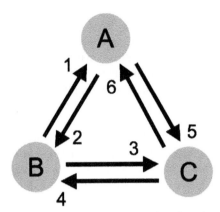

**FIGURE 7.9**   Diagram of the experimental design example 1 for a three-sample dye-swap loop design involving six 2-channel array slides. The three samples (condition groups) are $A$, $B$ and $C$, and the arrays are numbered 1–6. The back of the arrow points to a condition in one dye channel (e.g. red), whereas the front points to the condition in the other dye channel (e.g. green).

### 7.4.3.1 Example: Three-Sample Loop Design with Dye-swaps

*Example 1, Step 1: Specify the Experimental Design*

This first example is of a three-sample, dye-swap loop design involving six arrays. A diagrammatical interpretation[19] of the experiment and what is measured in each sample in relation to the other sample types is seen in Figure 7.9.

*Example 1, Step 2: Parameterization*

The effects measured by each sample type can be represented as follows:

$$\text{Sample } A = a$$

$$\text{Sample } B = b$$

$$\text{Sample } C = c$$

For two-channel arrays, the parameters in terms of their effects should be representative of log-ratio ($M$-value) comparisons and array channel (i.e. red or

---

19 The structure of the graph determines which main effects can be estimated and the precision of the estimates. For two-dye arrays, the mRNA samples can be compared directly or indirectly only if there is a path joining the corresponding two vertices. The precision of the estimated paths joining the two vertices is inversely related to the length of the path, i.e. direct comparisons made within slides yield more precise estimates than indirect ones (between slides) as the latter has a longer path.

green). This involves identifying which sample comparisons (i.e. differences between the samples) are important to the study. The comparisons of interest are to identify gene expression differences between all three pair-wise comparisons of each sample:

The direct comparison of $A_{red}$ and $B_{green}$ :   $A - B = a - b$

The direct comparison of $B_{red}$ and $A_{green}$ :   $B - A = b - a$

The direct comparison of $C_{red}$ and $B_{green}$ :   $C - B = c - b$

The direct comparison of $B_{red}$ and $C_{green}$ :   $B - C = b - c$

The direct comparison of $C_{red}$ and $A_{green}$ :   $C - A = c - a$

The direct comparison of $A_{red}$ and $C_{green}$ :   $A - C = a - c$

Next, any redundant parameters are removed. It does not matter which ones as long as each comparison type is included and there are no redundancies (i.e. there should be no other alternatives for that comparison type):

The direct comparison of $A$ and $B$ :   $A - B = a - b$

The direct comparison of $B$ and $C$ :   $B - C = b - c$

The direct comparison of $C$ and $A$ :   $C - A = c - a$

The parameters of interest are labeled to specify each array according to the parameterization so far and in this example, a 'loop' forming between the three samples is seen (Figure 7.10).

Only all the possible independent parameters are chosen, that is no combination of the parameters can equal any of the parameters. Therefore, it is important to ensure that the parameters involve every treatment type at least once and that they do not form a loop in the experimental design picture. If necessary, different parameterizations for the experiment could be explored, for example $d_{BC}$ and $d_{CA}$ instead of $d_{AB}$ and $d_{BC}$. For two-channel arrays, the rule is that the total number of independent parameters is the number of sample types minus 1. We have three parameters here so we arbitrarily remove one to have the following two parameters:

The direct comparison of $A$ and $B$ :   $A - B = a - b = d_{AB}$

The direct comparison of $B$ and $C$ :   $B - C = b - c = d_{BC}$

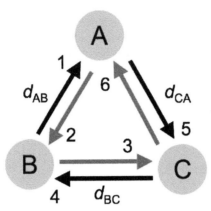

**FIGURE 7.10**  All possible nonredundant set of parameters for example 1: $d_{AB}$, $d_{BC}$ and $d_{CA}$ are labeled in the experimental design. However, the rule of thumb is that no loops in the experimental design should form. In this example, there seems to be a 'loop' forming between the three samples, $A$, $B$ and $C$.

Therefore, $a - b$ and $b - c$ are the parameters $\beta$ in the linear model $Y = X\beta + \varepsilon$ and these are labeled in the experimental design as $d_{AB}$ and $d_{BC}$ (Figure 7.11).

*Example 1, Step 3: Specify the Design Matrix*
The design matrix, $X$ in the linear model is specified using the two chosen parameters $d_{AB}$ and $d_{BC}$ and a 'matrix multiplication' procedure is used to calculate the multipliers. Each row in the matrix corresponds to an array in the experiment and each column corresponds to a coefficient. For each row, algebra is used to determine the 'multipliers' used to construct the matrix using

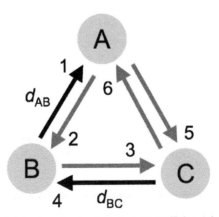

**FIGURE 7.11**  Example 1 has two chosen parameters (or coefficients) for the linear model, $d_{AB}$ and $d_{BC}$.

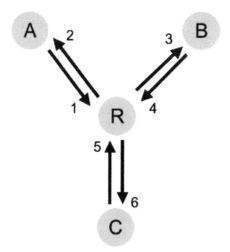

**FIGURE 7.12**   Example 2 is of a reference experimental design with dye-swaps to compare three sample types, $A$, $B$ and $C$ indirectly using the reference R. The arrays are numbered 1–6. The back of the arrow points to a condition in one dye channel (e.g. red), whereas the front points to the condition in the other dye channel (e.g. green). Here, the comparison of interest is to find gene expression differences indirectly between samples '$A$ and $B$', '$A$ and $C$', and '$B$ and $C$'.

information from $\beta$ and the effects from the direct comparison measurement from the corresponding array:

$$
\begin{aligned}
&\text{Array 1 comparison } A - B: &&1(a - b)^{d_{AB}} + 0(b - c)^{d_{BC}} = (a - b)\\
&\text{Array 2 comparison } B - A: &&-1(a - b) + 0(b - c) = (b - a)\\
&\text{Array 3 comparison } C - B: &&0(a - b) + -1(b - c) = (c - b)\\
&\text{Array 4 comparison } B - C: &&0(a - b) + 1(b - c) = (b - c)\\
&\text{Array 5 comparison } C - A: &&-1(a - b) + -1(b - c) = (c - a)\\
&\text{Array 6 comparison } A - C: &&1(a - b) + 1(b - c) = (a - c)
\end{aligned}
$$

The design matrix, $X$ is therefore represented in the linear model equation as follows:

$$
E\begin{bmatrix} y^1 \\ y^2 \\ y^3 \\ y^4 \\ y^5 \\ y^6 \end{bmatrix}
=
\begin{bmatrix} 1 & 0 \\ -1 & 0 \\ 0 & -1 \\ 0 & 1 \\ -1 & -1 \\ 1 & 1 \end{bmatrix}
\times
\begin{bmatrix} a - b \\ b - c \end{bmatrix}
=
\begin{bmatrix} a - b \\ b - a \\ c - b \\ b - c \\ c - a \\ a - c \end{bmatrix}
$$

After the design matrix is specified, the *contrast matrix* is specified and this allows the coefficients defined by the design matrix to be combined into contrasts of interest, where each contrast corresponds to a comparison of interest between the samples (condition groups). The next example (example 2), demonstrates how this is done.

### 7.4.3.2 Example 2: Reference Design With Dye-Swaps
*Example 2, Step 1: Specify the Experimental Design*
This next example is of a simple reference design involving six array slides and the experimental design is illustrated in Figure 7.12. There are three samples of interest, $A$, $B$ and $C$, each directly compared with a reference sample, $R$ and dye-swaps of the samples are also included in the design.

*Example 2, Step 2: Parameterization*
In this example, the comparison of interest is to find gene expression differences indirectly between samples '$A$ and $B$', '$A$ and $C$', and '$B$ and $C$'. There are a total of four sample types in the experiment and an interpretation of what are the effects is shown as follows:

$$A = r + a$$
$$B = r + b$$
$$C = r + c$$
$$R = r$$

All the possible parameters in terms of their 'effects' are represented in terms of log-ratio ($M$-value) direct or indirect comparisons and this requires some algebra:

The direct comparison of $R_{red}$ and $A_{green}$ :     $R - A = r - (r + a) = a$

The direct comparison of $A_{red}$ and $R_{green}$ :     $A - R = (r + a) - r = a$

The direct comparison of $B_{red}$ and $R_{green}$ :     $B - R = (r + b) - r = b$

The direct comparison of $R_{red}$ and $B_{green}$ :     $R - B = r - (r + b) = b$

The direct comparison of $R_{red}$ and $C_{green}$ :     $R - C = r - (r + c) = c$

The direct comparison of $C_{red}$ and $R_{green}$ :     $C - R = (r + c) - r = c$

The indirect comparison of $A_{red}$ and $B_{green}$ :     $A - B = (r + a) - (r + b) = a - b$

The indirect comparison of $C_{red}$ and $B_{green}$ :     $C - B = (r + c) - (r + b) = c - b$

The indirect comparison of $B_{red}$ and $A_{green}$ :     $B - A = (r + b) - (r + a) = b - a$

The indirect comparison of $A_{red}$ and $C_{green}$ :     $A - C = (r + a) - (r + c) = a - c$

... and many more, etc.

For reference designs with two-channel arrays, the number of chosen parameters equals the number of sample types (condition groups). As before, the chosen parameters must be independent, that is no combination of the chosen parameters can equal any of the other chosen parameters. To illustrate how to

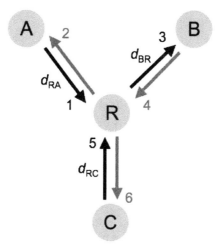

**FIGURE 7.13**   Example 2 has three chosen parameters (or coefficients) for the linear model, $d_{RA}$, $d_{BR}$ and $d_{RC}$.

determine this, take the following parameterization example where out of a selection of the following parameters, $A - R$, $A - B$, $R - C$ and $A - C$, the latter parameter can be considered not to be independent. The effects for these selection of parameters is as follows:

$$A - R = a$$
$$A - B = (R - A) - (B - R) = a - b$$
$$R - C = c$$
$$A - C = a - c$$

The parameter $A - C$ can be rewritten using the combination $A - R$ and $R - C$ as:

$$a - c = (a) - (c)$$

In other words the parameter $A - C$ is equal to the combination of the other two parameters $A - R$ and $R - C$ and can be regarded as not being independent so this cannot be selected. The final chosen parameters can thus be $A - R$, $A - B$ and $A - C$

Alternatively, the chosen three parameters could be chosen as $d_{R - A}$, $d_{B - R}$ and $d_{R - C}$ for the comparisons of interest (see Figure 7.13 to observe how this relates to the experimental design):

$$d_{R-A} = R - A = a$$
$$d_{B-R} = B - R = b$$
$$d_{R-C} = R - C = c$$

*Example 2, Step 3: Specify the Design Matrix*
The design matrix, $X$ in the linear model is specified using the three chosen parameters, $d_{R-A}$, $d_{B-R}$ and $d_{R-C}$ and the 'matrix multiplication' procedure:

$$
\begin{array}{ll}
\text{Array 1 comparison } R - A: & 1(a)^{d_{R-A}} + 0(b)^{d_{B-R}} + 0(c)^{d_{R-C}} = a \\
\text{Array 2 comparison } A - R: & 1(a) + 0(b) + 0(c) = a \\
\text{Array 3 comparison } B - R: & 0(a) + 1(b) + 0(c) = b \\
\text{Array 4 comparison } R - B: & 0(a) + 1(b) + 0(c) = b \\
\text{Array 5 comparison } R - C: & 0(a) + 0(b) + 1(c) = c \\
\text{Array 6 comparison } C - R: & 0(a) + 0(b) + 1(c) = c
\end{array}
$$

$$
E\begin{bmatrix} y_1 \\ y_2 \\ y_3 \\ y_4 \\ y_5 \\ y_6 \end{bmatrix}
=
\begin{bmatrix} 1 & 0 & 0 \\ 1 & 0 & 0 \\ 0 & 1 & 0 \\ 0 & 1 & 0 \\ 0 & 0 & 1 \\ 0 & 0 & 1 \end{bmatrix}
\times
\begin{bmatrix} a \\ b \\ c \end{bmatrix}
=
\begin{bmatrix} a \\ a \\ b \\ b \\ c \\ c \end{bmatrix}
$$

with $Y$, $X$, and $\beta$ labelling the respective matrices.

*Example 2, Step 4: Specify the Contrast Matrix*
For example 2 above, the contrasts of interest could be, for example, to compare $A - R$, $B - R$, to obtain the difference between $A - R$ and $B - R$ and to get an average of $A - R$ and $B - R$. Similarly with the design matrix, the contrast matrix is constructed using the two estimated parameters and a 'matrix multiplication' procedure: there can be as many rows as there are contrasts of interest.

$$
\begin{bmatrix}
1 & 0 & 0 \\
0 & 1 & 0 \\
0 & 0 & 1 \\
1 & -1 & 0 \\
1 & 0 & -1 \\
0 & 1 & -1 \\
0.5 & 0.5 & 0 \\
0.5 & 0 & 0.5
\end{bmatrix}
\times
\begin{bmatrix} a \\ b \\ c \end{bmatrix}
=
\begin{bmatrix}
a \\
b \\
c \\
a - b \\
a - c \\
b - c \\
0.5(a+b) \\
0.5(a+c)
\end{bmatrix}
\begin{array}{l}
\rightarrow A \\
\rightarrow B \\
\rightarrow C \\
\rightarrow A - B \\
\rightarrow A - C \\
\rightarrow B - C \\
\rightarrow 1/2(A+B) \\
\rightarrow 1/2(A+C)
\end{array}
$$

with $Z$, $P$, and $C$ labelling the respective matrices.

where $Z$ is the contrast matrix, $P$ is a vector of the parameter estimates and $C$ is a vector of the contrasts of interest (the contrast matrix[20]).

---

20 In Limma, the contrast matrix is represented as the transpose of what is shown here.

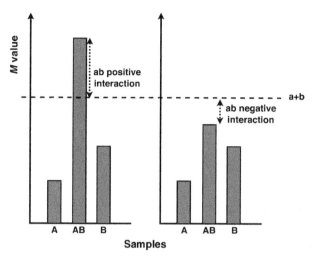

**FIGURE 7.14**  The Interaction effect of having two different drugs combined. The two plots illustrate how a combined effect of drug *A* and *B* (sample '*AB*') can result in either a positive or negative interaction in relation to the baseline of the gene expression *M*-values when adding the independent effects of drug *A* and drug *B* together ($a + b$).

### 7.4.3.3 *Example 3: Factorial Experiment With an Interaction Effect*

*Example 3, Step 1: Specify the Experimental Design*

This next example demonstrates how to create a design matrix for a factorial experiment with an interaction effect (e.g. for a drug or antibody, etc.). The primary question here is to determine whether there is a positive or negative effect of combing both drugs together compared to when the drugs are taken independently (Figure 7.14). As shown in Figure 7.15, the factorial experiment involves four different samples on six dual-channel arrays, using the control (untreated) sample as the common reference. The sample *A* is the sample treated with drug *A*; *B* is the sample treated with different drug *B*, *C* is the control or untreated sample (the 'baseline' effect), and *AB* is the sample treated with drugs *A* and *B* combined

*Example 3, Step 2: Parameterization*

An interpretation of the effects measured for each of the 4 sample types can be represented as follows:

$$
\begin{aligned}
\text{Sample untreated} : \quad & C = c \\
\text{Sample with drug } A : \quad & A = c + a \\
\text{Sample with drug } B : \quad & B = c + b \\
\text{Sample with } A \text{ and } B : \quad & AB = c + a + b + ab
\end{aligned}
$$

where $c$ is the base line effect, $a$ is the main effect of treatment with drug $A$, $b$ is the main effect of treatment with drug $B$, and $ab$ is the interaction effect between treatments $A$ and $B$.

In order to account for differences between the control, $C$ (untreated) and the samples with each of the two drugs on their own, including differences between the sample with the combination drug and the control, as well as differences between the combination drug and when the two different drugs are taken separately, the three chosen parameters are as follows:

$$A - C = (c + a) - (c)$$
$$B - C = (c + b) - (c)$$
$$AB - A - B + C = (c + a + b + ab) - (c + a) - (c + b) + (c) = ab$$

*Example 3, Step 3: Specify the Design Matrix*

$$E\begin{bmatrix} y^1 \\ y^2 \\ y^3 \\ y^4 \\ y^5 \\ y^6 \end{bmatrix} = \begin{matrix} X \\ \begin{bmatrix} 0 & 1 & 0 \\ 1 & 0 & 0 \\ 0 & 1 & 1 \\ 1 & 1 & 1 \\ 1 & -1 & 0 \\ 1 & 0 & 1 \end{bmatrix} \end{matrix} \times \begin{bmatrix} a \\ b \\ ab \end{bmatrix} = \begin{bmatrix} b \\ a \\ b + ab \\ a + b + ab \\ a - b \\ a + ab \end{bmatrix}$$

### 7.4.3.4 Example 4: Trend Analysis of a Single Time-Course Study

*Example 4, Step 1: Experimental Design*

This is an example of an experiment designed to identify a time-course trend over five points where mice have been treated with a drug. The five sample types being: $C$ (untreated or time = 0), $T_1$, $T_2$, $T_3$ and $T_4$ (a diagrammatical representation is shown in Figure 7.16). The design matrix should be constructed directly for the analysis of trends, as the `model.matrix` function of `limma` does not handle them. The latter formula approach only applies a linear model for various comparisons between the time-points and does not construct matrices for analyzing trends. This requires the construction of matrices with polynomial effects to account for 'quadratic' linear models.

*Example 4, Steps 2 and 3: Parameterization and Design Matrices*

As shown in Figure 7.17, there are three different possible models to fit in order to identify trends or particular shapes of gene expression profiles over time: a 'straight line' linear model fit to identify positive or negative 'straight' trends; a 'quadratic' linear model fit to identify any late response trends; and a 'quadratic' linear model fit to identify any early response trends. This kind of trend analysis

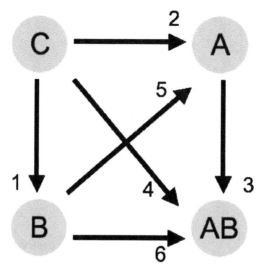

**FIGURE 7.15** Design for a factorial experiment to study the interaction ('joint') effect of two different drugs on the gene expression response of tumor cells. In this example there are four different samples $A$, $B$, $C$ and $AB$. The sample $A$ is the sample treated with drug $A$; $B$ is the sample treated with a different drug $B$, $C$ is the control or untreated sample, and $AB$ is the sample treated with both drug $A$ and $B$ combined. The combination of having both drugs may have an additional effect not observed when adding effects for both drugs alone (independently). The six arrays are labeled 1–6. The back of the arrow points to a condition in one dye channel (e.g. red), whereas the front points to the condition in the other dye channel (e.g. green).

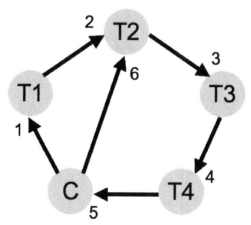

**FIGURE 7.16** Design for a single time-course study using six slides (labeled 1–6) for analyzing the effect of a drug over five time-points: $C$ (untreated or time = 0), $T_1$, $T_2$, $T_3$ and $T_4$. The back of the arrow points to a condition in one dye channel (e.g. red), whereas the front points to the condition in the other dye channel (e.g. green).

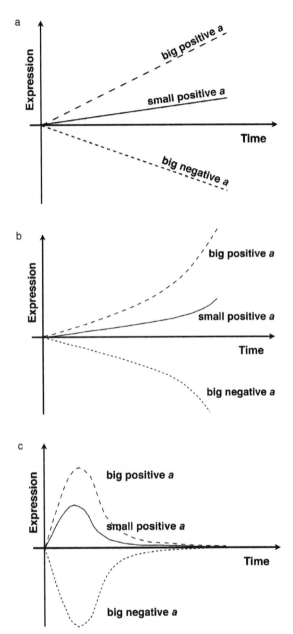

**FIGURE 7.17** Depending on what type of trend is being analyzed, three different models can be used to fit the data to identify gene expression profiles across the three different trend groups: (a) 'straight line' linear model; (b) 'quadratic linear model' to identify any late response trends of the drug; and (c) 'quadratic linear model' to identify any early response trends of the drug. With sufficient numbers of time points, the identification of both (b) and (c) trends can be very informative.

is complementary to the use of clustering methods such as $k$-means clustering, which can be used as an alternative approach to identify time-course trends (see Chapter 8 on 'Clustering'). The effects of interest, when looking at a particular trend is the control, $c$, the addition of a treatment, $a$, and the effect that time has on the treatment (i.e. $a$, $2a$, $3a$, etc. for a 'straight line' trend or $a$, $4a$, $9a$, $16a$, etc. for a 'quadratic' trend). If for example, there are two drugs being assessed in a multiple time-course, then an additional effect to look for is an interaction effect (i.e. if the two trend lines for each drug cross-over). The following are example parameterizations for all three models and the identification of the different effects to be measured from each of the five samples. For each model in the example, only one parameter is chosen. The commonality across the various time points is the effect $a$ and how it changes across the time points and the baseline effect, $c$. In each case the parameterization is followed by the design matrix:

Straight line

$$C = c$$
$$T1 = c + a$$
$$T2 = c + 2a$$
$$T3 = c + 3a$$
$$T4 = c + 4a$$

There are 4 possible different parameters, but only one parameter has been chosen for this particular example: $T2 - T1 = a$

Straight line design matrix

$$E \begin{bmatrix} y^1 \\ y^2 \\ y^3 \\ y^4 \\ y^5 \\ y^6 \end{bmatrix} = \begin{bmatrix} 1 \\ 1 \\ 1 \\ 1 \\ -4 \\ 2 \end{bmatrix} \times [a] = \begin{bmatrix} a \\ a \\ a \\ a \\ -4a \\ 2a \end{bmatrix}$$

Quadratic early response

$$C = c$$
$$T1 = c + a$$
$$T2 = c + 4a$$
$$T3 = c + 9a$$
$$T4 = c + 16a$$

There are 4 possible different parameters, but only one parameter has been chosen for this particular example: $T2 - T1 = a$

Quadratic early response design matrix

$$
E \begin{bmatrix} y^1 \\ y^2 \\ y^3 \\ y^4 \\ y^5 \\ y^6 \end{bmatrix} = \begin{bmatrix} 1 \\ 3 \\ 5 \\ 7 \\ -16 \\ 4 \end{bmatrix} \times [a] = \begin{bmatrix} a \\ 3a \\ 5a \\ 7a \\ -16a \\ 4a \end{bmatrix}
$$
$$
\overset{Y}{} \qquad \overset{X}{} \qquad \overset{\beta}{}
$$

Quadratic late response

$$
\begin{aligned}
C &= c \\
T1 &= c + 16a \\
T2 &= c + 9a \\
T3 &= c + 4a \\
T4 &= c + a
\end{aligned}
$$

There are 4 possible different parameters, but only one parameter has been chosen for this particular example: $T2 - T1 = a$

Quadratic late response design matrix

$$
E \begin{bmatrix} y^1 \\ y^2 \\ y^3 \\ y^4 \\ y^5 \\ y^6 \end{bmatrix} = \begin{bmatrix} 16 \\ -7 \\ -5 \\ -3 \\ -1 \\ 9 \end{bmatrix} \times [a] = \begin{bmatrix} 16a \\ -7a \\ -5a \\ -3a \\ -1a \\ 9a \end{bmatrix}
$$
$$
\overset{Y}{} \qquad \overset{X}{} \qquad \overset{\beta}{}
$$

### 7.4.3.5 Estimation of Coefficients Using Least Squares

Once the contrast matrix is constructed, the `limma` function `contrast.fit` is used to compute estimated coefficients and standard errors for these contrasts from the original model. This involves a mathematical procedure, *least squares*, to discriminate between the conditions or treatments of interest. Essentially, this is a regression approach that finds the estimated log expression ratios by fitting a linear equation to the observed data for each gene. In other words, it finds the closest fit (in terms of minimizing squared deviations) between the observed values and the estimated values (the linear model).

### 7.4.3.6 *The Moderated t-Statistic,* B-*Statistic and a p-Value*

A range of statistical methods can then be applied to the estimated coefficients using the empirical Bayes method (described earlier) and this approach moderates the standard errors of the fold changes (Smyth, 2004). This method also involves replacing the variance parameter with a posterior variance, which is estimated from the data. The function `eBayes` is used to calculate the moderated *t*-statistics.

Essentially, `limma` calculates and displays the moderated *t*-statistic, *B*-statistic and a *p*-value for every gene and these are used to provide an overall ranking of genes in order of evidence for differential expression (by default, ranking is done by the *B*-statistic). The *p*-value is also given after adjustment for multiple testing (i.e. the adjusted *p*-value). The most common form of adjustment is the Benjamini and Hochberg's approach to control the false discovery rate (Benjamini and Hochberg, 1995).

In order to calculate these statistics and decide which genes are differentially expressed, `limma` considers the following: the number of replicates for each gene (whether on the same slide or different slides); the variation in expression measurement values or *M* values; and the magnitude of expression measurement values or *M* values. All the genes in a replicate group are combined into estimates of the posterior distribution, and then at a gene level, these parameter estimates are then combined to form a *B*-statistic, *B* which is a Bayes log posterior odds (Lonnstedt and Speed, 2001) or the log-odds that the gene is differentially expressed. For example, if $B = 1.7$, the odds of differential expression is $\exp(1.7) = 5.47$, that is about five and a half to one. The probability that the gene is differentially expressed is $5.47/(1 + 5.47) = 0.84$, that is a 84% probability that this gene is differentially expressed. If the *B*-statistic is zero, the chance that the gene is differentially expressed is 50–50.

There is also an automatic adjustment of the *B*-statistic in `limma` for multiple testing by assuming that a particular user-defined percentage of genes are differentially expressed. Both the *p*-values and the *B*-statistics normally rank in the same order, and if there are no missing values or quality weights in the data, then the order will be exactly the same. The *p*-values may be preferred over *B*-statistics because unlike *B*-statistics, they do not require prior knowledge or estimation of the proportion of differentially expressed genes. Limma also computes the moderated *F*-statistic by combining the moderated *t*-statistics for all the contrasts into an overall significance test for a given gene. This statistic is used to identify whether a gene is differentially expressed on any of the contrasts. For a complex experiment with many comparisons, this may be desirable first before selecting genes on the contrasts of interests. This reduces the number of tests required and therefore on the amount of adjustment for multiple testing. For single-channel arrays that are used to compare a 'treatment against a control', a moderated paired *t*-test can also be computed using the linear model approach. Other useful features in `limma` include volcano plots, venn diagrams and the `geneSetTest` function for gene-set (or -class) analysis (discussed further in Section 7.6).

### 7.4.4 Handling of Complex Experimental Designs

On a more general note, the linear formulation model of `limma` allows the handling of experiments with arbitrary complexity including those with multiple factors, for example factorial designs to identify interaction effects and multiple time-series (or time-course) data. When looking for trends across single or multiple time-courses, a quadratic linear model can be applied to the data. In other words, `limma` is a framework for assessing differential expression in any microarray experimental design. However, it is essential to consider the objectives of any analysis at the experimental design stage, before collecting any samples for the array experiments, since some experimental designs are more precise or optimal than others at identifying differentially expressed genes. For example, for two-channel array studies with many related conditions, such as a time-course experiment, interwoven loop designs are considerably better than reference designs (Glonek and Solomon, 2004; Wit and McClure, 2004). When deciding which design is optimal during the experimental design stage, measures of optimality can be calculated using the *R* package `smida` (Wit and McClure, 2004). Understandably, optimal experimental designs can be confounded by limited material or financial resources and therefore the choice of a particular design can depend on the number of arrays available. Limma can also deal with 'unconnected designs' where there are no arrays linking a comparison that needs to be made. For such designs, the two-channel arrays can be analyzed as single-channel arrays, however, unconnected designs are not as precise as those that are connected with a direct comparison.

### 7.5 NONPARAMETRIC STATISTICS

Although parametric methods seem to work well in some situations, there is a huge drawback in their strong dependence on assuming normality as this is often violated in microarray data sets. It is sometimes difficult to verify normality considering the small sample size one often encounters in a microarray experiment. Several nonparametric methods, including Statistical Analysis of Microarrays (SAM; Tusher et al., 2001; Storey and Tibshirani, 2003a,b,c), an empirical Bayes method (Efron et al., 2001) and the mixture model method (Pan et al., 2003), have been proposed to alleviate the problem. The main feature of nonparametric statistical methods is that they do not assume that the data is normally distributed. The difference between parametric and nonparametric methods is the way the null distribution of the test statistic (Z) is derived. For example, in the two-sample *t*-test, the null distribution of Z is assumed to have

**TABLE 7.4** A Summary of Parametric Microarray Data Analysis Methods and Their Availability

| Package | Availability | Nonparametric method | Reference |
|---------|-------------|----------------------|-----------|
| WilcoxCV | CRAN | Wilcoxon rank sum test | Boulesteix (2007) |
| RankProd | Bioconductor | Rank Product | Hong et al. (2006) |
| siggenes | Bioconductor | Empirical Bayes and permutation | Tusher et al. (2001), Schwender et al. (2006) |
| Vbmp | Bioconductor | Regression with Gaussian process priors | Lama and Girolami (2008) |
| NPLR | NA | Nonparametric likelihood ratio | Bokka and Mathur (2006) |
| TMM | NA | Mixture model method | Jiao and Zhang (2008) |
| MMM | NA | Mixture model method | Pan et al. (2003) |

*Bioconductor*: http://www.bioconductor.org; CRAN (The Comprehensive *R* Archive Network): http://cran.r-project.org/.

the $t$-distribution, whereas in nonparametric methods, the null distribution of $Z$ is estimated directly from the repeated measurements of gene expression levels under each condition. Nonparametric methods are generally less powerful than the parametric equivalents but they are more robust to outliers since these tests usually rely on statistics based on the rank of the observations rather than the values themselves. As with classical parametric statistical methods, nonparametric methods generally require large sample sizes.

There are two types of nonparametric tests: classical and bootstrap. The former is equivalent to a parametric test and is used in place of either a paired or unpaired $t$-test when the data sets being compared are not normally distributed. However, these are less powerful than the parametric equivalents. The bootstrap test also assumes a nonnormal distribution but is more modern and applicable to a wide range of applications. Bootstrap methods also tend to be more powerful than the classical nonparametric methods and are generally the preferred nonparametric methods for microarray data analysis. A summary of packages available either in *Bioconductor* or as *R* code is provided in Table 7.4.

### 7.5.1 Classical Nonparametric Tests

The *Wilcoxon signed rank* test is the nonparametric equivalent to the paired $t$-test since it is used when each experimental subject is observed before and after a treatment. It involves sorting the absolute values of the differences from smallest to largest then replacing the true value with ranks regardless of sign. Next, the sum of the ranks of the positive differences (upregulated values) is determined and compared to that of the negative differences, and the smaller is taken and compared against a precomputed table to obtain a $p$-value. If the null hypothesis is true, the sum of the ranks of the positive differences should be similar to the sum of the ranks of the negative differences. However, if the test statistic is outside of this range, the null hypothesis can be rejected indicating that the treatment did have some effect. Troyanskaya et al. (2002) provide a reasonable assessment of this method as applied to microarray data.

The *Mann–WhitneyU-test* (also known as the *Wilcoxon rank sum* test) is the nonparametric equivalent to the unpaired test and is similar in method to the Wilcoxon signed rank test. The data from the two groups are combined into one sample of size $n_1 + n_2$ (i.e. the observations are ranked regardless to which sample group they are in), and given ranks from the smallest to the largest value. The sum of the ranks of the two groups, $U_1$ and $U_2$ are compared and the smaller of the two numbers is compared against a precalculated table to obtain a $p$-value for the null hypothesis that the two distributions are the same. The test can be defined as follows:

$$U_1 = \frac{n_1(n_1 + 1)}{2} - R_1 \quad U_2 = \frac{n_2(n_2 + 1)}{2} - R_2$$

where $n_1$ is the two sample size in group 1 and $R_1$ is the sum of the ranks for group 1. Both the Wilcoxon sign-rank and the Mann–Whitney tests are used when the distributions of the two samples have approximately the same shape, although the latter is quite robust against slight deviations. If this assumption is not met the null hypothesis is different.

The *Kruskal–Wallis* test is a nonparametric equivalent to an ANOVA. There are no restrictions on the number of conditions in the analysis, and in the special case where there are only two condition groups, the Kruskal–Wallis is equivalent to the Mann–Whitney. In the Kruskal–Wallis test, all observations are ranked without regard for the group in which they are found. The sums of each group's observations are first computed and then the distributions of the ranks are compared. As with the ANOVA, the Kruskal–Wallis is a statistic that examines overall variation among the groups. Information about which groups are significantly different is not offered, but these can be determined by performing pairwise comparisons after a significant result.

The Kruskal–Wallis statistic, $H$, is calculated by dividing the sum of squared differences (SSD) of ranks by $N(N + 1)/12$ as follows:

$$H = \frac{n_1(x_1 - x)^2 + \cdots n_k(x_k - x)^2}{N(N + 1)/12}$$

where $n_1, \ldots, n_k$ is the number of individuals in each group; $x_1, \ldots, x_k$ is the mean of each group's rank and $N$ is the total number of individuals in all groups. The $H$ statistic is used to generate a measurement that examines how the rank of each group compares with the average rank of all the groups. It is compared to the Chi-square distribution with degrees of freedom $k - 1$. If the sample size is

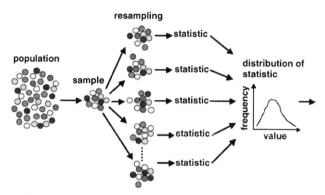

**FIGURE 7.18**   The basic idea behind bootstrapping. See text for details.

small, it may be necessary to prevent an overly conservative estimate using k degrees of freedom. Like the Mann–Whitney $U$-test, this method requires similar group distributions to test for a difference in medians.

## 7.5.2 Bootstrap Tests

Bootstrap methods are robust to noise and experimental artefacts and they are also more powerful than classical nonparametric methods. They tend to be used when the sample sizes are neither large nor are from a normal distribution. There are bootstrap equivalents for both paired and unpaired nonparametric methods and it is also possible to use them for more complex methods such as ANOVA models and data clustering approaches.

The use of the bootstrap term derives from the phrase 'to pull oneself up by one's bootstrap' from the legendary Baron Munchausen who grabbed at his bootstraps to pull himself out of a deep hole. It is another way of saying 'you succeed on your own despite limited resources'. The statistical bootstrap (Efron, 1979) does something similar in that it is a resampling method that involves reusing the data many times over. As illustrated in Figure 7.18, the original sample can be duplicated as many times as the computing resources allow, and then this expanded sample is treated as a virtual population. Samples are then drawn from this population to verify the estimators. The main objective of the test is to compare a particular property of the real data set with a distribution of the same property in random data sets, and the most common property to use is the $t$-statistic. Bootstrapping simulates alternative values for the $t$-statistic in order to calculate an empirical $p$-value.

With an unpaired analysis, for example, the bootstrap works by combining a large-data set from the two data groups to create a large number of alternative random data sets by resampling from the original data set with 'replacement'. With a replacement approach, different individuals in the bootstrap data could have an identical value from the real data set. The resulting bootstrap data resembles real data in that the values are similar but no biological sense can be made of it because the values have been randomized. Using this large randomized bootstrap data sample, an empirical distribution is generated using the $t$-statistics and this is compared with the $t$-statistic from the real data set. An empirical $p$-value is then calculated by computing the proportion of the bootstrap statistics that have a more extreme value than the $t$-statistic from the real data.

### 7.5.3 Permutation or Randomization Tests

Permutation or randomization methods are closely related to bootstrap analysis in that they are bootstrap methods that involve sampling without replacement. The main idea is to randomly shuffle the measurement from the two sample data sets several times to create a null distribution. Each of the bootstrap values is created by first resampling values from the real data set without replacement, so each of the real values is used only once, and an empirical $p$-value can be calculated using the shuffled samples. By permuting all genes in one microarray simultaneously, the covariance structure of the data can be preserved and this may also be considered to be a good way to get around assuming independence of the genes. Permutation can only be used if the two sample distributions are the same, except perhaps for the possible difference in mean. However, if the variances of the samples are not the same, the results may be inaccurate.

### 7.5.4 The Significance Analysis of Microarrays (SAM)

The SAM approach is available as `siggenes` from *Bioconductor* or as a Microsoft Excel plug-in. Further discussion and on usage details can be found from a variety of sources (Tusher et al., 2001; Storey and Tibshirani, 2003a,b,c; the `siggenes` *Biconductor* vignette).

When the number of replicates is small, variance estimation is more challenging as the estimated variance is unstable. Furthermore, low expression measurements lead to a low variance and this leads to a $t$-statistic that can be high. Penalized or regularized statistics add an extra term, $s_0$, a correction factor (constant) to increase the size of the standard deviation and prevent this small variance from overestimating t scores and inflating the statistic (i.e. producing a large false discovery probability) for weakly expressed genes:

$$d = \frac{x_1 - x_2}{\sqrt{(S_1^2/N^1) + (S_2^2/N^2)} + s_0}$$

This moderated $t$-statistic is used by SAM to determine the value of $s_0$ (Tusher et al., 2001). By adding a small correction factor to the denominator

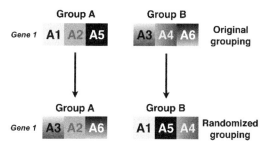

**FIGURE 7.19**   The basic idea behind permutation. Gene expression measurements are randomly shuffled across the two sample sets, Group A (arrays A1, A2, A5) and Group B (arrays A3, A4, A6). This is done several times to create a null distribution.

of the *t*-statistic (the gene specific variance estimate), the denominator is regularized and adjusted for the underestimated error. The constant is estimated from the sum of the global standard error of the genes and results in reducing the chance of detecting genes that have a low standard deviation by chance.

SAM also uses permutation to achieve greater statistical significance. This multiple testing approach involves computing a reasonably accurate estimate for the true FDR using permutations of the data, allowing the possibility of dependent tests. Therefore, as pointed out by the authors, it seems plausible that this estimated FDR approximates the strongly controlled FDR when any subset of null hypotheses is true. However, the authors also point out that due to the limited number of possible distinct permutations, the number of distinct values that the *p*-value can take is also limited. A drawback of the permutation method is that it is difficult to demonstrate low *p*-values. For example, showing that a *p*-value is lower than $10^{-7}$ requires at least $10^7$ permutations. Often, when the sample size is small, the total number of permutations is not large enough to attain very low significance levels.

Essentially, the permutation approach used by SAM involves the following steps:

   **i.** For each gene, a *d*-value (the modified test statistic) is computed and this is the *observed d-value* for that gene.
   **ii.** The genes are ranked in ascending order of their *d*-values.
   **iii.** The values of the genes are randomly shuffled between two condition groups, for example groups *A* and *B*, so that the reshuffled groups *A* and *B*, respectively have the same number of elements as the original groups *A* and *B* (Figure 7.19).
   **iv.** The *d*-value is computed for each randomized gene.
   **v.** The permuted *d*-values of the genes are ranked in ascending order.
   **vi.** Steps 3,4 and 5 are repeated many times, so that each gene has many randomized d-values corresponding to its rank from the observed

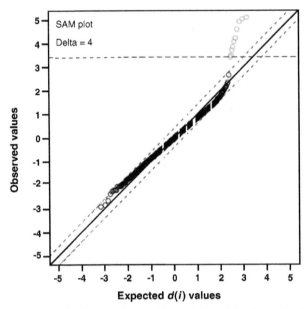

**FIGURE 7.20**   An example of a quantile–quantile ($q$–$q$) plot of the observed and the expected $d$-values produced by siggenes for a particular delta $\Delta$. The two broken parallel band of lines around the solid line marks the distance of delta = 0.4. The horizontal broken line marks a cutoff of 3.416 for the observed values (the $y$-axis). All genes above this point and to the left of the 45° solid line are called significant. Taken from the siggenes vignette from Bioconductor.

(unpermuted) $d$-value. The average of the randomized d-values are taken for each gene and this is the *expected d-value* of that gene.

vii. The observed $d$-values are plotted against the expected $d$-values on a quantile–quantile plot (an example is shown in Figure 7.20) to obtain global cutoffs.

viii. For a given cutoff or threshold, delta, the FDR is estimated by computing for each of the permutations, the number of genes with $d$-values above or below delta, and averaging this number over permutations.

ix. A threshold delta is chosen to control the expected number of false positives, PFER, under the complete null, at an acceptable nominal level.

In other words, for each permutation of the data, the number of differentially expressed genes that are statistically significant for a given delta threshold is determined. The median number of genes that are statistically significant from these permutations is the median FDR, therefore, any genes designated as statistically significant from the randomized data are being picked up purely by chance and are 'falsely' discovered. The median number picked up over many randomizations is a reasonably accurate estimate of the false discovery rate.

## 7.6 GENE-CLASS TESTING

Differential gene expression analysis can produce a long list of statistically significant differentially expressed genes and interpreting such a long list is usually a daunting task. This can be dealt with in several ways:

i. *Data integration and filtering*: By incorporating the normalized intensity values or the MAS5 present/absent calls, genes (probes or array features) lower than a certain threshold over all or a subset of samples can be filtered out or given a lower priority. Genes with lower fold changes between the condition groups of interest could be given lower priority over those with higher fold changes, but however small, these may still be of great importance and should not be removed. The `genefilter` package can be used to filter data before or after any statistical analysis. Adding gene annotations (e.g. Gene Ontology) to the data set and then filtering with these can help focus the attention on particular biological areas of interest. A well-known and popular online annotation and analysis tool, DAVID[21] (Dennis et al., 2003; Huang da et al., 2007a,b; Sherman et al., 2007) can be used for various simple and advanced annotation needs and there are numerous annotation packages in *Bioconductor*, including `annotate`, `AnnBuilder`, `hsa-homology` (for orthologues) and `AnnotationTools` (Kuhn et al., 2008). In addition to retrieving general gene and functional annotation, the latter tool can also be used to find both orthologues and integrate heterogeneous gene expression profiles from other datasets from different platforms. Integrating data from, for example, other relevant published data sets (see the next section on 'Within-and cross-platform meta-analysis') or data sets from other technologies (e.g. Chip–Chip) can also help draw attention to a smaller subset of genes and provide new insights into the data. Other tools also include the `geneplotter` package in *Bioconductor*, which can be used to visualize gene expression along genomic coordinates, and Cytoscape,[22] which can be used to integrate and visualize biological pathways or interaction networks (Shannon et al., 2003). Other similar data mining and visualization tools to Cytoscape include BioTapestry (Longabaugh et al., 2005) and StarNet (Jupiter and VanBuren, 2008).

ii. *Clustering*: Clustering can help confirm your findings and reveal some new ones (see Chapter 8 for more details). For example, hierarchical clustering of the selected gene list can be used to produce 'heatmaps', which can provide useful information on the overall structure of the data subset, that is how the different samples and condition groups relate to each other and partition given the list of selected genes, and how certain groups of genes

---

21 DAVID: http://david.abcc.ncifcrf.gov/home.jsp.
22 Cytoscape: http://www.cytoscape.org/.

expression profiles cluster together. Clustering can be based on gene, sample and/or functional annotation such as GO, biological pathway, etc.

iii. *Gene-set analysis*: Gene-set or category analysis methods are statistical tests that compare the number of genes in a class that are 'significant' with the number that are expected under a particular null hypothesis. Example categories include Gene ontology, biological pathway, chromosomal location or arm, clinical or disease phenotype. This section discusses gene-set analysis methods in further detail.

Understandably, there does seem to be a recent shift in emphasis from microarray data analysis methods for identifying differentially expressed genes to methods for identifying differentially expressed *gene categories* (sets or classes). As pointed out by Subramanian et al. (2005), this is because the former approach has several issues and limitations:

i. After adjusting $p$-values using a multiple testing procedure, genes may not meet the threshold for statistical significance because the relevant biological differences are modest in relation to the noise inherent in the microarray technology.

ii. Making a list long list of statistically significant genes biologically meaningful can be daunting. The process is often slow and ad hoc and relies on the biologists field of expertise.

iii. The focus on identifying large expression changes for particular genes can result in missing important effects on pathways. For example, an increase of 25% of all genes involved in a particular metabolic pathway may have dramatic effects on the dynamics of that pathway and this may be more important than a gene shown to have a 25-fold increase in expression.

iv. A weak effect in a group of genes may be missed when each gene is considered individually, but it may be captured when they are considered together. For example, when comparing results from different groups or array platforms, there may be a disappointingly low amount of overlap in the gene lists, but what is missed out is the potentially large overlap in pathway representations.

The main difference between these two approaches is that the gene category analysis approach uses a priori information about genes (i.e. genes known to be involved in the same metabolic pathway or are located on the same cytogenetic band) to group them into categories and this can enhance the interpretation of experiments aimed at identifying expression differences across conditions or treatments. Category data are data that map from specific entities (i.e. genes) to categories. Sometimes the mapping will be a partition where each gene maps to one category only, but more commonly, genes will map to multiple categories. Biological examples of category data include mappings from genes to chromosomes (which is nearly a partition), genes to pathways, genes to Gene Ontology

(GO) terms (Ashburner et al., 2000) or genes to protein complexes. The categories can come from various annotation sources, mainly from various public database resources, and there are two main ways to analyze gene categories: overrepresentation analysis (ORA) or gene set enrichment analysis (GSEA).

## 7.6.1 Overrepresentation Analysis

Earlier statistical methods for identifying gene categories that are overrepresented or underrepresented in statistically significant differentially expressed genes are reviewed in Khatri and Draghici (2005) and Allison et al. (2006) with more recent methods covered by Nettleton et al. (2008). The more popular earlier approaches used variations on the Fisher's Exact test or $\chi^2$-test from a contingency table (Berriz et al., 2003; Draghici et al., 2003; Leroux et al., 2003) including FatiGO (Al-Shahrour et al., 2004), Gostat (Beissbarth and Speed, 2004) and the NetAffx Gene Ontology Mining Tool (Cheng et al., 2004). Rivals et al. (2007) provide a comprehensive review of many of these methods, both early and more recent, and a selection of these is summarized in Table 7.5. Basically, most of them involve testing whether the statistically significant genes from a particular category are significantly larger (or smaller) than what would have been expected if they had been randomly and independently drawn without replacement, from the gene collection represented on the microarray.

However, as mentioned by the reviewers, there are certain drawbacks to these earlier methods that involve basic random sampling. In particular, these methods make the incorrect assumption that genes are independent of one another. Genes in a gene set are functionally related and are not independent; the complex structure of gene interactions within a gene set is not fully captured using univariate approaches. Under the random sampling assumption, the appearance of an unusually large (or small) proportion of genes from a particular category may be easily explained by positive correlation among genes in the category rather than 'enrichment'. Therefore, random sampling provides a poor probability model for the selection of statistically significant differentially expressed genes. Goeman and Buhlmann (2007) refer to them as *gene sampling* methods and use simulation to show that they can be quite liberal when genes are dependent.

Furthermore, the results of such methods can be sensitive to the significance level threshold used to produce the list of differentially expressed genes. In certain situations, all genes in a given category may exhibit small changes but when considered together, they may provide strong evidence of a condition or treatment effect. However, if many of the individual changes fail to reach the specified significance threshold level, such categories will go undetected. Newton et al. (2007) developed methods for addressing this problem, but their approach uses a gene sampling probability model, which incorrectly assumes all genes are independent from one another.

**TABLE 7.5** Selected examples of statistical methods for identifying gene categories that are either over- or under-represented

| Package | Availability | Gene Category Method | Reference |
|---|---|---|---|
| allez | R package[1] | Random set enrichment scoring | Newton et al. (2007) |
| BINGO | Stand-alone java tool[2] | Hypergeometric | Maere et al. (2005) |
| CLENCH | Stand-alone perl tool | Hypergeometric, binomial, $\chi^2$ | Shah & Fedoroff (2004) |
| DAVID | Web-based[3] | Fisher | Huang et al. (2007) |
| FatiGO | Web-based[4] | Fisher | Al-Shahrour et al. (2004) |
| FuncAssociate | Web-based[5] | Fisher | Berriz et al. (2003) |
| GeneMerge | Web-based[6] | Hypergeometric | Castillo-Davis & Hartl, (2003) |
| GeneTrail | Web-based[7] | Hypergeometric, $\chi^2$ | Backes et al. (2007) |
| GOEAST | Web-based[8] | Hypergeometric, Fisher, $\chi^2$ | Zheng & Wang (2008) |
| GoStat | Web-based[9] and Bioconductor | Fisher, $\chi^2$ | Beissbarth (2004) |
| NetAffx GO Mining | Web-based[10] | $\chi^2$ | Cheng et al. (2004) |
| Onto-Express | Web-based[11] | Binomial, Fisher, $\chi^2$ | Draghici et al. (2003) |
| STEM | Stand-alone java tool[12] | Hypergeometric | Ernst & Bar-Joseph (2006) |

[1] Download site for allez: http://www.stat.wisc.edu/~newton/.
[2] Download site for BiNGO: http://www.psb.ugent.be/cbd/papers/BiNGO/.
[3] DAVID url: http://david.abcc.ncifcrf.gov/.
[4] FatiGO url: http://www.fatigo.org/.
[5] FuncAssociate url: http://llama.med.harvard.edu/cgi/func/funcassociate.
[6] GeneMerge url: http://genemerge.bioteam.net/.
[7] GeneTrail url: http://genetrail.bioinf.uni-sb.de/.
[8] GOEAST url: http://omicslab.genetics.ac.cn/GOEAST/.
[9] GoStat url: http://gostat.wehi.edu.au/.
[10] NetAffx url: http://www.affymetric.com/analysis/index.affx.
[11] Onto-Express url: http://vortex.cs.wayne.edu/ontoexpress/.
[12] Download site for STEM: http://www.cs.cmu.edu/~jernst/stem/.

Another drawback with these *gene sampling* methods is that information is lost when continuous gene-specific measures of statistically significant differential expression (e.g. *p*-values) are split to produce genes that are declared differentially expressed and those that are not. The rank order of the evidence for differential expression is lost in these gene lists.

## 7.6.2 Gene Set Enrichment and Analysis (GSEA)

Two very similar methods were proposed to address some of the deficiencies found in Overrepresentation analysis or *gene sampling* methods. These methods are based on the gene set enrichment and analysis approach: GSEA-P (Mootha et al., 2003; Subramanian et al., 2005, 2007) and significance analysis of function and expression (SAFE; Barry et al., 2005). Essentially, these approaches evaluate the distribution of genes belonging to a biological category in a sorted list of genes by computing running sum statistics. The main difference between GSET-P and SAFE is that they calculate a different measure of differential expression at the start of the analysis. Goeman and Buhlmann (2007) refer to these methods as *subject sampling* methods for determining significance, since they permute treatment labels (*subject sampling)* rather than permuting gene labels (*gene sampling*).

GSEA-P is probably the most well-known gene category analysis approach and is freely available as *R* code or as a platform-independent package.[23] GSEA-P handles data sets from two condition groups and genes are ranked based on how correlated their expression is to the class distinction using various metrics. Given a list of a priori defined genes (i.e. a gene set or category) ranked by a sorted list, the main goal of GSEA-P is to determine whether these genes are randomly distributed throughout this list or found mainly at the top or bottom. The assumption is that gene sets related to phenotypic distinction usually shows the latter distribution. A schematic overview of the GSEA approach taken from Mootha et al. (2003) is shown in Figure 7.21. GSEA involves four main calculations:

i. *Calculation of the signal-to-noise statistic.* The signal-to-noise statistic is calculated for each gene and this quantifies the distance in expression between two phenotypic classes and any suitable metric can be used. Genes are ranked based on the correlation between their expression and the class distinction.

ii. *Calculation of an enrichment score (ES).* An enrichment score is calculated using the Kolmogorov–Smirnov statistics to reflect the degree to which a gene set is overrepresented at the top or bottom of the entire ranked list. The ES is the maximum deviation from zero and is calculated by walking down the gene set list, and increasing a running-sum statistic whenever a gene is encountered and decreasing it whenever a gene is not encountered in the list.

---

23 GSET-P is available from the following url: http://www.broad.mit.edu/gsea/index.jsp.

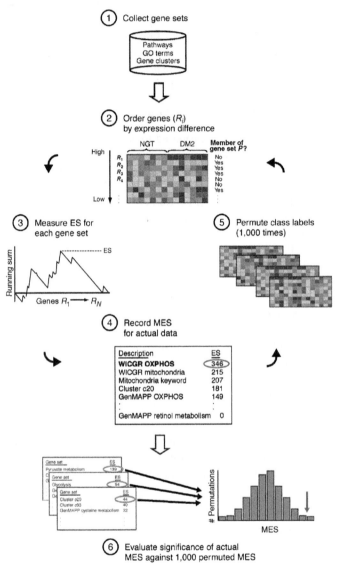

**FIGURE 7.21**  A schematic overview of GSEA taken from Mootha et al. (2003), see the text for details.

The magnitude of this increment depends on how correlated a gene is with the phenotype.

iii. *Estimation of significance level of ES.* The statistical significance (nominal *p*-value) of the ES is estimated by a permutation testing approach based on phenotype. This involves permuting the phenotype labels and recalculating the statistics for the ES of the gene set for each permuted data set. A thousand of these permutations generate a null distribution for the ES and the empirical, nominal $p$-value of the observed data is calculated relative to this distribution. The information is used for the multiple testing procedure below in step 4. Permutation approaches are popularly used because they retain gene correlation structure, and by permuting class labels rather than genes, it provides a more biologically meaningful assessment of significance.

iv. Adjustment for multiple hypothesis testing. An entire database of gene sets is evaluated to adjust the nominal $p$-value to account for multiple testing. This involves creating a normalized enrichment score (NES) by normalizing each gene set to account for set size before applying FDR control (Benjamini et al., 2001; Reiner et al., 2003) to each NES. The FDR of the resulting gene list is assessed by comparing the tail error from the null distribution of the statistic derived from the data permutations in step 3.

GSEA-P can also be used to analyze gene sets ranked by other setting such as differences or on how well their expression correlates with a given target pattern. For example, a GSEA-like procedure was used to show that enriched gene sets of Cyclin D1 targets ranked by correlation with the profile of Cyclin D1 in a compendium of tumor types (Lamb et al., 2003). In addition to this, GSEA-P can be used to extract the core members of high scoring gene sets that contribute to the ES since not all the members in a gene set will participate in a biological process. These core members are known as the 'leading edge subset' (LES) and can be interpreted as the core of a gene subset that accounts for the enrichment signal. Examination of this subset can reveal a biologically important subset within a gene set. High scoring subsets can be grouped based on shared LES of genes and this can reveal which of the gene sets correspond to either similar or distinct biological processes.

GSEA-P can also be used to evaluate a query microarray data set against a collection of gene sets in the database, MSigDB. The database (version 2.1) currently has a catalogue of 3337 gene sets belonging to four types: cytogenetic sets for human chromosomes and cytogenetic bands; functional sets for specific metabolic and signaling pathways; regulatory motif sets based on commonly conserved regulatory motifs in the promoter regions of human genes (Xie et al., 2005); and neighborhood sets centered on 380 cancer related genes. Many of these annotations originate from various publicly available manually curated databases (see Table 7.6 for details).

There are several problematic issues associated with approaches such as GSEA-P and these are mentioned in Efron and Tibshirani (2007), (Keller et al., 2007) and Nettleton et al. (2008). Nonparametric tests use ranks instead of measured values, therefore they tend to be less powerful, informative or flexible than corresponding parametric tests and they do not assume a Gaussian distribution of the data. Also, repeated runs of the permutation test algorithm may lead to

**TABLE 7.6** A List of Publicly Available and Manually Curated Database Annotation Resources Stored in MsigDB (Molecular Signature Database)

| Database | Website |
|---|---|
| BioCarta | http://www.biocarta.com |
| Signaling Pathway database | http://www.grt.kyushu-u.ac.jp/spad/menu.html |
| Signaling gateway | http://www.signaling-gateway.org |
| Signal transduction knowledge environment | http://stke.sciencemag.org |
| Human protein Reference Database (HPRD) (Peri et al., 2003) | http://www.hprd.org |
| GenMapp (Salomonis et al., 2007) | http://www.genmapp.org |
| Gene Ontology, (Ashburner et al., 2000) | http://www.geneontology.org |
| Sigma Aldrich pathways | http://www.sigmaaldrich.com/Area_of_Interest/Biochemicals/Enzyme_Explorer/Key_Resources.html |
| Cancer Genome Anatomy Project (Strausberg et al., 2000) | http://cgap.nci.nih.gov |

The database is used by GSEA-P to query a microarray data set against a collection of gene sets (Subramanian et al., 2005).

different significance values. They can also cause problems if the significance values are too small as these can make bad estimations of $p$-values, which are inappropriate for multiple testing procedures such as those controlling the FDR. It is also difficult to estimate how many permutations should be performed to obtain a reasonable sample size. Large numbers can be computationally intensive, especially if thousands of biological categories are being tested.

There are several suggested improvements or additional features to the GSEA-P approach and a summary of most of the various freely available GSEA methods are shown in Table 7.7. Among them, is the Gene Set Analysis (GSA) tool proposed by Efron and Tibshirani (2007), which uses the *maxmean* statistic in place of the Kolmogorov–Smirnov statistic for summarizing gene sets since it has superior power characteristics. They also proposed a *restandardization* approach, which combines both randomization and permutation ideas. Other improvements include an un-weighted approach involving dynamic programming utilized by an online tool, GeneTrail (Backes et al., 2007; Keller et al., 2007) to overcome computational running times, the problem of small $p$-values and other typical problems of nonparametric permutation tests. GeneTrail is based on the comprehensive integrative system BN++ (Küntzer et al., 2007), the biochemical network library,[24] which imports additional annotation not included in GSEA-P's MSigDB. Additional sources of data include; KEGG pathways (Kanehisa et al., 2006), TRANSFAC transcription factors

---

24 The comprehensive integrative system BN++ (the biochemical network library) is freely available at http://www.bnplusplus.org.

**TABLE 7.7** Selected Examples of Statistical Methods for Identifying Gene Categories

| Package | Availability | Gene category method | Reference |
|---|---|---|---|
| DEA | R code | Domain enhanced analysis using either principle components analysis (PCA) or partial least squares | Liu et al. (2007) |
| Eu.Gene Analyzer | Java stand alone[a] | Nonparametric permutation testing procedure | Cavalieri et al. (2007) |
| GeneTrail | Web-based tool[b] | Weighted GSEA involving dynamic programming | Backes et al. (2007), Keller et al. (2007) |
| GlobalTest | Bioconductor | Generalized mixed model | Goeman et al. (2005), Goeman and Mansmann (2008) |
| GSEA-P | R code | Nonparametric permutation testing procedure | Mootha et al. (2003); Subramanian et al. (2005, 2007) |
| GSA | R code | Randomization and permutation hybrid | Efron and Tibshirani (2007) |
| GSEAlm | Bioconductor | Linear modeling and posterior probabilities and PCA | Jiang and Gentleman (2007) |
| MRPP | | Nonparametric multivariate analysis | Nettleton et al (2008) |
| PAGE | Python stand alone | Parametric analysis of gene set enrichment involving z-scores. | Kim and Volsky (2005) |
| PGSEA | Bioconductor | Parametric gene set enrichment analysis | Furge et al. (2007) |
| SAFE | R code | Nonparametric permutation testing procedure | Barry et al. (2005) |
| SAM-GS | R code and Microsoft Excel | Significance analysis of microarray (SAM) for gene sets: nonparametric permutation testing | Dinu et al. (2007) |

[a] Eu.Gene Analyzer 1.0 url: http://ducciocavalieri.org/bio.
[b] GeneTrail url: http://genetrail.bioinf.uni-sb.de.

(Matys et al., 2006) and TRANSPATH pathways (Krull et al., 2006), protein–protein interactions from DIP (Salwinski et al., 2004), MINT (Zanzoni et al., 2002) and IntAct (Hermjakob et al., 2004).

### 7.6.3 Multivariate Gene-Set Analysis Methods

Methods such as GSEA-P, SAFE and GSA, do not have the ability to detect treatment effects on the multivariate expression distribution of genes in a category for two or more condition groups. There are many methods that have the ability to do this, including the following: the pathway level analysis of gene expression (PLAGE) method (Kong et al., 2006), which involves using the Hotelling's $T^2$ statistic, computes 'activity levels' of the genes in a given category using an approach involving singular value decomposition (SVD; Tomfohr et al., 2005). Liu et al. (2007) proposed the domain enhance analysis (DEA) method, which includes two different strategies involving either principal components (DEA-PCA) or partial least squares (DEA-PLS). The global test (GT) method addresses the problem of identifying gene categories whose expression can be used to predict a clinical outcome (e.g. cancer versus no cancer) using a generalized mixed model approach (Goeman et al., 2004). Finally, Nettleton et al. (2008) proposed the multiresponse permutation procedure (MRPP), which was demonstrated by the authors to perform well and in some cases, better than many of the other multivariate methods mentioned here. Some of these more esoteric methods are beyond the scope of our current discussion, however, readers should be aware that they are available and also get the idea that new methods to address the issues arising from assessing biological significance when faced with large gene sets are constantly being developed.

### 7.6.4 Within- and Cross-Platform Meta-Analysis

An increasing number of investigators are reusing and combining 'in house' or published data sets or studies from, for example supplementary information to support a publication, microarray data repositories or warehouses such as ArrayExpress and the Gene Expression Omnibus, GEO (see Chapter 9 for more information on microarray databases and warehouses) to strengthen and extend the results obtained from the individual studies. Meta-analysis has been shown to increase the statistical power of detecting small but consistent effects that might be false negatives in the individual study (Choi et al., 2003). There are several meta-analysis approaches available and they are based on what type of data is combined and whether the studies used arrays from within the same platform or from different platforms (cross-platforms).

The previous chapter so far has focused on normalizing methods for gene expression data within a single study, and much of that work can be applied to normalizing data from multiple studies that are based on the same array platform. The problem often encountered with normalizing combined data sets from the same platform is one of the processing power, since most combined data sets tend to be large. Depending on the available computational set-up, and combined study size, it may just be a case of using fast methods such as `justRMA` or `justRMALite` from the `AffyExtension`[25] package and the R package `aroma.affymetrix`[26] (Bengtsson et al. 2008) to normalize the large-data set, and/or adding more RAM to the computer. For even larger data sets, an appropriately configured cluster or farm of machines for processing may be necessary.

When normalizing cross-platform data sets, there are two main issues that need to be resolved: matching probes on different microarray formats, and preprocessing and normalizing data to address differences in platform and processing facilities. As mentioned already, the `annotationTools` package in *Bioconductor* can be used to both annotate and integrate data sets from multiple array platforms and cross-species platforms (Kuhn et al., 2008). For each independent data set, the appropriate preprocessing (to account for within array effects, etc.) and imputation missing value procedures are generally applied before matching any probes or applying any further normalization method. Then depending on the approach, the data set is merged on a common set of genes using a probe-matching tool to create one large-data matrix either before or after normalization. The various different normalization approaches that could be or have been applied for normalizing cross-platforms include:

1. Median centering (MC) each gene in each study and then on the combined studies (Shabalin et al., 2008)
2. The empirical Bayes method (Johnson et al., 2007).
3. Block linear model approach involving gene/sample clustering (Shabalin et al., 2008).
4. The distance weighted discrimination (DWD) method (Benito et al., 2004).
5. The metagene projection methodology (Tamayo et al., 2007).
6. Global or a local Z-score transformation (Teschendorff et al., 2006).
7. Median rank score transformation (Warnat et al., 2005), which is similar to the quantile normalization approach used by Bolstad et al. (2003) for normalizing single data sets.
8. Rankits transform based on ranked normal deviates (Royston and Sauerbrei 2004).
9. Singular value decomposition to discover and remove principal components that capture bias (Alter et al., 2000; Nielsen et al., 2002; Benito et al., 2004).

Shabalin et al. (2008) performed a comparative evaluation of methods 1, 2 and 3 on three different cancer data sets and found the block linear model approach

---

25 AffyExtensions url: http://bmbolstad.com/AffyExtensions/AffyExtensions.html.
26 More information on `aroma.affymetrix` can be found here: http://groups.google.com/group/aroma-affymetrix.

involving gene/sample clustering to perform the best. When assessing cross-plat-form prediction error rates of ER status of the breast cancer tumors (ER+ or ER−), the method was shown to successfully preserve biological information while removing systematic differences between platforms. On a heterogeneous set of array platform, comprised of Illumina, Agilent and Affymetrix, Stark compara-tively evaluated methods 5–8 above and found that the simplest method, the local $z$-score transformation performed the best (Stark, 2007).

At another level of meta-analysis, data can be combined at the statistical analysis stage, from primary statistics, such as $t$-statistics or $p$-values or from secondary statistics such as gene lists. A relatively simple method, is the Fisher's Inverse $\chi^2$-test (Fisher, 1925), which computes a combined statistic from the $p$-values obtained from the analysis of the individual studies, $S = -2\,log(\Pi_i P_i)$, where $S$ has a $\chi^2$ distribution with $2I$ degrees of freedom under the joint null hypothesis. There is no additional analysis required and is relatively easy to use, however, when working with $p$-values, it is impossible to estimate the average magnitude of differential expression. Choi et al. (2003) used a $t$-like statistic (defined as an effect size) as a summary statistic for each gene from each individual data set. To assess intrastudy and interstudy variation in the summary statistic across the data sets, they proposed a hierarchical modeling approach. This $t$-based hierarchical modeling approach, which has been implemented into a Bioconductor package as GeneMeta, estimates an overall effect size as the measurement of the magnitude of differential expression for each gene through parameter estimation and model fitting. Another alternative approach is the nonparametric rank product (RP) method, which is implemented in the *Biocon-ductor* RankProd package (Hong et al., 2006). Both the $t$-based and RP methods involve permutation tests to assess statistical significance, calculating FDR of the identification based on combined statistics, and the $p$-values gener-ated can also serve as input for the Fisher's inverse $\chi^2$-test. All three approaches were comparatively evaluated as meta-analysis methods for detecting differen-tially expressed genes recently by Hong and Breitling (2008). They showed that the RP method generally performed the best with higher sensitivity and selectivity than the $t$-based method in both individual and meta-analysis.

## 7.7 SUMMARY

Good *experimental design* practices are essential for meeting all the research objectives and to allow the application of the statistical methods appropriate for answering the relevant questions. It may be prudent to involve a statistician at the experimental design stage if you are unsure about the best methods to use. Before any statistical analyses, it is assumed that the data is of good quality as well as being appropriately transformed and normalized. *Filtering methods* to remove low intensity and variability across the data set may be necessary, when the latter is better than the former. However, when some filtering methods are applied, information may be lost rather than gained since deciding on which threshold to choose may be difficult to judge. Caution must be exercised regarding how to and when to apply filtering techniques. Other filtering methods focus on retaining a subset of array features associated with for instance, a particular Gene Ontology (GO) category of interest to the investigator. Focusing on a smaller list of array features can lead to a reduced number of multiple tests and this may result in many more significant genes.

*Statistical inference* involves extrapolating a representative sample taken from a whole population. When making a statistical inference, the interest lies in the variations between individuals in the sample population to which an extrapolation is made. It involves using unobserved parameters about a whole population to draw conclusions about the truth of hypotheses, based on statistics obtained from the samples. *Univariate* statistical tests (e.g. Student's *t*-test) are the most commonly used statistical methods, employed when there is only a single *dependent variable* and one or more *independent variables*. If the study has more than one dependent variable, then the procedures used are referred to as *multivariate*. Characteristics of the independent and dependent variables determine the statistical test and sampling distribution that will be used in order to assess significance.

Univariate statistical methods require a sufficient number of sample replicates to achieve statistical *power* and *sample size* is positively related to power. When comparing two groups of conditions to evaluate differential gene expression, evidence indicates that at least a minimum (and not the optimum) of five biological replicates per a condition group should be analyzed. Power can loosely be defined as the ability of a test statistic to detect change and depends on the number of replicates along with many other factors. Larger sample sizes have greater power to detect smaller differences in the population than smaller ones. Therefore, the researcher can increase the number of replicates used to increase the experimental power. For certain statistical methods involving medium to large-scale studies, *power analysis tools* can be used to estimate the number of replicates required for the research objectives.

For microarray studies, '*hypothesis testing*' approaches facilitate determining whether a gene is differentially expressed or not. There are two main schools of thought for hypothesis testing, the *classical* (frequentist) and the *Bayesian*. The classical approach tests, for each gene, the hypothesis that a particular gene is not differentially expressed and the negative hypothesis is the *null hypothesis*, $H_0$. Hypothesis testing requires a differential expression score, which is calculated for each gene and is also referred to as the *test statistic*, $t_s$. The location of $t_s$ in a *t-distribution curve* is determined in relation to the *p*-value, which is calculated as the area under the two tail ends of the curve for the cutoffs, $-t_s$ and $+t_s$. Essentially, the result of a classical hypothesis test is the *p*-value, a quantitative measure for reporting the result of a hypothesis test. There are two possible inferential errors when testing a single pair of hypotheses: a *type 1 error* (also known as a *false positive*) declares a gene to be differentially expressed at the $p$-value cutoff $p < \alpha$ when it is not, and occurs when the null hypothesis is wrongly rejected. A *type 2 error* (also known as a *false negative*) does not declare a gene to be differentially expressed at the $p$-value cutoff $p < \alpha$ when it is, and occurs when the null hypothesis is wrongly accepted.

When analyzing microarray data, statistical tests are performed on thousands of different genes in parallel and this causes a '*multiple testing*' problem. When many hypotheses are tested, the probability that a type I error is committed increases sharply with the number of hypotheses. There are several methods that address multiple testing by adjusting the $p$-value from a statistical test and these are based on the number of tests performed. There are two main categories of multiple testing procedures, depending on which error rate is being corrected: *Family Wise Error Rate (FWER)* corrections and *False Discovery Rate (FDR)* corrections. The FWER correction methods are more conservative than the FDR correction methods, but will give a higher number of false negatives (i.e. lower power). The FDR control is usually acceptable for situations where a small number of false positives is acceptable. The $p$-values *distribution* plot can provide useful information on whether to apply a multiple testing procedure. Good Bioconductor or CRAN packages that implement multiple testing approaches include `maanova`, `limma`, `siggenes`, and in particular, `multtest` and `qvalue`.

Classical and modern statistical approaches can be categorized whether or not they assume normality in the data: *parametric* methods assume *normality* (a Gaussian distribution) whereas *nonparametric* methods assume otherwise. Parametric statistics test the hypothesis that one or more treatments have no effect on the mean and variance of a chosen variable. Nonparametric are less powerful than the parametric equivalents but they are more robust to outliers since such tests usually rely on statistics based on the rank of the observations rather than the values themselves. Classical statistical methods generally require large sample sizes ($>30$) and the choice of statistical method depends on experimental setting, variable characteristics and sample distribution. For comparing two condition groups there are two versions of the classical parametric $t$-test, depending on whether the data is *paired* (one-sample $t$-test) or *unpaired* (two-sample $t$-test). The nonparametric equivalents to the one-sample $t$-test and the two-sample $t$-test are the *Wilcoxon signed rank* test and *Wilcoxon Rank Sum* (or the *Mann–Whitney U*) test respectively. For comparing multiple condition groups of two or more, the analysis of variance (**ANOVA**) is used and there are two versions. A nonparametric equivalent to an ANOVA is the *Kruskal–Wallis* test.

An issue with multifactor ANOVA is that a single model is used to fit all the genes and this cannot account for the variability between genes. To address this issue, there are methods that fit a general linear model (GLM) to each gene individually and can be used as an extension to ANOVA to allow for the inclusion of numeric variables. GLMs have been used to model nuisance technical effects associated with microarrays as well as biological effects. The bioconductor package, `maanova` (MicroArray ANalysis Of VAriance) uses mixed effect models to estimate variance components and either moderated *F*- and *t*-tests with multiple testing can be performed for differential gene expression analysis.

The simple 'gene by gene' testing approach of classical methods often fails. This is because many microarray experiments have small sample sizes (replicate numbers) and this reduces the power to detect differential gene expression. One solution, which is more powerful and robust and allows for gene-to-gene heterogeneity, involves '*shrinkage*' of the variance component estimates by pulling each individual estimate closer to the mean values across genes. *Bioconductor* packages that implement this type of approach include `maanova` and `limma` (*Linear Models for MicroArray data*). The linear formulation model of `limma` is used for both single and dual-channel arrays and allows the handling of experiments with arbitrary complexity including those with reference designs and factorial designs. In addition, quadratic linear models can be used to identify early or late response trends in single or multiple time-series studies. The linear model fit involves *parameterization* for the construction of the '*design matrix*' and the '*contrast matrix*' either directly or via a formula. The `limma` package uses the *empirical Bayes* approach for variance regularization and calculates a moderated *t*-statistic and a *B*-statistic, which is a Bayes log posterior odds (the log-odds that the gene is differentially expressed). The moderated *F*-statistic and moderated paired *t*-test can also be calculated.

A nonparametric approach that also addresses the issue of small sample sizes is the Bioconductor package, `siggenes` (based on the Significance Analysis of Microarrays; *SAM*). By adding a small correction factor to the denominator of the *t*-statistic (the gene specific variance estimate), the denominator is regularized and adjusted for the underestimated error. For each gene, the observed *d*-value (the moderated *t*-statistic) is calculated for each gene, and permutation is used to calculate the expected *d*-value and for FDR estimation.

Small sample sizes together with efforts to maintain low FDRs often result in low statistical power to detect differentially expressed genes, and interpreting a long list of genes is usually a daunting task. The gene list can be dealt with in one or more of the following different ways: filtering genes based on intensity level or fold change using, for example the package `genefilter`; integrating and filtering data in terms of gene and functional annotation, for example with tools such as DAVID, `annotate`, `AnnBuilder`, `hsahomology`; retrieving annotations and integrating results from other published studies from different array platforms, different species or studies using other array technologies such as chip–chip, for example using `AnnotationTools`. Data can also be integrated and visualized in terms of genomic coordinates using `geneplotter` or biological pathways and interaction networks using Cytoscape. Unsupervised clustering approaches can be used to produce, for example, heatmaps to visualize data structure and unsupervised clustering approaches can be used to identify genes with similar gene expression profiles.

Another approach involves using *gene-class (or category) testing* methods, which are statistical tests that compare the number of genes in a class or category (e.g. GO ontology category, biological pathway, chromosomal location, clinical or disease phenotype) that are 'significant' with the number that are expected under a particular null hypothesis. There are two main approaches: the overrepresentation or underrepresentation-based (*gene sampling*) methods such as GOstat and FatiGO and the gene set enrichment analysis (*GSEA*) (*subject sampling*) based tools such as GSEA-P and globalTest. The latter methods are generally considered to be better since they retain the gene correlation structure of microarray data and do not assume gene independence.

A meta-analysis approach of combing data sets from multiple data sets can lead to increased statistical power to detect small but consistent effects. Depending on whether the meta-analysis involves the comparison of multiple data sets from within the same platform or from different platforms or species even, there are a number of different approaches to handle and analyze the data and they are based on two main issues: merging probe sets from multiple platforms using tools such as annotationTools in Bioconductor, and normalizing the data to remove technical effects. Two promising methods include: a block linear model approach involving gene/sample clustering and a simple local *z*-transform approach. At another level of analysis, meta-analysis can involve combining and analyzing data sets of gene lists, *t*-statistics or *p*-values using for example, a nonparametric rank product (RP) method, which is implemented n the Rank-Prod package in Bioconductor.

## REFERENCES

Al-Shahrour F, Diaz-Uriarte R, Dopazo J. (2004) FatiGO: a web tool for finding significant associations of Gene Ontology terms with groups of genes. *Bioinformatics* **20:** 578–580.

Allison DB, Gadbury GL, Heo M, Fernandez JR, Lee C, Prolla TA, Weindrucha R. (2002) A mixture model approach for the analysis of microarray gene expression data. *Comput Stati Data Anal* **39:** 1–20.

Allison DB, Cui X, Page GP, Sabripour M. (2006) Microarray data analysis: from disarray to consolidation and consensus. *Nat Rev Genet* **7:** 55–65.

Alter O, Brown PO, Botstein D. (2000) Singular value decomposition for genome-wide expression data processing and modeling. *Proc Nat Acad Sci U S A* **97:** 10101–10106.

Ashburner M, Ball CA, Blake JA, Botstein D, Butler H, Cherry JM, Davis AP, Dolinski K, Dwight SS, Eppig JT, Harris MA, Hill DP, Issel-Tarver L, Kasarskis A, Lewis S, Matese JC, Richardson JE, Ringwald M, Rubin GM, Sherlock G. (2000) Gene ontology: tool for the unification of biology. The Gene Ontology Consortium. *Nat Genet* **25:** 25–29.

Backes C, Keller A, Kuentzer J, Kneissl B, Comtesse N, Elnakady YA, Muller R, Meese E, Lenhof HP. (2007) GeneTrail–advanced gene set enrichment analysis. *Nucleic Acids Res* **35**(Web Server issue): W186–W192.

Barry WT, Nobel AB, Wright FA. (2005) Significance analysis of functional categories in gene expression studies: a structured permutation approach. *Bioinformatics* **21**: 1943–1949.

Beissbarth T, Speed TP. (2004) GOstat: find statistically overrepresented Gene Ontologies within a group of genes. *Bioinformatics* **20**(9): 1464–1465.

Bengtsson H, Simpson K, Bullard J, Hansen K. (2008) aroma.affymetrix: A generic framework in R for analyzing small to very large Affymetrix data sets in bounded memory, Tech Report #745, Department of Statistics, University of California, Berkeley.

Benito M, Parker J, Du Q, Wu J, Xiang D, Perou CM, Marron JS. (2004) Adjustment of systematic microarray data biases. *Bioinformatics* **20**: 105–114.

Benjamini Y, Hochberg Y. (1995) Controlling the false discovery rate: a practical and powerful approach to multiple testing. *J R Stats Soc Ser B* **57**: 289–300.

Benjamini Y, Hochberg Y. (2000) On the adaptive control of the false discovery rate in multiple testing with independent statistics. *J Educ Behav Stats* **25**: 60–83.

Benjamini Y, Yekutieli D. (2001) The control of the false discovery rate in multiple testing under dependency. *Ann Stat* **29**: 1165–1188.

Benjamini Y, Krieger A, Yekutieli D. (2001) *Two-staged Linear Step-up FDR Controlling Procedure.* Technical Report, Department of Statistics and Operation Research, Tel-Aviv University, and Department of Statistics, Wharton School, University of Pennsylvania.

Berriz GF, King OD, Bryant B, Sander C, Roth FP. (2003) Characterizing gene sets with FuncAssociate. *Bioinformatics* **19**(18): 2502–2504.

Bokka S, Mathur SK. (2006) A nonparametric likelihood ratio test to identify differentially expressed genes from microarray data. *Appl Bioinformatics* **5**(4): 267–276.

Bolstad BM, Irizarry RA, Astrand M, Speed TP. (2003) A comparison of normalization methods for high density oligonucleotide array data based on variance and bias. *Bioinformatics* **19**: 2185–193.

Bonferroni CE. (1935) Il Calcolo delle assicurazioni su gruppi di teste. In: Studi. *Onore del Professore Salvatore Ortu Carboni.* Roma, Italia, pp. 13–60.

Boulesteix AL. (2007) WilcoxCV: an R package for fast variable selection in cross-validation. *Bioinformatics* **23**(13): 1702–1704.

Castillo-Davis CI, Hartl DL. (2003) GeneMerge–post-genomic analysis, data mining, and hypothesis testing. *Bioinformatics* **19**(7): 891–892.

Cavalieri D, Castagnini C, Toti S, Maciag K, Kelder T, Gambineri L, Angioli S, Dolara P. (2007) Eu. Gene Analyzer a tool for integrating gene expression data with pathway databases. *Bioinformatics* **23**(19): 2631–2632.

Cheng J, Sun S, Tracy A, Hubbell E, Morris J, Valmeekam V, Kimbrough A, Cline MS, Liu G, Shigeta R, Kulp D, Siani-Rose MA. (2004) NetAffx Gene Ontology Mining Tool: a visual approach for microarray data analysis. *Bioinformatics* **20**: 1462–1463.

Choi JK, Yu U, Kim S, Yoo OJ. (2003) Combining multiple microarray studies and modeling interstudy variation. *Bioinformatics* **19**: i84–90.

Churchill GA. (2004) Using ANOVA to analyze microarray data. *Biotechniques* **37**: 173–177.

Conesa A, Nueda MJ, Ferrer A, Talon M. (2006) maSigPro: a method to identify significantly differential expression profiles in time-course microarray experiments. *Bioinformatics* **22**(9): 1096–1102.

Cui X, Churchill GA. (2003) Statistical tests for differential expression in cDNA microarray experiments. *Genome Biol* **4**(4): 210.

Cui X, Kerr MK, Churchill GA. (2003) Transformations for cDNA microarray data. *Stat Appl Genet Mol Biol* **2**: A4.

Cui X, Hwang JT, Qiu J, Blades NJ, Churchill GA. (2005) Improved statistical tests for differential gene expression by shrinking variance components estimates. *Biostatistics* **6**: 59–75.

Dabney AR, Storey JD. (2006) A reanalysis of published Affymetrix GeneChip control dataset. *Genome Biol* **7**: 401.

Datta S. (2005) Empirical Bayes screening of many p-values with applications to microarray studies. *Bioinformatics* **21**(9): 1987–1994.

de Haan JR, Wehrens R, Bauerschmidt S, Piek E, van Schaik RC, Buydens LM. (2007) Interpretation of ANOVA models for microarray data using PCA. *Bioinformatics* **23**: 184–190.

Dennis G, Sherman BT, Hosack DA, Yang J, Gao W, Lane HC, Lempicki RA. (2003) DAVID: database for annotation, visualization, and integrated discovery. *Genome Biol* **4**: P3.

Dinu I, Potter JD, Mueller T, Liu Q, Adewale AJ, Jhangri GS, Einecke G, Famulski KS, Halloran P, Yasui Y. (2007) Improving gene set analysis of microarray data by SAM-GS. *BMC Bioinformatics* **8**: 242.

Do KA, Muller P, Tang F. (2005) A Bayesian mixture model for differential gene expression. *Journal of the Royal Statistical Society: Series C (Applied Statistics)* **54**: 627–644.

Dobbin K, Simon R. (2005) Sample size determination in microarray experiments for class comparison and prognostic classification. *Biostatistics* **6**: 27–38.

Dobson, A. (2008) Introduction to Generalized Linear Models. Chapman & Hall; 3 edition.

Draghici S, Khatri P, Martins RP, Ostermeier GC, Krawetz SA. (2003) Global functional profiling of gene expression. *Genomics* **81**(2): 98–104.

Dudoit S, Popper Shaffer J, Boldrick JC. (2003) Multiple hypothesis testing in microarray experiments. *Stat Sci* **18**: 71–103.

Dudoit S, van der Laan MJ, Pollard KS. (2004) Multiple testing. Part I. Single-step procedures for control of general type I error rates. *Stat Appl Genet Mol Biol* **3**: A13.

Dudoit S, Yang W, Callow M, Speed T. (2000) *Statistical Methods for Identifying Differentially Expressed Genes in Replicated cDNA Microarray Experiments.* Technical Report #578, Dept. of Statistics, University of California, Berkeley.

Edwards JW, Ghosh P. (2006) Bayesian analysis of microarray data. In: Allison DB, Page GP, Beasley TM, Edwards JW, editors. *DNA Microarrays and Related Genomics Techniques.* Chapman & Hall/CRC, pp. 267–288.

Efron B. (1979) Bootstrap methods: another look at the jacknife. *Ann Stats* **7**: 1–26.

Efron B, Tibshirani R. (2007) On testing the significance of sets of genes. *Ann Appl Statist* **1**(1): 107–129.

Efron B, Tibshirani R, Storey JD, Tusher V. (2001) Empirical Bayes analysis of a microarray experiment. *J Am Stats Assoc* **96**: 1151–1160.

Ernst J, Bar-Joseph Z. (2006) STEM: a tool for the analysis of short time series gene expression data. *BMC Bioinformatics* **7**: 191.

Everitt B, Palmer C. (2005) Encyclopaedic Companion to Medical Statistics. Hodder Arnold.

Fisher RA. (1925) *Statistical Methods for Research Worker.* Oliver and Boyd, Edinburgh.

Furge KA, Tan MH, Dykema K, Kort E, Stadler W, Yao X, Zhou M, Teh BT. (2007) Identification of deregulated oncogenic pathways in renal cell carcinoma: an integrated oncogenomic approach based on gene expression profiling. *Oncogene* **26**(9): 1346–1350.

Ge Y, Dudoit S, Speed T. (2003) *Resampling-based Multiple Testing for Microarray Data Analysis.* Technical Report, Department of Statistics, University of California, Berkeley Technical Report no. 633.

Gelman A, Carlin JB, Stern HS, Rubin DB. (2003) *Bayesian Data Analysis*, 2nd edition. CRC Press, Inc.

Gentleman R, Carey V, Huber W, Irizarry R, Dudoit S. (2005) *Bioinformatics and Computational Biology Solutions Using R and Bioconductor (Statistics for Biology and Health)*. Springer, Verlag New York Inc.

Glonek GF, Solomon PJ. (2004) Factorial and time course designs for cDNA microarray experiments. *Biostatistics* **5**: 89–111.

Goeman JJ, Buhlmann P. (2007) Analyzing gene expression data in terms of gene sets: methodological issues. *Bioinformatics* **23**: 980–987.

Goeman JJ, Mansmann U. (2008) Multiple testing on the directed acyclic graph of gene ontology. *Bioinformatics* **24**(4): 537–544.

Goeman JJ, van de Geer SA, de Kort F, van Houwelingen HC. (2004) A global test for groups of genes: testing association with a clinical outcome. *Bioinformatics* **20**(1): 93–99.

Hahne F, Huber W, Gentleman R, Falcon S. (2008) *Bioconductor Case Studies*. Springer, Verlag New York Inc.

Hair F, Black B, Babin B, Anderson E, Tatham RL. (2007) *Multivariate Data Analysis*, 6th edition. Pearson Education.

Hedenfalk I, Duggan D. (2001) Gene-expression profiles in hereditary breast cancer. *N Engl J Med* **344**(8): 539–548.

Hermjakob H, Montecchi-Palazzi L, Lewington C, Mudali S, Kerrien S, Orchard S, Vingron M, Roechert B, Roepstorff P, Valencia A, Margalit H, Armstrong J, Bairoch A, Cesareni G, Sherman D, Apweiler R. (2004) IntAct: an open source molecular interaction database. *Nucleic Acids Res* **32**: D452–455.

Holm S. (1979) A simple sequentially rejective multiple test procedure. *Scand J Stat* **6**: 65–70.

Hong F, Breitling R. (2008) A comparison of meta-analysis methods for detecting differentially expressed genes in microarray experiments. *Bioinformatics* **24**: 374–382.

Hong F, Breitling R, McEntee CW, Wittner BS, Nemhauser JL, Chory J. (2006) RankProd: a bioconductor package for detecting differentially expressed genes in meta-analysis. *Bioinformatics* **22**: 2825–2827.

Huang da W, Sherman BT, Tan Q, Collins JR, Alvord WG, Roayaei J, Stephens R, Baseler MW, Lane HC, Lempicki RA. (2007) The DAVID Gene Functional Classification Tool: a novel biological module-centric algorithm to functionally analyze large gene lists. *Genome Biol* **8**: R183.

Huang da W, Sherman BT, Tan Q, Kir J, Liu D, Bryant D, Guo Y, Stephens R, Baseler MW, Lane HC, Lempicki RA. (2007) DAVID Bioinformatics Resources: expanded annotation database and novel algorithms to better extract biology from large gene lists. *Nucleic Acids Res* **35**: W169–175.

Hummel M, Meister R, Mansmann U. (2008) GlobalANCOVA: exploration and assessment of gene group effects. *Bioinformatics* **24**(1): 78–85.

Jeffery IB, Higgins DG, Culhane AC. (2006) Comparison and evaluation of methods for generating differentially expressed gene lists from microarray data. *BMC Bioinform* **7**: 359.

Jiang Z, Gentleman R. (2007) Extensions to gene set enrichment. *Bioinformatics* **23**: 3306–313.

Jiao S, Zhang S. (2008) The t-mixture model approach for detecting differentially expressed genes in microarrays. *Funct Integr Genomics.*

Johnson WE, Li C, Rabinovic A. (2007) Adjusting batch effects in microarray expression data using empirical Bayes methods. *Biostatistics* **8**: 118–127.

Jørstad TS, Langaas M, Bones AM. (2007) Understanding sample size: what determines the required number of microarrays for an experiment? *Trends Plant Sci* **12:** 46–50.

Jupiter DC, VanBuren V. (2008) A visual data mining tool that facilitates reconstruction of transcription regulatory networks. *PLoS ONE* **3:** e1717.

Kanehisa M, Goto S, Hattori M, Aoki-Kinoshita KF, Itoh M, Kawashima S, Katayama T, Araki M, Hirakawa M. (2006) From genomics to chemical genomics: new developments in KEGG. *Nucleic Acids Res* **34:** D354–357.

Kanji G. (2006) *100 Statistical Tests*, 3rd edition. Sage Publications Ltd.

Keller A, Backes C, Lenhof HP. (2007) Computation of significance scores of unweighted gene set enrichment analyses. *BMC Bioinform* **8:** 290.

Kendziorski CM, Newton MA, Lan H, Gould MN. (2003) On parametric empirical Bayes methods for comparing multiple groups using replicated gene expression profiles. *Stat Med* **22**(24): 3899–3914.

Kerr KF, Churchill GA. (2001) Experimental design for gene expression microarrays. *Biostatistics* **2:** 183–201.

Kerr MK, Martin M, Churchill GA. (2000) Analysis of variance for gene expression microarray data. *J Comput Biol* **7:** 819–837.

Khatri P, Draghici S. (2005) Ontological analysis of gene expression data: current tools, limitations, and open problems. *Bioinformatics* **21:** 3587–3595.

Kim SY, Volsky DJ. (2005) PAGE: parametric analysis of gene set enrichment. *BMC Bioinformatics* **6:** 144.

Kong SW, Pu WT, Park PJ. (2006) A multivariate approach for integrating genome-wide expression data and biological knowledge. *Bioinformatics* **22:** 2373–2380.

Krull M, Pistor S, Voss N, Kel A, Reuter I, Kronenberg D, Michael H, Schwarzer K, Potapov A, Choi C, Kel-Margoulis O, Wingender E. (2006) TRANSPATH: an information resource for storing and visualizing signaling pathways and their pathological aberrations. *Nucleic Acids Res* **34:** D546–551.

Kuhn A, Luthi-Carter R, Delorenzi M. (2008) Cross-species and cross-platform gene expression studies with the Bioconductor-compliant R package 'annotationTools'. *BMC Bioinform* **9:** R26.

Küntzer J, Kneissl B, Kohlbacher O, Lenhof HP. (2007) Abstract analysis of pathways using the BN++ software framework. *BMC Syst Biol* **1:** P24.

Lama N, Girolami M. (2008) Vbmp: variational Bayesian Multinomial Probit Regression for multiclass classification in R. *Bioinformatics* **24:** 1135–136.

Lamb J, Ramaswamy S, Ford HL, Contreras B, Martinez RV, Kittrell FS, Zahnow CA, Patterson N, Golub TR, Ewen ME. (2003) A mechanism of cyclin D1 action encoded in the patterns of gene expression in human cancer. *Cell* **114:** 323–334.

Leroux L, Durel B, Autier V, Deltour L, Bucchini D, Jami J, Joshi RL. (2003) Ins1 gene up-regulated in a beta-cell line derived from Ins2 knockout mice. *Int J Exp Diabesity Res* **4**(1): 7–12.

Liu J, Hughes-Oliver JM, Menius JA. (2007) Domain-enhanced analysis of microarray data using GO annotations. *Bioinformatics* **23:** 1225–1234.

Lo K, Gottardo R. (2007) Flexible empirical Bayes models for differential gene expression. *Bioinformatics* **23**(3): 328–335.

Longabaugh WJ, Davidson EH, Bolouri H. (2005) Computational representation of developmental genetic regulatory networks. *Dev Biol* **283:** 1–16.

Lonnstedt I, Speed T. (2001) *Replicated Microarray Data*. Technical Report, Dept. of Statistics, University of California, Berkeley.

Maere S, Heymans K, Kuiper M. (2005) BiNGO: a Cytoscape plugin to assess overrepresentation of gene ontology categories in biological networks. *Bioinformatics* **21**(16): 3448–3449.

Matys V, Kel-Margoulis OV, Fricke E, Liebich I, Land S, Barre-Dirrie A, Reuter I, Chekmenev D, Krull M, Hornischer K, Voss N, Stegmaier P, Lewicki-Potapov B, Saxel H, Kel AE, Wingender E. (2006) TRANSFAC and its module TRANSCompel: transcriptional gene regulation in eukaryotes. *Nucleic Acids Res* **34**(Database issue): D108–D110.

Mootha VK, Lindgren CM, Eriksson KF, Subramanian A, Sihag S, Lehar J, Puigserver P, Carlsson E, Ridderstrale M, Laurila E, Houstis N, Daly MJ, Patterson N, Mesirov JP, Golub TR, Tamayo P, Spiegelman B, Lander ES, Hirschhorn JN, Altshuler D, Groop LC. (2003) PGC-1alpha-responsive genes involved in oxidative phosphorylation are coordinately downregulated in human diabetes. *Nat Genet* **34**(3): 267–273.

Nettleton D, Recknor J, Reecy JM. (2008) Identification of differentially expressed gene categories in microarray studies using nonparametric multivariate analysis. *Bioinformatics* **24**: 192–201.

Newton MA, Kendziorski CM, Richmond CS, Blattner FR, Tsui KW. (2001) On differential variability of expression ratios: improving statistical inference about gene expression changes from microarray data. *J Comput Biol* **8**(1): 37–52.

Newton MA, Noueiry A, Sarkar D, Ahlquist P. (2004) Detecting differential gene expression with a semiparametric hierarchical mixture method. *Biostatistics* **5**: 155–176.

Newton MA, Quintana FA, den Boon JA, Sengupta S, Ahlquist P. (2007) Random-set Methods identify aspects of the enrichment signal in gene-set analysis. *Ann Appl Stat* **1**: 85–106.

Nielsen TO, West RB, Linn SC, Alter O, Knowling MA, O'Connell JX, Zhu S, Fero M, Sherlock G, Pollack JR, Brown PO, Botstein D, van de Rijn M. (2002) Molecular characterization of soft tissue tumors: a gene expression study. *Lancet* **359**: 1301–1307.

Nueda MJ, Conesa A, Westerhuis JA, Hoefsloot HC, Smilde AK, Talon M, Ferrer A. (2007) Discovering gene expression patterns in time course microarray experiments by ANOVA-SCA. *Bioinformatics* **23**(14): 1792–1800.

Pan W, Lin J, Le CT. (2003) A mixture model approach to detecting differentially expressed genes with microarray data. *Funct Integr Genomics* **3**: 117–124.

Pavlidis P, Li Q, Noble WS. (2003) The effect of replication on gene expression microarray experiments. *Bioinformatics* **19**(13): 1620–1627.

Peri S, Navarro JD, Amanchy R, Kristiansen TZ, Jonnalagadda CK, Surendranath V, Niranjan V, Muthusamy B, Gandhi TK, Gronborg M, Ibarrola N, Deshpande N, Shanker K, Shivashankar HN, Rashmi BP, Ramya MA, Zhao Z, Chandrika KN, Padma N, Harsha HC, Yatish AJ, Kavitha MP, Menezes M, Choudhury DR, Suresh S, Ghosh N, Saravana R, Chandran S, Krishna S, Joy M, Anand SK, Madavan V, Joseph A, Wong GW, Schiemann WP, Constantinescu SN, Huang L, Khosravi-Far R, Steen H, Tewari M, Ghaffari S, Blobe GC, Dang CV, Garcia JG, Pevsner J, Jensen ON, Roepstorff P, Deshpande KS, Chinnaiyan AM, Hamosh A, Chakravarti A, Pandey A. (2003) Development of human protein reference database as an initial platform for approaching systems biology in humans. *Genome Res* **13**(10): 2363–2371.

Pollard KS, Dudoit S, van der Laan MJ. (2005) Multiple testing procedures: the multiset package and application to genomics. In: Gentleman R, Carey V, Huber W, Irizarry R, Dudoit S, editors. *Bioinformatics and Computational Biology Solutions Using R and Bioconductor.* Springer, Verlag New York Inc, pp. 249–271.

Popper Shaffer J. (2006) Recent developments towards optimality in multiple hypothesis testing. *IMS Lecture Notes-Monograph Series* **49**: 16–32.

Pounds S, Morris SW. (2003) Estimating the occurrence of false positives and false negatives in microarray studies by approximating and partitioning the empirical distribution of *p*-values. *Bioinformatics,* **19**(10): 1236–1242.

Reiner A, Yekutieli D, Benjamini Y. (2003) Identifying differentially expressed genes using false discovery rate controlling procedures. *Bioinformatics* **19**: 368–375.

Rivals I, Perssonaz L, Taing L, Potier MC. (2007) Enrichment of Depletion of a GO category within a class of genes: which test? *Bioinformatics* **23**(4): 401–407.

Royston P, Sauerbrei W. (2004) A new measure of prognostic separation in survival data. *Stat Med* **23**: 723–748.

Salomonis N, Hanspers K, Zambon AC, Vranizan K, Lawlor SC, Dahlquist KD, Doniger SW, Stuart J, Conklin BR, Pico AR. (2007) GenMAPP 2: new features and resources for pathway analysis. *BMC Bioinformatics* **8**: 217.

Salwinski L, Miller CS, Smith AJ, Pettit FK, Bowie JU, Eisenberg D. (2004) The database of interacting proteins: 2004 update. *Nucleic Acids Res* **32**: D449–451.

Scholtens D, von Heydebreck A. (2005) Analysis of differential gene expression analysis. In: Gentleman R, Carey V, Huber W, Irizarry R, Dudoit S, editors. *Bioinformatics and Computational biology solutions using R and Bioconductor.* Springer, Verlag New York Inc, pp. 229–248.

Schwender H, Krause A, Ickstadt K. (2006) Identifying interesting genes with siggenes. *RNews* **6**(5): 45–50.

Searle S, Casella G, McCulloch C. (1992) *Variance Components.* John Wiley and sons, Inc., New York, NY.

Shabalin AA, Tjelmeland H, Fan C, Perou CM, Nobel AB. (2008) Merging two gene-expression studies via cross-platform normalization. *Bioinformatics* **24**: 1154–1160.

Shah NH, Fedoroff NV. (2004) CLENCH: a program for calculating Cluster ENriCHment using the Gene Ontology. *Bioinformatics* **20**(7): 1196–1197.

Shannon P, Markiel A, Ozier O, Baliga NS, Wang JT, Ramage D, Amin N, Schwikowski B, Ideker T. (2003) Cytoscape: a software environment for integrated models of biomolecular interaction networks. *Genome Res* **13**: 2498–2504.

Sherman BT, Huang da W, Tan Q, Guo Y, Bour S, Liu D, Stephens R, Baseler MW, Lane HC, Lempicki RA. (2007) DAVID Knowledgebase: a gene-centered database integrating heterogeneous gene annotation resources to facilitate high-throughput gene functional analysis. *BMC Bioinform* **8**: R426.

Sivia D, Skilling J. (2006) *Data Analysis: A Bayesian Tutorial,* 2nd edition. Oxford University Press.

Smyth GK. (2004) Linear models and empirical Bayes methods for assessing differential expression in microarray experiments. *Stat Appl Genet Mol Biol* **3**: A3.

Smyth GK. (2005) Limma: linear models for microarray data. In: Gentleman R, Carey V, Huber W, Irizarry R, Dudoit S, editors. *Bioinformatics and Computational Biology Solutions Using R and Bioconductor.* Springer, Verlag New York Inc, pp. 397–420.

Stark RJ. (2007) *Integrating Gene Expression Datasets for Cancer Transcriptomics.* MPhil Computational Biology, University of Cambridge, UK.

Storey JD. (2001) *The Positive False Discovery Rates: A Bayesian Interpretation and the q-Value.* Technical Report 2001-12, Department of Statistics, Stanford University.

Storey JD, Tibshirani R. (2003) SAM thresholding and false discovery rates for detecting differential gene expression in DNA microarrays. In: Parmigiani G, Garrett ES, Irizarry R, Zeger SL, editors. *The Analysis of Gene Expression Data: Methods and Software.* Springer, New York.

Storey JD, Tibshirani R. (2003) Statistical Significance for Genomewide Studies. *Proc Natl Acad Sci U S A* **100:** 9440–9445.

Storey JD, Tibshirani R. (2003) SAM thresholding and false discovery rates for detecting differential gene expression in DNA microarrays. In: Parmigiani G, Garrett ES, Irizarry R, Zeger SL, editors. *The Analysis of Gene Expression Data.* Springer, pp. 272–290.

Strausberg RL, Buetow KH, Emmert-Buck MR, Klausner RD. (2000) The cancer genome anatomy project: building an annotated gene index. *Trends Genet* **16**(3): 103–106.

Subramanian A, Tamayo P, Mootha VK, Mukherjee S, Ebert BL, Gillette MA, Paulovich A, Pomeroy SL, Golub TR, Lander ES, Mesirov JP. (2005) Gene set enrichment analysis: a knowledge-based approach for interpreting genome-wide expression profiles. *Proc Natl Acad Sci U S A* **102**(43): 15545–15550.

Sykacek P, Furlong RA, Micklem G. (2005) A friendly statistics package for microarray analysis. *Bioinformatics* **21**(21): 4069–4070.

Tai YC, Speed T. (2004) *A multivariate empirical bayes statistic for replicated microarray time course data.* University of California, Techreports 667 Berkeley.

Tamayo P, Scanfeld D, Ebert BL, Gillette MA, Roberts CW, Mesirov JP. (2007) Metagene projection for cross-platform, cross-species characterization of global transcriptional states. *Proc Nat Acad Sci U S A* **104:** 5959–5964.

Teschendorff AE, Naderi A, Barbosa-Morais NL, Pinder SE, Ellis IO, Aparicio S, Brenton JD, Caldas C. (2006) A consensus prognostic gene expression classifier for ER positive breast cancer. *Genome Biol* **7:** R1901.

Thomas JG, Olson JM, Tapscott SJ, Zhao LP. (2001) An efficient and robust statistical modeling approach to discover differentially expressed genes using genomic expression profiles. *Genome Res* **11:** 1227–1236.

Tomfohr J, Lu J, Kepler TB. (2005) Pathway level analysis of gene expression using singular value decomposition. *BMC Bioinformatics* **12**(6): 225.

Troyanskaya OG, Garber ME, Brown PO, Botstein D, Altman RB. (2002) Nonparametric methods for identifying differentially expressed genes in microarray data. *Bioinformatics* **18:** 1454–1461.

Tsai C, Hsueh H, Chen JJ. (2003) Estimation of false discovery rates in multiple testing: application to gene microarray data. *Biometrics* **59:** 1071–1081.

Tusher VG, Tibshirani R, Chu G. (2001) Significance analysis of microarrays applied to the ionizing radiation response. *Proc Natl Acad Sci U S A* **98:** 5116–5121.

van der Laan MJ, Dudoit S, Pollard KS. (2004) Multiple testing. Part II. Step-down procedures for control of the family-wise error rate. *Stat Appl Genet Mol Biol* **3:** A14.

Venables WN, Smith DM. (2002) An Introduction to R. *Network Theory Ltd.*

Warnat P, Eils R, Brors B. (2005) Cross-platform analysis of cancer microarray data improves gene expression based classification of phenotypes. *BMC Bioinform* **4:** R265.

Welch BL. (1947) The generalization of student's problem when several different population variances are involved. *Biometrika* **34:** 28–35.

Westfall PH, Young SS. (1993) *Resampling-based Multiple Testing.* Wiley, New York.

Wettenhall JM, Smyth GK. (2004) limmaGUI: a graphical user interface for linear modeling of microarray data. *Bioinformatics* **20**(18): 3705–3706.

Wit E, McClure J. (2004) *Statistics for Microarrays.* John Wiley and Sons Ltd.

Witmer JA, Samuels ML. (2003) *Statistics for the Life Sciences,* International edition. Pearson Education.

Wu H, Kerr KF, Cui X, Churchill GA. (2003) MAANOVA: a software package for the analysis of spotted cDNA microarray experiments. In: Parmigiani G, Garett E, Irizarry R, Zeger SL, editors. *The Analysis of Gene Expression Data.* Springer, New York, pp. 313–341.

Xie X, Lu J, Kulbokas EJ, Golub TR, Mootha V, Lindblad-Toh K, Lander ES, Kellis M. (2005) Systematic discovery of regulatory motifs in human promoters and 3' UTRs by comparison of several mammals. *Nature* **434**(7031): 338–345.

Yang X. (2004) Qvalue methods may not always control false discovery rate in genomic applications. *Computational Systems Bioinformatics Conference.*IEEE 2004, pp. 556–557.

Yekutieli D, Benjamini Y. (1999) Resampling-based false discovery rate controlling procedure for dependent test statistics. *J Stat Plan Inference* **82:** 171–196.

Zanzoni A, Montecchi-Palazzi L, Quondam M, Ausiello G, Helmer-Citterich M, Cesareni G. (2002) MINT: a Molecular Interaction database. *FEBS Lett* **513:** 135–140.

Zheng Q, Wang XJ. (2008) GOEAST: a web-based software toolkit for Gene Ontology enrichment analysis. *Nucleic Acids Res* **36:** W358–W363.

# Clustering and Classification

After the application of appropriate quality control metrics, data normalization and statistical evaluation, the experimenter may be left with lists of hundreds or even thousands of genes that are relevant to the biology under investigation. How can the biologist, who is more normally used to dealing with a few observations, makes sense of these data? Fortunately there are a variety of analysis tools that can assist in the organization and presentation of high dimensionality data. While such tools are without doubt incredibly useful, it must be remembered that they are not necessarily revealing 'biological truth'. Rather the analysis methods provide views of the data that may guide subsequent investigations or suggest new hypotheses. The choice of analysis tools should, of course, be determined by the type of experiment being carried out and the objectives of that experiment. For example, if the objective is to find out how a drug effects gene expression over time then an unsupervised analysis method that does not make assumptions about the data is the most likely choice. Conversely, if normal tissues are being compared with cancer biopsies to identify gene expression signatures that uniquely identify cancers, a supervised classification method is a more appropriate analytical tool.

## 8.1 INTRODUCTION

In the previous chapter we described the statistical methods that underpin the evaluation of gene expression data and facilitate the selection of differentially expressed genes. In this chapter we introduce some of the methods underpinning the classification and organization of sets of differentially expressed genes. We discuss some of the basic principles of classification and describe some of the more common approaches for supervised and unsupervised data analysis. The discussion is by no means exhaustive and is intended as an entry into the literature as well as an introduction to some of the tools provided in the freely available BioConductor platform and $R$ environment. Some of these tools may also be available in commercial analysis platforms but we feel the BioConductor packages offer the most comprehensive and cost effective (i.e. free) solutions available to the microarray community. In the sections that follow, we precede each discussion with an indication of the relevant Bio-Conductor packages. Interested readers are directed to the comprehensive vignettes that accompany each package available from the BioConductor web site.

The principles underlying most gene expression analysis tools are fairly similar since, broadly speaking, the objectives of most expression studies are the same: find the differences, or conversely the similarities, between the samples under investigation. To do this, measured intensities from expression experiments are represented in a data matrix. The matrix can be considered as having rows corresponding to genes and columns corresponding to experiments. Thus, each row represents the complete set of intensities for a given gene from a set of experiments. Each row can then be considered as a single point in a high dimensional space. Once data is considered in this way there are various methods for determining whether individual rows of the data matrix are similar to each other or different. In particular, a variety of distance metrics, including Euclidean and Manhattan distance, or correlation metrics such as the Pearson correlation, are commonly used to measure the relationship between gene expression profiles. In some situations it may be useful to abandon the actual gene expression intensities and simply use the rank of the genes in the experiment. For example, the gene with the 10th lowest intensity has a rank of 10 while the gene with the 1000th lowest intensity has a rank of 1000. Such rank-based approaches are often useful when only a small number of genes or probes are expected to show significant differences, such as in ChIP-on-chip experiments, and quantitative results are not important.

Microarray analysis tools fall broadly in two categories, using supervised or unsupervised approaches. If prior knowledge is available for the samples under consideration, then it is possible to group or classify samples into different classes. For example, one may have a set of tissue samples from cancer patients and a set of normal samples. In such cases supervised approaches, so called because the algorithm is guided to discriminate gene expression profiles based on differences between the samples, are frequently used to identify gene expression patterns or features specific to each class. Such supervised approaches can also be used to predict what class a new sample belongs to. In contrast, unsupervised, sometimes referred to as hypothesis-free, approaches are important for discovering novel biological mechanisms, for revealing genetic regulatory networks and for analyzing large datasets for which little prior knowledge is available. Examples here include clustering genes based on expression over a time-course, comparing experimental and control samples such as wild type and mutant or experiments examining drug effects.

Unsupervised analysis methods for microarray data can be divided into three categories: clustering approaches, model-based approaches and projection methods. Clustering approaches group genes and experiments with similar behavior, making the data simpler to analyze. These methods group genes that behave similarly under similar experimental conditions, with the underlying assumption that coexpressed genes are functionally related. Most clustering methods do not attempt to model the underlying biology. A disadvantage of such methods is that they partition genes and experiments into mutually exclusive clusters, whereas in reality a gene or an experiment may be part of several biological processes.

Model-based approaches first generate a model that explains the interactions among biological entities participating in genetic regulatory networks and then train the parameters of the model with the gene expression datasets. One challenge associated with model-based approaches, depending on the complexity of the model, is a lack of sufficient data to train the parameters. A second challenge is the prohibitive computational requirement most training algorithms have, though clearly the increasing power of desktop computers continues to reduce the impact of this concern.

Projection methods linearly decompose expression data into a dataset with a reduced number of dimensions, enabling the sets of genes that maximally discriminate the samples under analysis to be grouped and easily visualized. There are various kinds of projection methods but the most commonly used are principal component analysis (PCA) and independent component analysis (ICA). PCA projects the data into a new space spanned by the principal components. Each successive principal component is selected to be orthonormal to the previous ones and to capture the maximum information that is not already present in the previous components. Applied to expression data, PCA finds principal components, or eigenarrays, which can be used to reduce the dimensionality of expression data for visualization, noise filtering and for simplifying the subsequent computational analyzes. In contrast to PCA, ICA reduces the dimensions of an input dataset into a lower dimension set of components so that each component is statistically as independent from the others as possible. Liebermeister (2002) first proposed using linear ICA for microarray analysis to extract expression modes, where each mode represents a linear influence of a hidden cellular variable. However, there has been no systematic analysis of the applicability of ICA as an analysis tool in diverse datasets, or comparison of its performance with other analysis methods.

A word of caution: microarray experiments typically analyze few samples and these experiments typically have few, if any, replicates. Conversely, each microarray generates thousands of data points for each sample. This is sometimes referred to as the low N, high P problem and is a particular challenge for most statistical techniques that generally operate best with high N, low P data. Thus statistical tools for analyzing microarray data must be chosen with care.

## 8.2 SIMILARITY METRICS

BioConductor packages: `dist`, `bioDist`, `daisy`

The first choice to be made in terms of most comparative microarray analyzes is how similarity between the gene expression data is to be defined, essentially computing how similar two series of numbers are. It should be fairly obvious that measures of similarity are the basis underpinning most expression analysis techniques and therefore it is pertinent that we briefly review the more common measures. There are many ways of calculating how similar sets of numbers are and we are not exhaustive in this overview, however the different methods described below suffice to highlight the general principals that can be applied.

### 8.2.1 A Distance Metrics

Distance metrics are conceptually the simplest methods for determining similarity or difference: points that are close together in the high dimensional space can be considered as having similar properties, while points that are distant from each other represent points or vectors with different properties. When gene expression vectors are compared, points close in space should have similar expression patterns and points that are distant should show differential expression. Similarly, when experiment profiles (samples) are compared, proximal points represent samples with similar overall patterns of gene expression, while distant points represent samples with different overall patterns of gene expression. Various methods for calculating distance may be used with the following being the simplest:

**(i)** *Euclidean distance*: this is simply the length of the line connecting two points in space and is calculated using a multidimensional version of Pythagoras theorem:

$$d = \sqrt{\sum_{i=1}^{n} (x_i - y_i)^2}$$

**(ii)** *Manhattan distance*: this is the distance that is traveled if only right angle turns are made to traverse the space between two points rather than the sum of squares of Euclidean distance. This measure of distance is less sensitive to outlier values than the Euclidean distance:

$$d = \sum_{i=1}^{n} |x_i - y_i|$$

## 8.2.2 Correlation Metrics

Correlation measures the strength and direction of a linear relationship between two variables. Various methods for measuring a linear relationship can be used and in the context of gene expression analysis the following are the most commonly used:

**(i)** *Pearson correlation*: The Pearson correlation coefficient is obtained by dividing the covariance of two variables by the product of their standard deviations:

$$r = \frac{\sum_{i=1}^{n}(x_i - \bar{x})(y_i - \bar{y})}{(n - 1)S_x S_y}$$

where $S_x$ is the standard deviation of $x$ and $S_y$ is the standard deviation of $y$. The Pearson correlation is insensitive to the magnitude of the gene expression vectors being compared. For example, if the pattern of expression of a gene is a sine wave with an intensity of 3 and the pattern for a second gene is a sine wave in the same phase with an intensity of 10, the genes will have a perfect correlation coefficient of 1. Thus, the Pearson coefficient is sensitive to the similarity of the shape of an expression pattern but not the overall intensity.

**(ii)** *Spearman rank correlation*: in some situations it is preferable to use the rank of expression values rather than their intensity as a measure of change. The Spearman rank correlation is calculated as:

$$r_s = \frac{\sum_i (R_i - \bar{R})(S_i - \bar{S})}{\sqrt{\sum_i (R_i - \bar{R})^2} \sqrt{\sum_i (S_i - \bar{S})^2}}$$

where $R_i$ is the rank of gene expression intensity $x_i$ and $S_i$ is the rank of gene expression intensity $y_i$.

**(iii)** *Jackknife correlation*: this method makes correlation metrics less sensitive to outliers. If a gene expression matrix has gene expression vectors of length $i$ (with $i$ different intensity measures), then the jackknife correlation of two vectors $x$ and $y$ is defined as the minimum correlation found when a series of correlation is determined for $x$ and $y$ dropping the $j$th intensity measurement (for $j − 1$ to $i$). In other words, the correlation coefficient is calculated repeatedly dropping one corresponding intensity from each vector. Finally, the lowest coefficient is selected to represent the correlation of $x$ and $y$. Therefore, when a single outlier is present in the dataset, artificially increasing correlation coefficients, then application of the jackknife process will eliminate their effect. The process can be generalized to account for any number of outliers, but in this more general case the method becomes computationally very intensive and is probably not worthwhile unless many outliers are actually expected.

## 8.3 UNSUPERVISED ANALYSIS METHODS: CLUSTERING AND PARTITIONING

### 8.3.1 Hierarchical Clustering

*R* packages: `cclust, cluster, hclust`

The first point of entry into data analysis is often hierarchical clustering and although the method has its peculiarities, it is quite often helpful to use hierarchical clustering to provide an initial exploration or visualization of a dataset. The technique organizes gene expression vectors in a strict hierarchy of nested subsets in the same way phylogenetic trees are often rendered. During the clustering, the closest pair of points is grouped and replaced by a single expression value representing the features of the set, the next closest pair of points is treated similarly, and so on. The data points are thus fashioned into a tree with the objective that the most similar gene expression vectors are closest in the tree hierarchy and dissimilar vectors are distant. The *R* package `cluster` provides the `agnes` tool (Kaufman and Rousseeuw, 1990) for performing agglomerative clustering.

In principle, hierarchical clustering can be performed by 'agglomeration' or by 'division'. Agglomerative methods work 'bottom up' and start by placing individual gene expression vectors (or profiles when experiments are compared) from a gene expression matrix into their own clusters or nodes, where the nodes are assigned the same value as the expression vector. These nodes are compared with all others, using an appropriate similarity metric, to find the most similar nodes. Similar nodes are merged into higher nodes. Each higher node is then assigned a modified expression vector that represents the subnode vectors, which

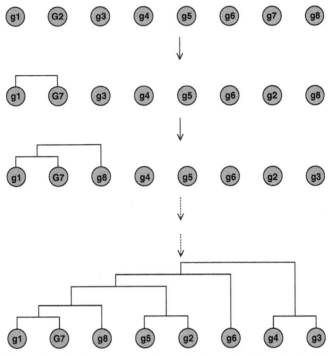

**FIGURE 8.1**  A schematic illustration of agglomerative hierarchical clustering. In this example a distance metric determines that $g1$ and $g7$ have the closest expression profiles and groups these. A measure of the combined $g1$–$g7$ expression values is used in subsequent comparisons, in this case indicating similarity with the expression profile of $g8$. The process continues until all genes (or samples) are linked in a single tree with the length of the branches an indication of the distance between each gene (or sample).

we will refer to as the representative vector (Figure 8.1). The method for calculating the representative vector is referred to as the 'linkage rule'. The process is then repeated with the new nodes to build higher nodes until finally a single top node is obtained. In contrast, divisive methods work 'top–down' and start by placing the complete data matrix into a top node. This node is then split into subnodes, which are iteratively subdivided until individual expression vectors are in their own nodes. Top–down approaches will be discussed in more detail in Section 8.3.5.

The behavior of agglomerative hierarchical clustering methods is determined to a large extent by the nature of the similarity metrics and linkage rules used to build the tree. The most common linkage rules are as follows:

**(i)** *Single linkage/nearest neighbor*: the representative vector of a node is taken as the highest similarity value from all the pairwise comparisons of all the members of the node.

**(ii)** *Complete linkage*: the representative vector of a node is recorded as the lowest similarity value from all the pairwise comparisons of all the members of the node.

**(iii)** *Mean linkage*: the representative vector is calculated as the mean of the similarity values from all the pairwise comparisons of all the members of the node.

**(iv)** *Centroid linkage*: The representative vector is calculated as the mean of the actual values of the members of the node.

Single linkage tends to produce loose clusters with a 'chaining effect' as nodes may be linked by similarity between only one of the members of each node, but it is computationally efficient as only one similarity matrix with all pairwise comparisons needs to be calculated. Complete linkage is similarly computationally efficient, but tends to produce tight clusters. Average linkage tends to produce looser clusters than complete linkage but tighter than single linkage. It is computationally more intensive than either single or complete linkage. Centroid linkage is also quite computationally intensive and the meaning of the representative vectors is somewhat less clear than with the other methods. In addition, the representative vectors can become quite different from their component vectors, particularly if the component vectors are anticorrelated or as the subhierarchy becomes large.

Hierarchical clustering has a number of issues that may be considered short-comings for the study of gene expression. Strict phylogenetic trees are best suited to situations dealing with true hierarchical descent (such as in the evolution of species) and are not really designed to reflect the multiple distinct ways in which expression patterns can be similar. This problem is exacerbated as the size and complexity of the data set grows. In addition, most hierarchical clustering is performed using 'greedy algorithms', which make local evaluations about similarity between gene expression vectors. This causes points to be grouped based on local decisions, with no opportunity to reevaluate the clustering. It is known that the resulting trees can lock in accidental features, reflecting idiosyncrasies of the similarity metrics and linkage rules used. These local effects lead to nonuniqueness, inversion problems and a general lack of robustness, which can complicate the interpretation of the clusters. Bearing these limitations in mind, hierarchical clustering should probably be considered more as an exploratory or visualization tool than as a rigorous analysis tool. Even when using hierarchical clustering as a visualization tool, it should be remembered that for just a single clustering there are a large number of different dendrograms that can represent the same hierarchy and some consistent representation should be used.

These issues notwithstanding, hierarchical clustering has been used successfully used to identify expression signatures in many different circumstances. For example, classifying lymphomas (Alizadeh et al., 2000), finding melanoma specific expression signatures (Ryu et al., 2007), finding patterns in DNA copy number changes in breast cancer comparative genomic hybridization data (Yau et al., 2007), classify responders and nonresponders with a rheumatoid arthritis treatment (Lequerre et al., 2006) and mapping expression patterns in the central nervous system (Stansberg et al., 2007) to name but a few.

## 8.3.2 Partitioning Methods

Partitioning approaches attempt to cluster data into a predetermined number of sets and can be thought of as a form of dimensionality reduction. Various approaches have been developed but probably the most widely used are $k$-means and self-organizing maps.

### 8.3.2.1 k-Means Clustering

$R$ package: mva

The objective of the $k$-means algorithm is to partition $n$ gene expression vectors into $k$ user-defined sets or clusters, where $k \ll n$ and the mean squared cluster distance from the $k$ cluster centers (intracluster error) is minimized. It is conceptually similar to the self-organizing map method discussed below. In practice, the global minimum intracluster error is computationally impractical to determine as it is an NP-hard problem. The most common form of the algorithm uses an iterative refinement heuristic known as Lloyd's algorithm. Lloyd's algorithm starts by initializing each of $k$ different candidate cluster centers with a different randomized representative vector of the same dimension as the gene expression matrix. Each gene vector from the gene expression matrix is then assigned to the cluster centre to which the gene vector is most similar. The representative vector of each of these sets is then recalculated as the mean point (or centroid) of the gene expression vectors assigned to each set. Each gene vector is then reassigned to the set to which it is now most similar. The process of recalculating the representative vectors and then reassigning the gene vectors is repeated iteratively until the sets no longer change, that is $k$ cluster centers have been found (Figure 8.2). Another related approach that is sometimes used is the $k$-medioid technique, which forces one of the points in the data to be a cluster centre. This is implemented in the Partitioning About Medioids algorithm (Tibshirani et al., 2001). PAM is similar to the QT clustering approach discussed below.

The $k$-means method using Lloyd's algorithm is sensitive to the starting values of the representative set and is thus not guaranteed to return a global optimum, although it normally gets close. Since the algorithm is extremely fast, the process is typically repeated many times to determine whether a stable solution is found. A drawback of the $k$-means algorithm is that the number of clusters $k$ is an input parameter and an inappropriate choice of $k$ may yield poor results. This can be addressed in a rough and ready way by rerunning the algorithm with increasing numbers of clusters. If the number of clusters is plotted

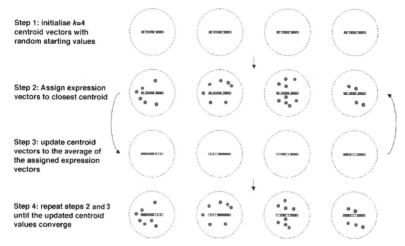

Step 1: initialise *k*=4
centroid vectors with
random starting values

Step 2: Assign expression
vectors to closest centroid

Step 3: update centroid
vectors to the average of
the assigned expression
vectors

Step 4: repeat steps 2 and 3
until the updated centroid
values converge

**FIGURE 8.2**   A schematic representation of the $k$-means algorithm when $k = 4$.

against the mean squared difference in position between points in the cluster and the centroid for that number of clusters, it will be apparent that initially, increases in the number of clusters will result in tighter and tighter clusters while further increases will lead to diminishing returns. When plotted this should be apparent as an 'elbow' on the graph as the amount of variance explained by increasing numbers of clusters starts to diminish. Silhouette width can also be used to choose the appropriate value for $k$ (see Box 8.1). An attempt to formalize this notion is the Gap statistic (Tibshirani et al., 2001). Clusters should also be reviewed for biological meaning to determine whether the results look sensible.

The $k$-means approach has been used for a wide range of applications including gene selection for further studies (Leung et al., 2006) and analysis of transcription factors (Ishida et al., 2002). A criticism of both $k$-means and hierarchical clustering is that genes or experiments can only be assigned to one cluster. Since genes may be involved in more than one pathway, this limitation is significant. A variation on $k$-means clustering is the fuzzy $c$-means (FCM) approach, which operates in a similar fashion to $k$-means in that a predetermined number of clusters are used and cluster centroids are found using more or less the same algorithm. The difference is that with FCM all genes are considered members of all clusters but the membership is weighted by their distance from cluster centroids, so that genes are more likely to be in clusters whose centroids are closer (Dembele and Kastner, 2003). An $R$-packages, *c-means*, is available.

### 8.3.2.2 Self-Organizing Map (SOM)

$R$ packages: som

The self-organizing map (SOM) method partitions gene expression vectors into a userdefined grid of cells, typically two-dimensional ($a \times b$), where each

---

**Box 8.1** Silhouette width (Rousseeuw, 1987) is a measure of quality and the Silhouette width for a clustered data point $i$ is defined as:

$$Sil_i = \frac{(b_i - a_i)}{\max(a_i, b_i)}$$

where $a_i$ is the average distance between point $i$ and all the other points in the cluster, $b_i$ is the minimum of the average distances between $i$ and points in other clusters. A different average is calculated for each cluster and $b_i$ is defined to be the lowest of these. The silhouette width is a value between 1 and $-1$.

If $Sil_i$ is close to 1, it means that the sample is considered to be 'well-clustered' and point $i$ was assigned to the correct cluster.

If $Sil_i$ is near zero, it means that that point $i$ could equally well be assigned to another nearby cluster, and the sample lies equally far away from both clusters.

If $Sil_i$ is close to $-1$, it means that sample is 'misclassified' and is merely somewhere in between the clusters.

Averages of $Sil_i$ can be calculated for individual clusters and for a complete clustering to give an indication of the quality of the clustering. The clustering average can be used to assess algorithms that require userdefined parameters, such as $k$-means or QT clustering, by plotting the average silhouette width against the parameter to identify the optimal value.

---

cell on the grid is intended to represent a cluster of genes with similar expression (or experiments with similar profiles). The fundamental difference between an SOM and a $k$-means partitioning is that the SOM is supposed not only to partition the data into $(a \times b) = k$ clusters but also the grid defined by the user is intended to represent the relationships between the clusters that are found. Thus a large gene expression matrix is clustered into $(a \times b)$ cells and each gene expression vector is assigned to a single cell. The contents of each cell are most similar to each other and each cell on the grid is more similar to neighboring cells on the grid than to distant cells on the grid. It is this similarity of the neighboring cells on the SOM grid that distinguishes the SOM method from other partitioning approaches, particularly the $k$-means method. Furthermore, the clustering results are displayed using the grid that was chosen to be fitted to the data. Depending on the software used the resulting 'map' of the data may display various features of the clustering, such as the number of points associated with each cluster and the distance between the clusters. In the context of gene expression profiling, the maps can also be used to display average expression intensities of the clusters for each experiment or time point. Thus SOMs are very useful for creating low-dimensional views of high-dimensional data.

As with the $k$-means method, each cell of an SOM grid is typically initialized with a different randomized vector of the same dimension as the gene expression matrix. The cells are then compared with real expression values from the gene expression matrix. At each iteration of the algorithm a gene is assigned to the cell to which it is most similar (using one of the distance measures described above). The value of the cell vector is then modified so that it becomes more similar to the value that has been assigned to it. In addition, the values of the neighboring cells are also modified to be more similar, but to a lesser extent. The precise weightings used will vary depending on the particular implementation of the algorithm. Geometrically speaking, the most similar cell moves toward the point defined by the gene assigned to the cell. As more genes are tested, the cells on the grid can be thought of as moving closer and closer to the genes to which they are most similar and that neighboring cells will move towards adjacent clusters. At the beginning of the process the cells of the SOM grid are randomly located in the expression space and initially will move rapidly toward nearby gene clusters, but as the process continues movements will become smaller and by the end of the process, the grid will be spread through the space where each cell of the grid converges to a final position that is hopefully the centre of a cluster of closely related genes (Figure 8.3). The SOM thus attempts to preserve the topology of the input data space but the method is sensitive to the size and topology of the SOM grid used to generate the map in the first place. SOMs have been applied to the analysis of haematopoiesis (Tamayo et al., 1999), cancer

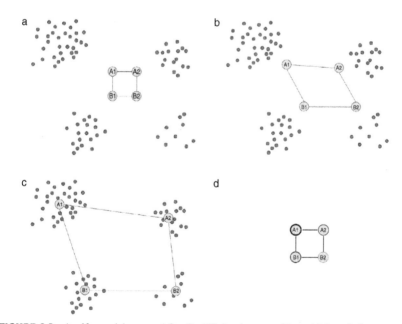

**FIGURE 8.3**   A self-organizing map. A 2 × 2 grid is fitted to a set of data with four obvious clusters of different sizes. Part (a) shows the starting configuration. Part (b) shows the progress of the clustering as the grid points move toward their nearest clusters. Part (c) shows the final clustering positions of the grid. Part (d) shows the resulting map with the cluster sizes of the final clusters indicated by the thickness of the lines surrounding the circles.

(Covell et al., 2003) and to the more general problem of determining how many clusters are present in a dataset (Costa and Netto, 1999). Variations on the approach include growing self-organizing maps, which attempt to overcome the problem of choosing an appropriate grid size and topology a priori (Hsu et al., 2003).

### 8.3.3 Quality-Based Clustering Algorithms

*R* package: `flexclust`.

   The quality threshold-clustering (QT-clustering) algorithm (Heyer et al., 1999) is a partitioning algorithm that does not require the user to define the number of partitions at the outset. Instead, the user defines a 'maximum diameter' that defines a 'hypersphere' in expression space. For every expression vector (or experiment vector) from a gene expression matrix, a candidate cluster is built from the starting point by including the closest point, then the next closest, and so on, until the diameter of the cluster surpasses the maximum diameter. Various distance measures can be used, including Euclidean, Manhattan and maximum distance. The following two steps are applied:

(i) The candidate cluster with the most points is saved as the first true cluster, and the contents of thus cluster is removed from further consideration.
(ii) The process is then repeated recursively with the reduced set of expression vectors remaining after removal of the first cluster until all the gene vectors are assigned to a cluster.

   This method always returns the same result when repeated, but the result is obviously dependent on the maximum radius that is defined. By setting the radius parameter of the hyperspheres, the QT method imposes a quality threshold on the inclusion of data points in a cluster. A further feature of the QT method is the use of jackknife correlation to assess similarity between expression vectors or experiments. While similar in principle to *k*-means or rather to *k*-medioid clustering, a crucial difference is that outliers are not added to clusters due to the use of jacknife correlation. With *k*-means, all genes are assigned to a cluster and consequently outliers are always included in a cluster somewhere. Furthermore, the QT method does not globally minimize the within cluster distances – it is only locally optimal. Some variants on the QT approach have also been proposed, for example: Adaptive QT clustering (De Smet et al., 2002) and stochastic QT clustering (Scharl and Leisch, 2006). The adaptive method addresses the issue of determining the correct diameter for the 'hypersphere' that defines each cluster by attempting to find the optimum radius for each cluster. The process of finding an optimum radius uses a threshold setting for significance. This is a useful feature of adaptive QT clustering since very few algorithms implicitly implement a significance test and so the significance of the clusters that are produced is known. SOTA (discussed below) also has a similar feature.

Stochastic QT sacrifices the deterministic nature of the original QT approach to try and achieve tighter clustering. In the original QT approach, the algorithm tests every gene expression vector as a nucleation point for cluster construction. In stochastic QT, a parameter *ntry* is specified that limits the number of nucleation points tested in each iteration of the algorithm and the gene expression vectors to be used as nucleation points are picked randomly. The algorithm is run several times to determine whether reproducible or stable clusters are being produced. Obviously with this approach, the reproducibility of the deterministic version of the algorithm is lost but it does find tighter (if smaller) clusters, as measured by within cluster distance.

## 8.3.4 Divisive Hierarchical Clustering

### 8.3.4.1 *Diana* (*Divisive Analysis*)

*R* package: `cluster`

*Diana* (Kaufman and Rousseeuw, 1990) produces a hierarchical clustering using a top–down divisive approach, starting from the complete data matrix. The data are split using binary recursive partitioning. The term 'binary' means that each group of samples, represented by a 'node' in a decision tree, is only split into two child nodes at each iteration of the algorithm. Thus, each parent node can give rise to two child nodes and, in turn, each of these child nodes may themselves be split, forming additional children. Diana splits data by identifying the point in the current cluster that is most different from the mean and using this point to form a splinter cluster. Points that are more similar to the splinter cluster point than the old cluster mean are assigned to the new cluster. The process continues until each expression vector is resolved.

### 8.3.4.2 *Hopach* (*Hierarchical Ordered Partitioning and Collapsing Hybrid*)

*R* package: `hopach`

Hopach (van der Laan and Pollard, 2003) may be regarded as a hybrid of Partitioning About Medioids and a top–down hierarchical clustering. A criticism of agglomerative clustering is the binary nature of the clustering, as only pairwise comparisons are made at each stage. The Hopach method applies PAM on the whole data set using silhouette widths (Box 8.1) to choose the optimum number of partitions. The algorithm then successively applies PAM to the clusters formed in the first partitioning. In this way, the optimum number of clusters is selected at each level of the hierarchy avoiding the binary partitioning. After each partitioning stage, the clusters are assessed for similarity and closely related clusters may be merged or 'collapsed'. Thus, Hopach avoids another of the issues with conventional agglomerative clustering: that incorrect associations become fixed in the hierarchy. The depth of the hierarchy can be selected by the user or allowed to continue until individual expression vectors are resolved or other criteria are met.

### 8.3.4.3 *Self-Organizing Tree Algorithm*

*R* package: `clValid`; SOTA is also available through the GEPAS (Gene Expression Profile Analysis Suite, Tárraga et al., 2008): http://www.gepas.org/.

SOTA is a top–down hierarchical clustering approach (Herrero et al., 2001). The approach may be regarded as a hybrid of hierarchical clustering and self-organizing maps. SOTA starts with the complete data set and performs a series of binary splits on the data segregating the data into successively smaller pairwise partitions based on a predefined criterion for determining differences in the data. SOTA transforms the data matrix into a binary tree representation using a self-organizing neural network that is similar to SOM methods. The algorithm starts with a root node and two terminal neurones branching from the root. The terminal cells are comprised of vectors of the same size as the gene expression vectors. These are initialized with the mean values for the gene expression vectors of the gene expression matrix. All the gene expression vectors in the data set are assigned to the root node. Every gene expression vector is then 'presented' to the offspring cells. Genes are assigned to the cell to which they are most similar (using any of the similarity measures discussed above). As each gene is assigned to a cell, the cell vector, which is essentially a representative vector as defined for hierarchical clustering, is updated with a weighted average of the assigned expression vector and the cell vector. Similarly the parent and sister nodes receive updates with smaller weightings. Although the starting values of the first two cells of the network are the same, the gene expression vectors are tested against one of the cells first and, if sufficiently similar, are assigned to it and then not tested further. The cell values are then immediately updated leading to asymmetry. After the gene expression vectors have been assigned to cells, a value called the 'Resource' is calculated for each cell, which is essentially the mean of the 'distance' of each gene expression vector assigned to a cell from that cell's vector. Two daughter cells are then created for the cell with the highest resource, which then becomes a node. Each daughter is assigned the same value as the parent node. All the expression vectors assigned to the parent node are then presented to the offspring cells as for the starting cells (Figure 8.4).

This process continues until the distance between the gene expression vectors assigned to a cell and the cell vector falls below a predetermined threshold or until a predetermined amount of branching or number of clusters has been reached. The threshold for termination can be measured in various ways:

- Determining the total resource value of all of the terminal cells of the tree and comparing this with some predetermined threshold value.
- Setting a threshold that terminal nodes must reach before they will no longer produce daughter cells.

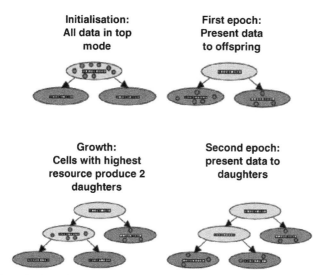

**FIGURE 8.4**   A schematic representation of the self-organizing tree algorithm.

- Continuing until the maximum of the distances between the expression vector pairs assigned to the cell falls below a predefined threshold.

Alternatively the algorithm can run until only single genes/experiments are assigned to each terminal cell. This flexibility in the SOTA approach allows the user to direct the clustering process to achieve specific goals. In the simplest instance, the data can be split into a hierarchy that ends with individual genes, much like a conventional agglomerative hierarchical clustering. Alternatively, a threshold for a given number of clusters can be selected, in which case SOTA behaves somewhat like a SOM. Furthermore, SOTA allows a confidence level to be defined for clustering, that is a value that determines the proportion of assignments to a cluster that occur through chance rather than real correlation.

Although SOTA notionally uses binary partitioning, the hierarchy of SOTA does not necessarily correspond to the hierarchy in *Diana* due to the nature of the stopping criteria employed. Termination of tree growth may happen at different points in the hierarchy on different branches. Thus, it is the contents of the terminal nodes that have met the stopping criterion that determines the resulting partitioning of the data, rather than the depth of the tree. Consequently, SOTA is more similar to dynamic self-organizing maps and other partitioning approaches in its end result.

SOTA has not been widely applied but some recent studies comparing this tool with other algorithms suggest that it is more robust than either of agglomerative hierarchical clustering or self-organizing maps (Handl et al., 2005; Yin et al., 2006). The hierarchical nature of the results makes visualization and interpretation of the clustering simple, similar to the views generated by agglomerative hierarchical clustering. Moreover, since the algorithm allows for the definition of confidence threshold values used for assignment of gene or experiment vectors to clusters, the significance of clusters can be quantitatively assessed. This is a very useful feature that most clustering algorithms do not provide. Similarly, the nonterminal nodes are effectively weighted averages of the expression or experiment vectors assigned to them. These node values can be used to assess correlation between clusters, that is to find clusters that are similar.

## 8.4 ASSESSING CLUSTER QUALITY

*R* package: `clValid`

From the preceding discussion of unsupervised analysis tools and a cursory glance at PubMed, it should be clear that there is a bewildering array of tools available for clustering expression data. It is important to recognize that many of the more novel algorithms that appear in the literature on a regular basis have often not been very thoroughly evaluated in a wide range of applications and weaknesses are often not readily apparent. Impartial assessment of the success of different clustering methods is essential both to assess new tools, as the field evolves, and to ensure that results produced are significant and meaningful. The discussion here is fairly cursory with the intention of introducing the reader to some of the issues to consider and some of the techniques available to assess cluster quality. Wider reading is definitely encouraged and the following are some good entry points into the literature (Chen et al., 2002; McShane et al., 2002; Handl et al., 2005). The clvalid *R* package implements most of the validation measures discussed here and is recommended as a good starting point for any validation of clustering (Brock et al., 2008).

It is important to understand certain key issues that arise with clustering:

**(i)** Most clustering methods identify related data in different ways and as a consequence perform better with certain types of data structure (Figure 8.5)

**(ii)** Almost all clustering methods will return some kind of clustering result even in the absence of any real structure in the data and most tools do not provide any assessment of the significance of the results. SOTA (Herrero et al., 2001) and Adaptive QT clustering (De Smet et al., 2002) are among the few exceptions where confidence levels can be predefined.

There are numerous measures for determining the quality of a cluster and results from a typical `clValid` run are shown in Figure 8.6. Simplistically, a good clustering algorithm should minimize the average distance between cluster members while maximizing distance between separate clusters. Various simple statistics can be calculated to assess this, such as the sum of squared distance of points from cluster centers (this is what *k*-means attempts to minimize) to determine cluster 'tightness' or 'compactness' and the average of between-cluster distances to assess the effectiveness of clustering. More formal

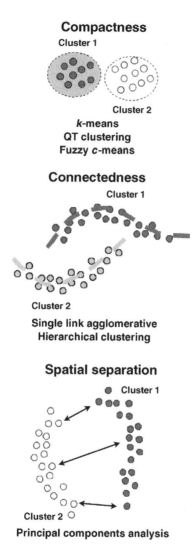

**FIGURE 8.5**   Different algorithms apply different discrimination criteria to effect a separation of data into clusters and this leads to them being better able to deal with certain kinds of structure in data. Algorithms that work using spatial separation or connectedness can deal with clusters that are not compact but methods like $k$-means should, in principle deal with clusters that are close to each other spatially as long as there is sufficient intracluster distances are smaller than the intercluster distances.

assessments can be made using tools such as Dunn indices, Silhouette widths and Connectivity (Boxes 8.1–8.3).

It is important to remember that measures of clustering exhibit biases in terms of what types of cluster they will score highly (Handl et al., 2005). Ideally

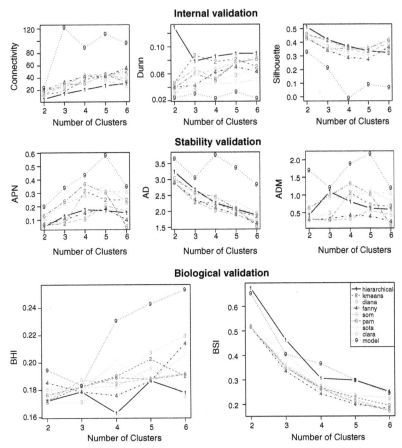

**FIGURE 8.6**   Typical results from clValid. Various clustering algorithms were applied to a published dataset. The plots show how the various metrics vary with the number of clusters sought by each of the algorithms. See color plate section.
*Source*: Taken from Brock et al. (2008).

multiple measures should be used in the assessment of clustering results and in comparing algorithms. Clustering algorithms used should also be 'stable', that is they should produce essentially the same clusters from similar datasets. Stability of algorithms can be tested using resampling, if sufficient data is available. Resampling methods (Levine and Domany, 2001; Gana Dresen et al., 2008) are used to create simulated data, which are then subjected to a given clustering. These clusters can be compared to see if different but related data sets give similar clusters using some form of Rand index (Box 8.4), or other preferred measure. Ideally, the algorithm should also be relatively insensitive to outliers.

Some further useful stability measures can be determined by performing clustering on a dataset and comparing the clustering on the whole dataset with the same algorithm on the same dataset with single columns removed. Various metrics can be determined from the comparisons of the clusterings on complete data and on incomplete data (Yeung et al., 2001; Datta and Datta, 2003), these include:

**(i)** Average proportion of nonoverlap (APN) measures the average number of vectors not assigned to the same cluster in the full and different incomplete data sets.

**(ii)** Average distance (AD) measures the average distance between observations placed in the same cluster by clustering based on the full data and clustering based on the data with single columns removed.

**(iii)** Average distance between means (ADM) measures the average distance between cluster centers for expression vectors placed in the same cluster by a clustering on the full data and clustering based on the data with a single column removed.

**(iv)** Figure of merit (FOM) measures the average intracluster variance of the observations in the deleted column, where the clustering is based on the remaining (undeleted) samples.

Stable clustering algorithms should give APN values that approach zero while the other measures should be maximized. Furthermore, it is worth assessing the probability of clusters arising by chance alone. This can be done analytically in some situations but, more generally, it can be done by simulation using Monte Carlo methods to generate numerous clusterings on random or resampled data with the same number of partitions as a putatively meaningful clustering to give an estimate of the probability of chance clustering (Handl et al., 2005).

While these sorts of measures of algorithm behavior are useful, clustering should also be biologically meaningful and some efforts have been made to develop automated tools for assessing the biological meaning of clusters. For example, Datta and Datta (2006) have proposed two measures of biological significance:

**(i)** The biological homogeneity index (BHI), which as the name suggests, is a measure of the biological homogeneity of clusters and assesses how well genes are clustered based on known functional annotations.

**(ii)** Biological stability index (BSI), which measures the consistency of the clustering algorithm's ability to produce biologically meaningful clusters when applied repeatedly to similar data sets.

The BHI measure is calculated by assigning a functional class to each gene in an experiment and then determining an index of how often genes with the same class turn up in the same cluster. The class assignments can be made using

---

**Box 8.2** The Dunn index (Dunn, 1974) is an attempt to formalize the idea that clusters should be compact and well separated, that is maximizing intracluster separation and minimizing intercluster distances. For a given clustering C, the Dunn index, D, is calculated as follows:

$$D = min_{1 \leq i \leq n} min_{1 \leq j \leq n, i \neq j} \frac{d(c_i, c_j)}{max_{1 \leq k \leq n}\{d'(c_k)\}}$$

where $c_i$, $c_j$ and $c_k$ represents the $i$th, $j$th and $k$th cluster of the clustering, $d(c_i, c_j)$ is the distance between clusters $c_i$, and $c_j$, that is intercluster distance, $d'(c_k)$ is the intracluster distance of cluster $c_k$, $n$ is number of clusters in C. A good clustering should maximize D. Like the Silhouette width, the Dunn index can be used to help select the optimum values for clustering tools that require user specified parameters.

---

**Box 8.3** Connectivity (Handl and Knowles, 2005) assumes that points that are nearest neighbors in expression space are most likely to end up in the same cluster. The connectivity measure counts violations of this principle in clustering (C) with N data points:

$$Conn(C) = \sum_{i=1}^{N} \sum_{j=1}^{L} x_{i,nn_{i(j)}}$$

where $nn_{i(j)}$ is the $j$th nearest neighbor of observation $i$, $x_{i;nni(j)}$ is defined to be zero if $i$ and $nn_{i(j)}$ are in the same cluster and $1/j$ otherwise. L is a parameter that determines the number of neighbors that contribute to the connectivity measure. The connectivity has a value between zero and infinity and should be minimized.

---

sources such as the Gene Ontology. The BSI measure determines how often genes with the same functional class turn up when a clustering algorithm is repeatedly applied to subsets of the data. The rational here being that if some of the data is removed from an experiment, robust clusters should still have the same levels of significance and the BSI attempts to measure this. Obviously this approach is limited by the reliability of the functional assignments used and is also based on the assumption that functionally related genes should turn up in the same cluster. Various related approaches have also been developed for assessing the biological significance of clusters (Clare and King, 2002) but all suffer from the same limitation that they ultimately rely on the quality of the underlying biological annotations.

## 8.5 DIMENSIONAL REDUCTION

Dimensional reduction of data can be regarded as the process of finding a function that will transform an $m$-dimensional space to a lower $n$-dimensional space so that the transformed variables give information on the data that is

> **Box 8.4** One of the simplest and most well-known tools for quantifying similarity between different clusterings, whether by the same algorithm on different data or different algorithms on the same data, is the Rand index. The basic Rand index is a measure of the similarity between clusterings based on pairwise agreement between clusterings (Rand, 1971):
>
> $$\text{Rand} = \frac{a+b}{a+b+c+d}$$
>
> where $a$ is the number of pairs of elements in a dataset that are in the same cluster in a clustering by a first algorithm and in the same cluster in a clustering by a second algorithm, $b$ is the number of pairs of elements in a dataset that are in different clusters in a clustering by a first algorithm and in different clusters in a clustering by a second algorithm, $c$ is the number of pairs of elements in a dataset that are in the same cluster in a clustering by a first algorithm and in different clusters in a clustering by a second algorithm and $d$ is the number of pairs of elements in a dataset that are in different clusters in a clustering by a first algorithm and in the same cluster in a clustering by a second algorithm. This index captures the similarity between two different algorithms or between application of the same algorithm on different but related sets of data. However, the basic Rand index does not take into account the fact that two points might be in the same cluster by chance alone so an adjusted Rand index is more commonly used as a measure for comparing different clusterings. The adjusted Rand index introduces a correction to take into account chance associations (Hubert and Arabie, 1985). More recently, a further modification of the Rand index was introduced, which attempts to give a higher similarity score to two clusterings if points that are in different clusters but the clusters are very close to each other. This rank adjusted Rand index (Pinto et al., 2007) should give a better assessment of whether two clusterings are similar even if some of the data that is grouped in one cluster in one clustering has been split into two or more neighboring clusters in a second clustering.

otherwise hidden in the large data set, that is the transformed variables should be the underlying *factors* or *components* that describe the essential structure of the data. It is hoped that these components correspond to some physical process involved in the generation of the data in the first place. In the lower dimensional space the components represent some combination of the observed variables.

Typically, only linear functions are considered, because the interpretation of the representation is simpler and the computational effort is relatively smaller. Thus, every component is expressed as a linear combination of the observed variables:

$$y_i(t) = \sum_j w_{ij} x_j(t), \text{ for } i = 1, \ldots, n \text{ and } j = 1, \ldots, m$$

where the variables $w_{ij}$ are coefficients that determine the transformed representation $y_i$ in terms of $x_j$. This can be rephrased as the problem of determining the coefficients $w_{ij}$. We can also express the linear transformation of $x$ to $y$ as a matrix multiplication, collecting the coefficients $w_{ij}$ in a matrix $(W)$. The equation becomes:

$$
\begin{array}{cc}
y_1(t) & x_1(t) \\
y_2(t) & x_2(t) \\
- & = W - \\
- & - \\
y_n(t) & x_m(t)
\end{array}
$$

Various methods for the determination of $W$ can be considered. Principal components analysis (PCA) for example, attempts to project the data in a new basis so that the orthogonal basis vectors have as little correlation with each other as possible while maximizing the signal as well. Another approach is to attempt to make sure that the components $y_i$ are as statistically independent as possible, that is finding the underlying variables whose effects are independent of each other. In the case where data is Gaussian, independence is found by determining uncorrelated components. This is factor analysis. In the more general case where data is non-Gaussian statistical independence is determined using independent components analysis (ICA).

## 8.5.1 Principle Components Analysis

BioConductor package: `pcaMethods`

PCA is used for the analysis of data sets that are known to be redundant and that may also be noisy (Stacklies et al., 2007). Technically, PCA does not actually reduce the number of dimensions of the data, but reexpresses the data in the same number of dimensions or components but projected so that as much of the variance of the data is captured in a small subset of the new components, referred to as the principal components. Thus, most of the components of the projected data may be ignored, hence reducing the dimensionality in a practical sense. The components of the projected data, as discussed above, are linear combinations of the source data. Redundant data, in the context of genes (or samples) that show similar patterns of expression, are thus reexpressed as new components that are weighted sums of the source genes or samples. The principal components found by PCA correspond to clusters of genes showing similar expression. However, if the expression differences are small in a dataset, that is there is low variance in the data, then the principal components can be difficult to interpret. Technically, the principal components are found by determining the eigenvectors and eigenvalues of the covariance matrix of the source data or by performing singular value decomposition (SVD). Either method is appropriate. PCA has been used for the analysis of breast cancer progression (Alexe et al., 2007), colorectal cancer (Hong et al., 2007) and Crohn's disease (Csillag et al., 2007) to give just a few examples. It is worth noting that $k$-means and PCA are closely related and it has been shown that PCA can be used to find near optimal cluster centers for $k$-means (Ding and He, 2004).

## 8.5.2 Gene Shaving

$R$ package: GeneClust

Gene shaving (Hastie et al., 2000) is a clustering algorithm based on PCA. PCA is used to identify the principal components in a gene expression matrix. The correlation values of the genes with the eigengene defined by the first principal component is determined. A fraction of the genes with low correlation to the eigengene are eliminated from the dataset, typically ten percent. The process is then repeated on the resulting subset of genes. Thus the gene shaving procedure generates a sequence of nested clusters, in a top–down manner, starting with $N$, the total number of genes, and typically decreasing down to a single gene. The optimum number of genes in the cluster associated with the first principal component can then be determined by comparing the nested clusters with each other. The process can then be repeated on other principal components. Due to the nested nature of the gene-shaving process, cluster membership is graded. An important difference between gene shaving and $k$-means or hierarchical clustering is that genes can be in more than one cluster.

## 8.5.3 Independent Components Analysis

$R$ package: fastICA

ICA decomposes an input dataset into components so that each component is statistically as independent from the others as possible. ICA is normally used to decompose a mixture of signals into their component source signals, assuming that the measured signal is a mixture of underlying sources. Liebermeister (2002) first proposed using linear ICA for microarray analysis. In this implementation, the independent components are regarded as linear influences of hidden variables on gene expression, which are referred to as 'expression modes', with each mode representing a linear influence of a hidden variable. Alternatively, the independent components can be regarded as putative biological processes (Lee and Batzoglou, 2003). For example, it has been shown, with data from well-characterized breast cancer studies, that independent components do indeed correspond to known cancer related pathways (Teschendorff et al., 2007). Further it is claimed that ICA is more reliable than PCA, analysis of variance (ANOVA) and the partial least squares methods for finding meaningful associations in microarray data (Carpentier et al., 2004). ICA has also been successfully used for development of molecular diagnosis tools for prostate and colon cancer (Zhang et al., 2005), analysis of endometrial cancer (Saidi et al., 2004, 2006) and for the analysis of regulatory signals in *Saccharomyces cerevisiae* (Liao et al., 2003).

## 8.6 SUPERVISED METHODS

Generally speaking, supervised methods of data analysis are used when the objective is to create a tool that can classify new gene expression patterns based on a set of samples with known expression patterns. The classifier is built using known data, such as a series of microarray results from tumor biopsies and corresponding normal tissue samples. This data is then fed into a suitable supervised classification tool to 'train' it. The trained tool can then be used to classify new samples to determine whether they are cancerous or not. A large number of such supervised methods have been developed for applications outside of gene expression profiling and many of these have now been validated for use in the classification of gene expression data. The following is a brief introduction to a small cross section of available methods.

### 8.6.1 K-Nearest Neighbor (KNN) Classification

*R* package: `class`

The KNN algorithm is one of the simplest of all machine-learning algorithms for discriminating between classes of data (Figure 8.7). In the context of expression data, an expression data space is trained simply by populating it with data (the unmodified gene expression vectors) from known samples such as cancer and normal profiles, with the profiles flagged as such. A new profile or experiment vector can then be classified by determining the distance of the new profile from its neighbors in the trained data space. It is usual to use the Euclidean distance, though other distance measures such as the Manhattan distance can be used instead. The classification is determined as a 'majority vote' of the $k$ (user defined) nearest neighbors in the data space, with the object being assigned to the class (cancer or normal) that is most common among these $k$ nearest neighbors. $k$ is typically a small but positive integer. If $k = 1$, then the object is simply assigned to the class of its nearest neighbor. In binary (two class) classification problems, it is helpful to choose $k$ to be an odd number since this avoids tied votes. More sophisticated 'voting' can be implemented, such as weighting the contributions of the neighbors, so that the nearer neighbors contribute more to the average than the more distant ones.

KNN has been used to build reliable classifiers for microbial blood pathogens based on genomic DNA microarray hybridizations (Wiesinger-Mayr et al., 2007). Since the KNN algorithm is sensitive to the local structure of the data the approach is often combined with other techniques. Genetic algorithms have been used to identify critical genes from breast cancer samples, that were then used to construct a KNN classifier (Jain and Mazumdar, 2003). KNN has also been combined with a regression method to predict gene function from array data (Yao and Ruzzo, 2006). Incidentally, the sensitivity of KNN to local data structure has been harnessed to impute missing values from gene expression matrices (Chapter 6; Hastie et al., 2000).

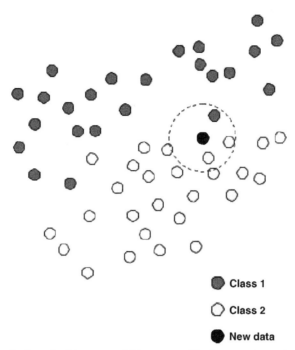

● **Class 1**

○ **Class 2**

● **New data**

**FIGURE 8.7**   A schematic of *k*-nearest neighbor process. The data space is populated with points with known class labels. New points are classified by a vote of the nearest neighbors. In this figure the new data point is assigned a class 2 label since more of the nearest neighbors are of class 2.

## 8.6.2 Support Vector Machines (SVM)

*R* package: `svmpath`

A support vector machine attempts to discriminate between clusters of data-points in large data sets by fitting a 'hyperplane', or high dimensional surface, between the points. For example, if the experiment vectors from a gene expression matrix of cancer biopsies and normal control tissue samples is analyzed, the SVM attempts to intercalate a surface in the intensity space between the cancer expression vectors and the normal tissue vectors to separate them. The points that are nearest the surface of the hyperplane are referred to as support vectors. Only these data points actually determine the topology of the plane. A new data point can then be tested to see whether it falls on one side of the hyperplane or the other: in this example, on the cancer side or the normal side (Figure 8.8). Thus, the fitted hyperplane acts as a classifier separating normal biopsies from cancerous biopsies based on gene expression. SVMs have been used to identify multidrug resistance in pancreatic cancer samples (Zhao et al., 2007), identify expression signatures in melanoma corresponding to disease progression and tumor subtype (Jaeger et al., 2007) and in ovarian cancer prognosis (Wei et al., 2006).

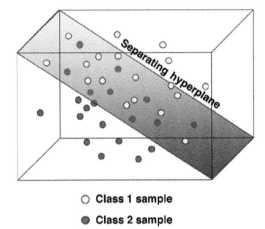

○ **Class 1 sample**

● **Class 2 sample**

**FIGURE 8.8**   A schematic of the support vector machine approach. Expression vectors distributed through a three-dimensional volume are separated by a hyperplane calculated by a support vector machine.

### 8.6.3 Linear Discriminant Analysis (LDA)

*R* package: sma

LDA attempts to find a line in expression space so that when points in the expression space are projected onto the line, points from one class fall on one side of a discrimination point and points from the other class fall on the other side of the discrimination point. Figure 8.9 illustrates this concept.

In some recent examples, LDA has been used to classify TB infections (Jacobsen et al., 2007), to determine the prognosis of stage II colon cancer (Barrier et al., 2007) and to discriminate oral leukoplakia subtypes (Kondoh et al., 2007). As with the *k*-nearest neighbor method, LDA is often used to produce a classifier once subsets of significant genes have been identified from microarray expression studies using other tools, for example, LDA has been used with 'Random Forests' to act as a classifier of breast cancer prognosis (Ma et al., 2007) or as a preliminary classification tool leading to a more sophisticated tool, for initializing a neural network for molecular diagnostics (Wang et al., 2006).

### 8.6.4 Decision Tree Classification Methods

*R* package: tree, randomForest

Decision trees methods take a top–down divisive approach to data partitioning much like *Diana* for unsupervised analysis, starting from the complete data matrix. As with *Diana*, the data are split using binary recursive partitioning. Decision tree methods differ from unsupervised divisive hierarchical clustering in that class labels are used to perform the classification. In the decision tree

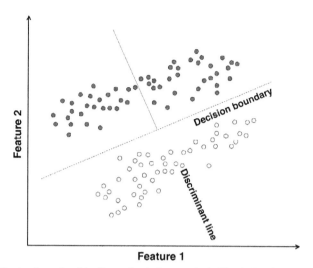

**FIGURE 8.9**  A schematic of the linear discriminate analysis. The dark points represent a set of samples of type A and the light points represent samples of type B. The solid line indicates the linear discriminant that distinguishes the two sample types, while the dotted line represents the discrimination boundary.

approach, nodes are characterized by 'purity'. For example, in the case of two classes: cancer versus normal, with each iteration of a typical algorithm, the data points will be split using the criterion that creates largest and purest nodes, that is the nodes that have greatest separation of cancer and normal samples. The process continues until all data points are in pure nodes and the classification labels are all the same. Tree depth may be limited by the user or tree growth stops when all data points are in their own nodes or they are in 'pure' nodes. Data points are most commonly split using the two following methods to assess node purity:

**(i)** Gini index (Breiman et al., 1984)

$$\text{gini} p_1, p_2, \ldots, p_n = \sum_{i \neq j} p_i p_j$$

where $p$ is the probability of each class in the node of the tree, that if a node is splitting samples into cancer or normal based on a particular criteria, then $p_1$ is the probability of the sample being diseased if the criterion for the given node is true and $p_2$ is the probability of the sample being healthy if the criterion is true.

**(ii)** Information or entropy

$$\text{node entropy} = -p_1 \, log \, p_1 - p_2 \, log \, p_2 - \cdots - p_n \, log \, p_n$$

Again $p$ is the probability of each class in the node of the tree.

The initial phase of tree construction generally results in an overfitted model, as all of the noise and variation in the training set is fitted along with relevant information. As a result, trees need to be 'pruned'. Various methods are known but two commonly used approaches are:

**(i)** Minimal cost complexity (MCC) (Breiman et al. 1984), which prunes the tree from its outermost leaves using the criterion that the pruned tree should minimize an ad hoc function that is the sum of the misclassification rate of the new tree with a measure of the tree complexity. Cross-validation is typically used to determine the misclassification rate (see Section 8.7).

**(ii)** Minimal description length (MDL) is an information or entropy measure. The underlying principle is the assertion that the best model of a set of data is the model that enables the greatest compression of the data. This principle can be used to test for overfitting of data by defining a coding scheme for a decision tree and comparing this to the source data. Pruned trees can be tested for their encoding efficiency.

Pruned trees can then be used as practical classifiers. The decision tree approach has a number of advantages. It is inherently nonparametric, in other words, no assumptions are made regarding the underlying distribution of values of the predictor variables. Thus, the decision tree approach can handle numerical data that are highly skewed or multimodal, as well as categorical predictors with either ordinal or nonordinal structure. With cancer data for example, decision trees allow you to include additional data alongside expression data, such as age, whether patients smoke, whether they have been treated, etc. Furthermore, data such as mutational status may also be included. The decision tree approach is a classification tool that copes well with the low N, high P problem characteristic of most microarray experiments.

The resulting models are inherently visualizable and are thus simple to understand and interpret. As a consequence of the clear logical hierarchy that is generated, decision trees may be regarded as 'white box models' whereas neural networks, in contrast, are 'black box' models since the decision-making process with neural networks is generally not easy for humans to interpret. The technique also requires less data preparation than other statistical techniques that must use dummy variables to represent categorical data and that need blank values to be removed.

Decision tree approaches have been used to model gene-networks (Soinov et al., 2003), that is taking a gene-centric rather than experiment-centric approach. The authors of this study report a number of advantages for such a supervised classification approach. First, it allows the identification of genes affecting the target gene directly from the classifier. Second, there is no requirement to assume any arbitrary discretization thresholds. Third, each data sample is treated as an example and classification algorithms are constructed in such a way that they learn from these examples. Decision trees have also been used to identify 'expression signatures' that are significant in microarray data (Boulesteix et al., 2003) and for the more usual problem of cancer classification (Guo et al., 2005).

'Random Forests' are a variation of the basic decision tree approach that addresses the problem of finding the optimum tree without overfitting the data by generating a number of trees for a dataset that are typically not pruned. New instances can then be classified by a 'vote', that is the classification of the new instance by the forest is the most common classification of the various trees that make up the forest. Alternatively, the best tree can be selected. The trees that make up the forest are generated by splitting the starting dataset into a number of random subsets for which individual trees are generated. In addition, trees may be made using subsets of the variables that make up each data point. In practice this means that the random forest approach can find the most significant variables (genes) in the data (Diaz-Uriarte and Alvarez de Andres, 2006). Random forest approaches have been used for diagnostic testing of childhood leukemias (Hoffmann et al., 2006), classification of renal cell carcinoma subtypes (Shi et al., 2005) and for more general analysis of expression data in conjunction with wider clinical metadata (Eichler et al., 2007).

### 8.6.5 Neural Networks

Artificial neural networks encompass a range of tools that exploit a virtual network of simple processing elements (neurons or nodes), which can model complex data by varying the connections between the processing elements and by varying what each element does to the signals in the network. Neural networks are typically adaptive, that is they come with algorithms designed to alter the strength (weights) of the connections in the network. Genetic algorithms are particularly good for this purpose (Chiang and Chao, 2007). In principle, neural networks can model data that is difficult to deal with using linear fitting (LDA) or fitting hyperplanes (SVMs). However, depending on the implementation of the network, the resulting models can be difficult to interpret.

### 8.7 ASSESSING MODEL QUALITY

A good classification tool should have a good true positive detection rate, in other words it should have high sensitivity. In addition, one would hope for both low false positive and false negative rates: the classification should have high specificity. Statisticians refer to false positives as type I error and false negatives as type II error and good classifiers minimize both types of error. Various standard statistical tools for assessing the strength of a classifier exist including the F-statistic (see Box 8.5 for further details). Another commonly used tool is the receiver operating characteristic (ROC) curve. An ROC curve

> **Box 8.5** The F-statistic or F-measure (van Rijsbergen, 1979) is a useful metric to quantify the success of a classifier
>
> $$F = \frac{2PR}{P + R} \text{ where } P = \frac{TP}{TP + FP} \text{ and } R = \frac{TP}{TP + FN}$$
>
> where TP is the true positive rate, FP is the false positive rate and FN is the false negative rate.

simply plots the true positive rate against the false positive rate. For a discrete classifier like a decision tree, this represents a point on the graph but probabilistic models, such as neural networks, give a curve dependent on the threshold values used (for examples see: De Smet et al., 2004; Tsai and Chen, 2004; Reid et al., 2005).

Cross-validation should also be used, if possible, to test the robustness of any classifier (Molinaro et al., 2005). In cross-validation the training data should be split into at least two subsets. One subset is used to create the classifier and the second is used to test that the classifier gives the correct predictions. In more sophisticated approaches, the data is split into $k$ subsets. One subset is retained for testing while the remaining $k - 1$ subsets of samples are used to create the classifier. This process is repeated $k$ times retaining a different sample for validation each time. The mean error rate can then be calculated. A number of other variations on the theme are possible but the premise is the same, a subset of data is held back to check the correctness of the classifier. However, it should be remembered that cross-validation works best when there are a substantial number of samples in the training data (Braga-Neto and Dougherty, 2004). Cross-validation for dealing with small sample sizes are being developed though (Fu et al., 2005).

## 8.8 SUMMARY

To conclude, there is a bewildering array of tools available now for clustering. The appropriate analysis tool needs to be chosen to suit the task, that is supervised or unsupervised, whether the analysis is a first pass for visualization or whether a detailed analysis of significance is to be assessed.

- Unsupervised clustering attempts to find similarities in the data analyzed but for complex data like expression matrices there is almost never a 'correct' clustering or structure to the data (except when data is prestructured for testing purposes). Thus the choice of tool to use is somewhat arbitrary, but it is important to understand the tool and how it finds similarities in the raw data so that it is clearly understood why data is grouped in the way that it is.

- Agglomerative hierarchical clustering is a typical starting point for unsupervised analysis but should be regarded more as a visualization tools to be used prior to more rigorous analysis. That is not to say it should not be used: it is essential to have a good look at your data before embarking on detailed analysis but it is important to remember the peculiarities of agglomerative clustering and also the nature of the dendrograms that are produced.
- Care should be taken in applying new algorithms. Many tools are developed by researchers whose primary interest is development of novel algorithms rather than application of the algorithm and many tools are not thoroughly assessed in a wide range of contexts. If a new algorithm is to be employed, it should be assessed carefully first.
- The choice of tool is perhaps less important than checking the results carefully for significance and there are numerous tools and metrics available to help, but again these tools and metrics measure success according to particular criteria and it is essential to bear these in mind when interpreting results.
- There are advantages in using tools that allow control over the significance of the results such as quality threshold clustering or SOTA.
- It is common practice to try more than one approach and then choose the tool that seems most appropriate. However, this can lead to very subjective assessments based on preconceptions about the results and it is important to use impartial cluster validation tools to check that the chosen clustering tool is giving tight, stable and significant clusters and just as importantly, that they are biologically meaningful. Checking results by resampling, cross-validation, Monte Carlo methods to assess probabilities of chance results and other tools is essential
- For supervised clustering, it is essential that any classifiers generated is tested properly, ideally by cross-validation, ROC curves, $F$-measures and other tools.
- As a final point, it is important to bear in mind that a full statistical validation of clustering requires a meaningful number of samples, the more the better. In particular, this issue needs to be considered at the design stage before embarking on the experimental work.

## REFERENCES

Alexe G, Dalgin GS, Ganesan S, Delisi C, Bhanot G. (2007) Analysis of breast cancer progression using principal component analysis and clustering. *J Biosci* **32:** 1027–1039.

Alizadeh AA, Eisen MB, Davis RE, Ma C, Lossos IS, Rosenwald A, Boldrick JC, Sabet H, Tran T, Yu X, Powell JI, Yang L, Marti GE, Moore T, Hudson J, Lu L, Lewis DB, Tibshirani R, Sherlock G, Chan WC, Greiner TC, Weisenburger DD, Armitage JO, Warnke R, Levy R, Wilson W, Grever MR, Byrd JC, Botstein D, Brown PO, Staudt LM. (2000) Distinct types of diffuse large B-cell lymphoma identified by gene expression profiling. *Nature* **403:** 503–511.

Barrier A, Roser F, Boelle PY, Franc B, Tse C, Brault D, Lacaine F, Houry S, Callard P, Penna C, Debuire B, Flahault A, Dudoit S, Lemoine A. (2007) Prognosis of stage II colon cancer by non-neoplastic mucosa gene expression profiling. *Oncogene* **26**: 2642–2648.

Boulesteix AL, Tutz G, Strimmer K. (2003) A CART-based approach to discover emerging patterns in microarray data. *Bioinformatics* **19**: 2465–2472.

Braga-Neto UM, Dougherty ER. (2004) Is cross-validation valid for small-sample microarray classification?. *Bioinformatics* **20**: 374–380.

Breiman L, Friedman JH, Olshen RA, Stone CJ. (1984) *Classification and Regression Trees*. Wadsworth, Monterey, CA.

Brock B, Pihur V, Datta S, Datta S. (2008) clValid: An *R* package for cluster validation. *J Stat Software* **25**: 4.

Carpentier AS, Riva A, Tisseur P, Didier G, Henaut A. (2004) The operons, a criterion to compare the reliability of transcriptome analysis tools: ICA is more reliable than ANOVA, PLS and PCA. *Comput Biol Chem* **28**: 3–10.

Chen G, Jaradat SA, Bannerjee N. (2002) Evaluation and comparison of clustering algorithms in analyzing ES cell gene expression data. *Stat Sinica* **12**: 241–262.

Chiang JH, Chao SY. (2007) Modeling human cancer-related regulatory modules by GA-RNN hybrid algorithms. *BMC Bioinform* **8**: 91.

Clare A, King RD. (2002) How well do we understand the clusters found in microarray data?. *In Silico Biol* **2**: 511–522.

Costa JA, Netto ML. (1999) Estimating the number of clusters in multivariate data by self-organizing maps. *Int J Neural Syst* **9**: 195–202.

Covell DG, Wallqvist A, Rabow AA, Thanki N. (2003) Molecular classification of cancer: unsupervised self-organizing map analysis of gene expression microarray data. *Mol Cancer Ther* **2**: 317–332.

Csillag C, Nielsen OH, Borup R, Nielsen FC, Olsen J. (2007) Clinical phenotype and gene expression profile in Crohn's disease. *Am J Physiol Gastrointest Liver Physiol* **292**: G298–304.

Datta S, Datta S. (2003) Comparisons and validation of statistical clustering techniques for microarray gene expression data. *Bioinformatics* **19**: 459–466.

Datta S, Datta S. (2006) Methods for evaluating clustering algorithms for gene expression data using a reference set of functional classes. *BMC Bioinformatics* **7**: 397.

De Smet F, Mathys J, Marchal K, Thijs G, De Moor B, Moreau Y. (2002) Adaptive quality-based clustering of gene expression profiles. *Bioinformatics* **18**: 735–746.

De Smet F, Moreau Y, Engelen K, Timmerman D, Vergote I, De Moor B. (2004) Balancing false positives and false negatives for the detection of differential expression in malignancies. *Br J Cancer* **91**: 1160–1165.

Dembele D, Kastner P. (2003) Fuzzy C-means method for clustering microarray data. *Bioinformatics* **19**: 973–980.

Diaz-Uriarte R, Alvarez de Andres S. (2006) Gene selection and classification of microarray data using random forest. *BMC Bioinform* **7**: 3.

Ding C, He X. (2004) K-means Clustering via Principal Component Analysis. *ICML'04: Proceedings of the Twenty-first International Conference on Machine Learning.*, pp. 225–232.

Dunn JC. (1974) Well separated clusters and optimal fuzzy partitions. *J Cybern* **4**: 95–104.

Eichler GS, Reimers M, Kane D, Weinstein JN. (2007) The LeFE algorithm: embracing the complexity of gene expression in the interpretation of microarray data. *Genome Biol* **8**: R187.

Fu WJ, Carroll RJ, Wang S. (2005) Estimating misclassification error with small samples via bootstrap cross-validation. *Bioinformatics* **21**: 1979–1986.

Gana Dresen IM, Boes T, Huesing J, Neuhaeuser M, Joeckel KH. (2008) New resampling method for evaluating stability of clusters. *BMC Bioinform* **9**: 42.

Guo Z, Zhang T, Li X, Wang Q, Xu J, Yu H, Zhu J, Wang H, Wang C, Topol EJ, Wang Q, Rao S. (2005) Towards precise classification of cancers based on robust gene functional expression profiles. *BMC Bioinform* **6**: 58.

Handl J, Knowles J. (2005) Exploiting the trade-off – the benefits of multiple objectives in data clustering. *Proceeding of the Third International Conference on Evolutionary Multi-criterion Optimization.*Lecture Notes in Computer Science 3410 Springer, pp. 547–560.

Handl J, Knowles J, Kell DB. (2005) Computational cluster validation in post-genomic data analysis. *Bioinformatics* **21**: 3201–3212.

Hastie T, Tibshirani R, Eisen MB, Alizadeh A, Levy R, Staudt L, Chan WC, Botstein D, Brown P. (2000) 'Gene shaving' as a method for identifying distinct sets of genes with similar expression patterns. *Genome Biol* **1**: R0003.

Herrero J, Valencia A, Dopazo J. (2001) A hierarchical unsupervised growing neural network for clustering gene expression patterns. *Bioinformatics* **17**: 126–136.

Heyer LJ, Kruglyak S, Yooseph S. (1999) Exploring expression data: identification and analysis of coexpressed genes. *Genome Res* **9**: 1106–1115.

Hoffmann K, Firth MJ, Beesley AH, de Klerk NH, Kees UR. (2006) Translating microarray data for diagnostic testing in childhood leukaemia. *BMC Cancer* **6**: 229.

Hong Y, Ho KS, Eu KW, Cheah PY. (2007) A susceptibility gene set for early onset colorectal cancer that integrates diverse signaling pathways: implication for tumorigenesis. *Clin Cancer Res* **13**: 1107–1114.

Hsu AL, Tang SL, Halgamuge SK. (2003) An unsupervised hierarchical dynamic self-organizing approach to cancer class discovery and marker gene identification in microarray data. *Bioinformatics* **19**: 2131–2140.

Hubert L, Arabie P. (1985) Comparing partitions. *J Classification* **2**: 193–218.

Ishida N, Hayashi K, Hoshijima M, Ogawa T, Koga S, Miyatake Y, Kumegawa M, Kimura T, Takeya T. (2002) Large scale gene expression analysis of osteoclastogenesis in vitro and elucidation of NFAT2 as a key regulator. *J Biol Chem* **277**: 41147–41156.

Jacobsen M, Repsilber D, Gutschmidt A, Neher A, Feldmann K, Mollenkopf HJ, Ziegler A, Kaufmann SH. (2007) Candidate biomarkers for discrimination between infection and disease caused by *Mycobacterium tuberculosis*. *J Mol Med* **85**: 613–621.

Jaeger J, Koczan D, Thiesen HJ, Ibrahim SM, Gross G, Spang R, Kunz M. (2007) Gene expression signatures for tumor progression, tumor subtype, and tumor thickness in laser-microdissected melanoma tissues. *Clin Cancer Res* **13**: 806–815.

Jain R, Mazumdar J. (2003) A genetic algorithm based nearest neighbor classification to breast cancer diagnosis. *Austral Phys Eng Sci Med* **26**: 6–11.

Kaufman L, Rousseeuw PJ. (1990) *Finding Groups in Data: An Introduction to Cluster Analysis.* John Wiley & Sons, New York.

Kondoh N, Ohkura S, Arai M, Hada A, Ishikawa T, Yamazaki Y, Shindoh M, Takahashi M, Kitagawa Y, Matsubara O, Yamamoto M. (2007) Gene expression signatures that can discriminate oral leukoplakia subtypes and squamous cell carcinoma. *Oral Oncol* **43**: 455–462.

Lee SI, Batzoglou S. (2003) Application of independent component analysis to microarrays. *Genome Biol* **4**: R76.

Lequerre T, Gauthier-Jauneau AC, Bansard C, Derambure C, Hiron M, Vittecoq O, Daveau M, Mejjad O, Daragon A, Tron F, Le Loet X, Salier JP. (2006) Gene profiling in white blood cells predicts infliximab responsiveness in rheumatoid arthritis. *Arthritis Res Ther* **8**: R105.

Leung YY, Chang CQ, Hung YS, Fung PW. (2006) Gene selection for brain cancer classification. *Conf Proc IEEE Eng Med Biol Soc* **1:** 5846–5849.

Levine E, Domany E. (2001) Resampling method for unsupervised estimation of cluster validity. *Neural Comput* **13:** 2573–2593.

Liao JC, Boscolo R, Yang YL, Tran LM, Sabatti C, Roychowdhury VP. (2003) Network component analysis: reconstruction of regulatory signals in biological systems. *Proc Natl Acad Sci U S A* **100:** 15522–15527.

Liebermeister W. (2002) Linear modes of gene expression determined by independent component analysis. *Bioinformatics* **18:** 51–60.

Ma Y, Qian Y, Wei L, Abraham J, Shi X, Castranova V, Harner EJ, Flynn DC, Guo L. (2007) Population-based molecular prognosis of breast cancer by transcriptional profiling. *Clin Cancer Res* **13:** 2014–2022.

McShane LM, Radmacher MD, Freidlin B, Yu R, Li MC, Simon R. (2002) Methods for assessing reproducibility of clustering patterns observed in analyzes of microarray data. *Bioinformatics* **18:** 1462–1469.

Molinaro AM, Simon R, Pfeiffer RM. (2005) Prediction error estimation: a comparison of resampling methods. *Bioinformatics* **21:** 3301–3307.

Pinto FR, Carrico JA, Ramirez M, Almeida JS. (2007) Ranked Adjusted Rand: integrating distance and partition information in a measure of clustering agreement. *BMC Bioinform* **8:** 44.

Rand WM. (1971) Objective criteria for the evaluation of clustering methods. *J Am Stats Assoc* **66:** 846–850.

Reid JF, Lusa L, De Cecco L, Coradini D, Veneroni S, Daidone MG, Gariboldi M, Pierotti MA. (2005) Limits of predictive models using microarray data for breast cancer clinical treatment outcome. *J Natl Cancer Inst* **97:** 927–930.

Rousseeuw PJ. (1987) Silhouettes: a graphical aid to the interpretation and validation of cluster analysis. *J Comput Appl Math* **20:** 53–65.

Ryu B, Kim DS, Deluca AM, Alani RM. (2007) Comprehensive expression profiling of tumor cell lines identifies molecular signatures of melanoma progression. *PLoS ONE* **2:** e594.

Saidi SA, Holland CM, Charnock-Jones DS, Smith SK. (2006) In vitro and in vivo effects of the PPAR-alpha agonists fenofibrate and retinoic acid in endometrial cancer. *Mol Cancer* **5:** 13.

Saidi SA, Holland CM, Kreil DP, MacKay DJ, Charnock-Jones DS, Print CG, Smith SK. (2004) Independent component analysis of microarray data in the study of endometrial cancer. *Oncogene* **23:** 6677–6683.

Scharl T, Leisch F. (2006) The stochastic QT–clust algorithm: evaluation of stability and variance on time–course microarray data. In: Rizzi A, Vichi M, editors. *COMPSTAT 2006 – Proceedings in Computational Statistics*. Physica-Verlag, Heidelberg, pp. 1015–1022.

Shi T, Seligson D, Belldegrun AS, Palotie A, Horvath S. (2005) Tumor classification by tissue microarray profiling: random forest clustering applied to renal cell carcinoma. *Mod Pathol* **18:** 547–557.

Soinov LA, Krestyaninova MA, Brazma A. (2003) Towards reconstruction of gene networks from expression data by supervised learning. *Genome Biol* **4:** R6.

Stacklies W, Redestig H, Scholz M, Walther D, Selbig J. (2007) PCAMethods – a bioconductor package providing PCA methods for incomplete data. *Bioinformatics* **23:** 1164–1167.

Stansberg C, Vik-Mo AO, Holdhus R, Breilid H, Srebro B, Petersen K, Jorgensen HA, Jonassen I, Steen VM. (2007) Gene expression profiles in rat brain disclose CNS signature genes and regional patterns of functional specialisation. *BMC Genomics* **8:** 94.

Tamayo P, Slonim D, Mesirov J, Zhu Q, Kitareewan S, Dmitrovsky E, Lander ES, Golub TR. (1999) Interpreting patterns of gene expression with self-organizing maps: methods and application to hematopoietic differentiation. *Proc Natl Acad Sci U S A* **96:** 2907–2912.

Tárraga J, Medina I, Carbonell J, Huerta-Cepas J, Minguez P, Alloza E, Al-Shahrour F, Vegas-Azcárate S, Goetz S, Escobar P, Garcia-Garcia F, Conesa A, Montaner D, Dopazo J. (2008) GEPAS, a web-based tool for microarray data analysis and interpretation. *Nucleic Acids Res* **36:** W308–W314.

Teschendorff AE, Journee M, Absil PA, Sepulchre R, Caldas C. (2007) Elucidating the altered transcriptional programs in breast cancer using independent component analysis. *PLoS Comput Biol* **3:** e161.

Tibshirani R, Walther G, Hastie T. (2001) Estimating the number of clusters in a data set via the gap statistic. *J R S Soc B* **63:** 411–423.

Tsai CA, Chen JJ. (2004) Significance analysis of ROC indices for comparing diagnostic markers: applications to gene microarray data. *J Biopharm Stat* **14:** 985–1003.

van der Laan MJ, Pollard KS. (2003) A new algorithm for hybrid hierarchical clustering with visualization and the bootstrap. *J Stat Plan Inference* **117:** 275–303.

van Rijsbergen C. (1979) *Information Retrieval*, 2nd edition. Butterworths, London.

Wang Z, Wang Y, Xuan J, Dong Y, Bakay M, Feng Y, Clarke R, Hoffman EP. (2006) Optimized multilayer perceptrons for molecular classification and diagnosis using genomic data. *Bioinformatics* **22:** 755–761.

Wei SH, Balch C, Paik HH, Kim YS, Baldwin RL, Liyanarachchi S, Li L, Wang Z, Wan JC, Davuluri RV, Karlan BY, Gifford G, Brown R, Kim S, Huang TH, Nephew KP. (2006) Prognostic DNA methylation biomarkers in ovarian cancer. *Clin Cancer Res* **12:** 2788–2794.

Wiesinger-Mayr H, Vierlinger K, Pichler R, Kriegner A, Hirschl AM, Presterl E, Bodrossy L, Noehammer C. (2007) Identification of human pathogens isolated from blood using microarray hybridisation and signal pattern recognition. *BMC Microbiol* **7:** 78.

Yao Z, Ruzzo WL. (2006) A regression-based *K* nearest neighbor algorithm for gene function prediction from heterogeneous data. *BMC Bioinform* **7:** S11.

Yau C, Fedele V, Roydasgupta R, Fridlyand J, Hubbard A, Gray JW, Chew K, Dairkee SH, Moore DH, Schittulli F, Tommasi S, Paradiso A, Albertson DG, Benz CC. (2007) Aging impacts transcriptomes but not genomes of hormone-dependent breast cancers. *Breast Cancer Res* **9:** R59.

Yeung KY, Haynor DR, Ruzzo WL. (2001) Validating clustering for gene expression data. *Bioinformatics* **17:** 309–318.

Yin L, Huang CH, Ni J. (2006) Clustering of gene expression data: performance and similarity analysis. *BMC Bioinform* **7:** S19.

Zhang XW, Yap YL, Wei D, Chen F, Danchin A. (2005) Molecular diagnosis of human cancer type by gene expression profiles and independent component analysis. *Eur J Hum Genet* **13:** 1303–1311.

Zhao YP, Chen G, Feng B, Zhang TP, Ma EL, Wu YD. (2007) Microarray analysis of gene expression profile of multidrug resistance in pancreatic cancer. *Chin Med J (Eng)* **120:** 1743–1752.

# Microarray Data Repositories and Warehouses

Microarray technology produces enormous amounts of complex data and, as discussed elsewhere in this book, there are many different commercial or in house microarray platforms as well as a wide range of analytical protocols in current use. If we are to fully capitalize on this wealth of data it is important that the scientific community can store and communicate their experimental results to each other in a common language. There are several obvious motivations for this including (i) the verification, comparison and recycling of published data using identical or alternative methods of analysis (ii) facilitating the development of novel algorithms for analyzing data from single or multiple data sets that are from the same or different array platforms (iii) helping integration and analysis of expression data with CHIP-array or array-CGH data or with data derived from other sources such as, for example, genetic interaction, protein interaction or proteomics studies. Here we briefly review the data capture and annotation standards being developed for microarray data as well as the public repositories being built to store these data.

## 9.1 INTRODUCTION

When sharing data between scientists or computer applications, it is helpful to have common standards in terminology for both the production and presentation of the data. For example, both the terms *probe* and *target* have been used to mean either the DNA on the array or the labeled sample in solution and this can be potentially misleading. In order to design and implement computer software or databases to handle and integrate data from a variety of sources, it is important to have an agreed set of standards from which to work. There are three main areas where microarray data standards are required:

i.  *Recording data.* Effective storage and retrieval of microarray data requires a formalized way for capturing details of array design, probe sequence, experimental process, and the gene expression data itself. Developing such a formal language is the main focus of the MIAME initiative (minimal information about a microarray experiment; Brazma et al., 2001). When publishing and submitting microarray data to a public data repository, it is now generally accepted that the data and experimental records be MIAME compliant.

ii. *Describing data.* If microarray data are to be broadly accessible, experimental details need to be described using widely understood terminology. This involves the use of standardized ontologies that provide controlled vocabularies and formal relationships describing, for example, microarray experiments, array features, experimental and analytical protocols, genes, samples, etc. Developing such an ontology is one of the goals of the Microarray Gene Expression Data (MGED) Society (Ball et al., 2004), which was founded in 1999 to establish standards for microarray data annotation and enable the development of public database repositories. The Gene Ontology (GO) Consortium (www.geneontology.org) is a broader initiative, established in 1999 to develop a uniform terminology for describing gene products and their functions that is independent of species or biological system. There are three categories of GO annotation: molecular function, biological process and cellular component (Ashburner et al., 2000, 2003; Harris et al., 2004; Consortium, 2008).

iii. *Implementing software.* The main aim of the MAGE (MicroArray Gene Expression) group is the development of a system for unifying the storage and exchange of microarray data between different data systems. Using MIAME and standard ontologies, the MAGE project provides object models, for example MAGE-OM (MAGE Object Model), exchange languages, for example MAGE-ML (MAGE mark-up language) and software modules, for example MAGE-stk (MAGE software toolkit) a collection of packages that act as converters between MAGE-OM and MAGE-ML) for a variety of programming platforms. Due to the complexity of MAGE-ML, a simplified system, MAGE-TAB (Rayner et al., 2006) was developed to enable laboratories with little bioinformatics experience or support to manage, exchange and submit well-annotated microarray data in a standard format using a simple spreadsheet.

There are three major databases recommended by MGED for the deposition of microarray data: ArrayExpress (Brazma et al., 2003), Gene Expression Omnibus (GEO) (Edgar et al., 2002) and Center for Information Biology gene EXpression database (CiBEX) (Tateno and Ikeo, 2004). In this chapter, we briefly review the former two, since these are the most popular, describing how data can be submitted and how they provide and present data to the research community. These databases serve as data repositories for publicly available archived experimental data as well as warehouses of gene expression profiles. The difference between these functions is that data repositories typically store raw data and information about experiments, whereas warehouses store normalized and analyzed experimental data to facilitate gene-based queries of expression profiles across a selection of different data sets. We will also touch upon other databases containing microarray-based data. However, many of these are integrated data resources focused on particular biological areas.

## 9.2 ARRAYEXPRESS

In 2003, the European Bioinformatics Institute (EBI) established ArrayExpress (Brazma et al., 2003; Sarkans et al., 2005; Parkinson et al., 2007), a MIAME supportive public database (http://www.ebi.ac.uk/arrayexpress). Since then, it has grown quite rapidly in size, especially since journals are becoming more forceful in requiring submission of micorarray data to public repositories. To date, (April 2008), 3415 experiments and 101 092 hybridizations are available in ArrayExpress. The stored experiments cover studies involving over 200 different organisms, the largest datasets representing experiments with human, mouse, Arabidopsis, yeast and rat samples. Although the majority of the data is related to gene expression studies, there are an increasing number of studies from CGH array studies and CHIP-on chip experiments.

### 9.2.1 Data Submission

Data can be submitted to the ArrayExpress repository by one of the following routes:

i. *Web-based submission*: this involves completing a series of online forms describing samples and protocols, then uploading data files for each hybridization via MIAMExpress. This route is best suited to small and medium sized experiments since a maximum of 50 hybridizations can be uploaded at one time using a batch upload tool.

ii. *Spreadsheet submission*: in this route a spreadsheet is downloaded, completed and uploaded together with data files. The spreadsheet uses a simple 'one-row-per-channel' model and is suitable for a wide-range of array designs and technologies. It is recommended for large experiments

(50+ hybridizations) or if large amounts of sample annotation are already in spreadsheet format (e.g. a table of clinical history information).

**iii.** *External microarray database submission*: this route requires that the submitter has their own MIAME compliant database. Data is exported automatically using the MAGE-ML or MAGE-TAB submission pipelines.

## 9.2.2 Data Curation

Following submission, the curation team at ArrayExpress process the data before it is uploaded into the database repository. This involves checking the data for MIAME compliance in terms of the completeness of the information required and for data consistency, for example, checks are made to see if the data files are complete and uncorrupted and whether data files match the specified array designs. The process typically takes a few days and ArrayExpress provides an accession number for the experiment as well as a reviewer login account to allow prerelease access to the data in the repository. Password-protected access to private data can also be set up for data that is not yet published. Many journals require accession numbers for microarray data before acceptance of a paper for publication and it is recommended that data are submitted well in advance of this.

After loading into ArrayExpress all experiments are given an automatically generated MIAME compliance score, between 0 and 5, where 0 means that none of the MIAME requirements were met and 5 means that all requirements were met. This is an assessment score for reviewers and other users to check how much MIAME-required information was provided for an experiment.

## 9.2.3 Data Retrieval

ArrayExpress organizes data by experiments, a collection of array hybridizations related to a particular study, that are usually linked to a publication. The experiments can be queried to find out more information about particular experiments or to retrieve data using the query interface to the ArrayExpress Repository (Figures 9.1 and 9.2). Data files can be downloaded as either tab-delimited text files for two-color experiments or the standard raw data file formats associated with the array technology, for example Affymetrix CEL format files provided within zip archives. Graphical representation of each experiment can also be retrieved, as can spreadsheets containing sample annotations and MAGE-ML format files from the ArrayExpress ftp site.

## 9.2.4 The Data Warehouse

The ArrayExpress Warehouse contains 7 921 771 gene expression profiles from 350 experiments comprising a total of 10 096 hybridizations populated from the ArrayExpress Repository (February 2008 statistics). The warehouse allows the user to search for one or more gene expression profiles by the gene name or

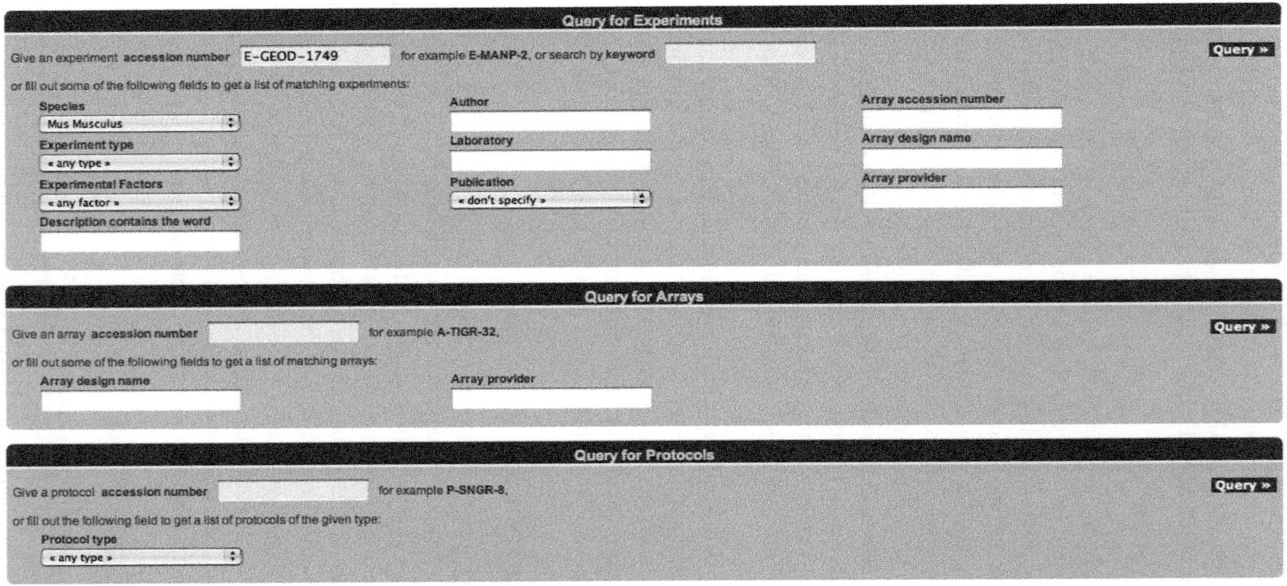

**FIGURE 9.1**   ArrayExpress query interface. Data may be retrieved using a variety of queries. In this example, a particular experiment with the *accession number* E-GEOD-1749. Other ways to query experimental data include searches using *keywords* such as 'mouse development', 'embryonic stem cell' or 'leukemia' or by species, author, etc. Queries can also be made for *Arrays* or for *Protocols* via an accession number or other fields as indicated on the form.

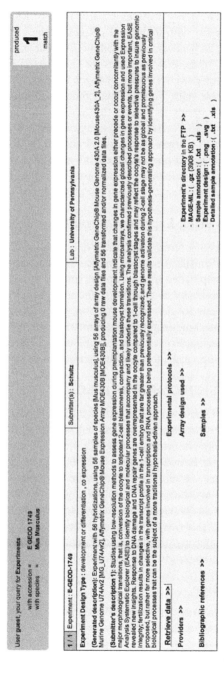

**FIGURE 9.2**    ArrayExpress data retrieval. The result returned when querying for an experiment using the *accession number*, E-GEOD-1749. A brief summary of the record is displayed, further information can be retrieved for the experiment; including array design, experimental protocols, experimental design, sample annotation, raw data, etc., using the links at the bottom right.

identifier, database accession number (e.g. UniProt) or other annotation such as Gene Ontology terms. The query involves a search for profiles across the entire dataset and generates a list with thumbnail images of the expression profiles (Figures 9.3 and 9.4). The lists can be ranked by different statistical criteria to measure the 'relevance' of these experiments for the selected genes. For example, genes can be selected to view any changes in particular experimental conditions such as disease state. Further developments are reported to include the implementation of queries to search for genes with similar expression profiles (Parkinson et al., 2007). Additional array annotations in ArrayExpress include data from Ensembl (Birney et al., 2006; Flicek et al., 2008) and UniProt (Hull et al., 2006), which also provide access to gene expression data in the ArrayExpress warehouse from their own databases via a DAS track and cross-references, respectively.

### 9.2.5 Data Analysis using ExpressionProfiler

From the EBI website an online microarray data analysis tool, Expression Profiler, can be accessed and used either to analyze data retrieved from ArrayExpress or to analyze data uploaded from any other source, such as the user's own local private data (Kapushesky et al., 2004). The tool provides a workflow of several linked components for data normalization, statistical analysis, clustering, pattern discovery, machine-learning algorithms and visualization. The most commonly used Bioconductor packages are gradually being integrated with ExpressionProfiler, thereby providing a web interface to these tools.

### 9.3 GENE EXPRESSION OMNIBUS (GEO)

The Gene Expression Omnibus (GEO), established in 1999, is currently the largest fully public gene expression resource and is accessible at the National Center for Biotechnology Information (NCBI): http://www.ncbi.nlm.nih.gov/geo/ (Edgar et al., 2002; Barrett and Edgar, 2006; Barrett et al., 2007). The repository and warehouse contains data from single- and dual-channel microarray-based experiments measuring mRNA, miRNA, genomic DNA (CGH, ChIP-chip, and SNP) and protein abundance, as well as non-array technologies such as serial analysis of gene expression (SAGE), mass spectrometry peptide profiling, and various forms of massively parallel sequencing technologies. To date (April 2008), there are 8308 series records and 214 400 sample records stored in GEO.

### 9.3.1 Data Submission

There are five different ways to submit data and these are summarized in Table 9.1. Data submitted to GEO are stored as three main entity types:

i. *Platform record*: describes the list of elements on the array (cDNAs, oligonucleotide probes, ORFs, antibodies) or the list of elements that

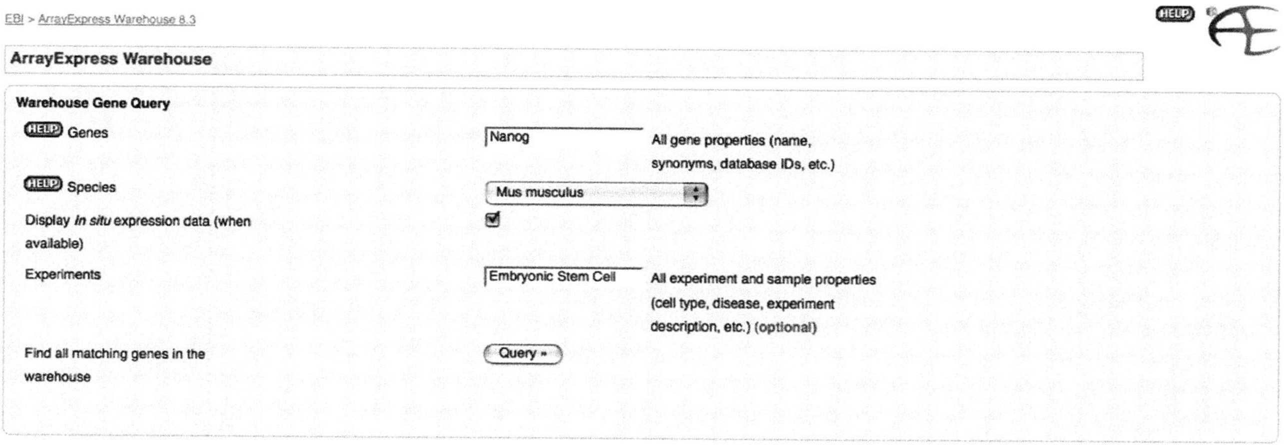

EBI > ArrayExpress Warehouse 8.3

## ArrayExpress Warehouse

**Warehouse Gene Query**

HELP **Genes**

Nanog

All gene properties (name, synonyms, database IDs, etc.)

HELP **Species**

Mus musculus

Display *in situ* expression data (when available)

Experiments

Embryonic Stem Cell

All experiment and sample properties (cell type, disease, experiment description, etc.) (optional)

Query »

Find all matching genes in the warehouse

HELP

**FIGURE 9.3** ArrayExpress warehouse. An example of the query interface searching for the Nanog gene in any mouse experimental data associated with the keyword *embryonic stem cell*. There is also the option to display any in situ expression data if it is available.

**FIGURE 9.4**   ArrayExpress warehouse data example of the data retrieved from the data warehouse when querying for Nanog expression in mouse embryonic stem cell experiments. A total of 36 relevant experiments are retrieved, which are ranked from the lowest *p*-value. In the graph, the *y*-axis represents the expression value while the *x*-axis shows all the arrays or samples in the experiment (one bar for each array), which are grouped by condition or treatment type.

**TABLE 9.1** Summary of GEO Data Submission Method Options

| Option | Format | Comments |
|---|---|---|
| Web deposit | Web forms | A step-by-step interactive web form used for submitting individual records for a small number of samples |
| GEOarchive or SOFTmatrix | Spreadsheets, for example Excel | A batch submission for a large number of samples; particularly Affymetrix submissions. If you use RMA or another program that generates value matrices, the GEOarchive deposit option is recommended |
| SOFT (Simple Omnibus Format in Test): SOFTtext and SOFTmatrix | Plain text | A rapid batch submission for a large number of samples, particularly if your data is already in a database or spreadsheet, or if you have any code to format all your data into one large SOFT formatted file. There are two versions of SOFT: SOFT*text*, where SOFT-formatted data are organized as concatenated records and SOFT*matrix*, where SOFT-formatted data are organized side-by-side as a matrix table, usually in a spreadsheet |
| MINiML (MIAME Notation in Markup Language) | XML | A data exchange format for a rapid batch submission for samples already in a MIAME-compliant database. Database must be able to export XML MINiML format |
| MAGE-ML | XML | A data exchange format for rapid batch submission of a large number of samples already in a database. Records must be exported as MIAME compliant MAGE-ML dump |

may be detected and quantified in that experiment (e.g., SAGE tags, peptides). Each Platform record is given a unique GEO accession number (GPLxxx) and may reference many Sample records submitted by multiple submitters.

**ii.** *Sample record*: describes the conditions under which an individual sample was handled, the manipulations it underwent and the abundance measurement of each element derived from it. Each Sample record, which is given a unique GEO accession number (GSMxxx), references only one Platform record and may be included in multiple Series.

**iii.** *Series record*: summarizes an experiment and is defined as a record that links together a group of related samples and provides a description of the whole study. A Series record is given a unique GEO accession number (GSExxx) and may also contain tables describing extracted data, summary conclusions, or analyses.

## 9.3.2 Data Curation

Once series records are submitted, the data are both automatically and manually parsed by the curation team, who check the data for MIAME compliance in terms of the completeness of the information required as well as for data consistency. The validation process typically takes a few days. Note that processing times for MAGE-ML submission can be substantially longer than for the other deposit routes. GEO provides an accession number for each experiment and a reviewer login account to view the data in the repository. Through GEO accounts, researchers can update or edit their records at any time and password-protected access to private data can also be set up if data is yet to be published. Any accompanying supplementary data and file types are linked from each record and stored on an FTP server.

GEO curators do not score the degree of MIAME compliant data as it is thought to be impractical given the huge diversity of biological themes, forms of technology, processing techniques and statistical transformations applied to microarray data (Barrett et al., 2007). The GEO philosophy is that the completeness, quality and accuracy of the submitted data is ultimately the responsibility of the researcher and this type of validation process is thought to benefit from journal editorial reviewer feedback or funding agency enforcement.

For certain submissions, the GEO staff reassemble and organize the data into an upper-level object called a GEO dataset (GDSxxx). A GDS record is essentially a curated set of GEO sample data and represents a collection of biologically and statistically comparable GEO samples. Samples within a GDS record refer to the same Platform record and therefore share a common set of probe elements. Datasets allow for the transformation and normalization of expression measurements from a diverse range of incoming data from multiple unrelated projects into a relatively standardized format for further downstream data analyses. Datasets provide two different views on the data:

i. An *experiment-centered representation* that reflects the entire study and is presented as a *dataset record*. This provides an overview of the experiment, a breakdown of the experimental variables, access to auxiliary objects, several data visualization and analysis tools as well as various download options.

ii. A *gene-centered representation* of quantitative gene expression measurements for one gene across a dataset presented as a *GEO Profile*. The profile provides gene annotation, dataset title, links to auxiliary information and a chart showing the expression level and rank of that gene across each sample in the dataset. All gene annotations are derived from querying sequence identifiers (e.g. GenBank accessions, clone IDs) with the latest versions of the NCBI Entrez Gene and UniGene databases.

### 9.3.3 The GEO Query Interface

NCBI's Entrez search system serves as the basis for most GEO queries and is routinely used by researchers to search other NCBI databases such as UniGene, Gene, PubMed, MapViewer, OMIM and GenBank (Wheeler et al., 2008). In addition, Entrez's powerful linking capabilities are used for intradatabase links to connect GEO gene profiles to annotations from various NCBI databases and, where possible, reciprocal links from various NCBI databases to GEO data. The Entrez interface is straightforward and is used to locate information and data by typing in keywords, various experiment categories or Boolean phrases restricted to supported attribute fields. Advanced Entrez features (shown as a toolbar at the head of all Entrez query and retrieval pages) allow for the generation of complex multipart queries or combination of multiple queries that find common overlaps in retrievals. Entrez query facilities also have a spell-check function as well as automatic term mapping using MeSH translation tables.

As shown in Figure 9.5 there are four main GEO query tools that can be used to locate, analyze and visualize data.

i. *GEO accession* query tool can be used to retrieve data using a valid GEO accession as the search term, for example GPLxxx, GSMxxx, GSExxx and GDSxxx.

ii. *GEO BLAST* tool can be used to query for gene expression profiles based on nucleotide sequence similarity.

iii. *Entrez GEO DataSets* contain experiment-centered data.

iv. *Entrez GEO Profiles* contain gene-centered data. Whereas the datasets query tool is used to locate GEO series, DataSets, or platforms of interest using search terms such as experiment keywords, authors, etc., the profiles query is used to locate and view individual and GEO series, DataSet-specific gene expression profiles of interest with search terms such as gene name, symbol, experiment keywords, etc.

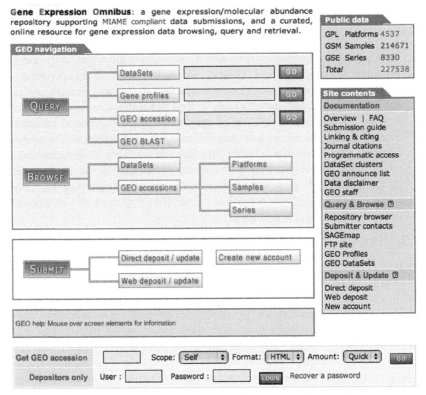

**FIGURE 9.5**   Gene Expression Omnibus. The main GEO query interface providing access to the four main Entrez-base query tools: DataSets, gene profile, GEO accession and GEO BLAST. The main window also provides entry points for browsing the database and for initiating a data submission.

As shown in Figure 9.6, a thumbnail list of expression profiles from various DataSets can be retrieved for a particular gene using the Entrez GEO profiles tool. There are four different links at the top right side of each retrieval:

i. *Profile neighbors link* provides thumbnail list of other genes with similar expression profiles (calculated by Pearson correlation coefficients) within a particular DataSet, suggesting that those genes have a coordinated transcriptional response.

ii. *Homolog neighbors link* connects groups of genes related by HomoloGene group across all DataSets. HomoloGene is an Entrez system that automatically detects homologous genes within sequenced eukaryotic genomes.

iii. *Sequence neighbor link* connects groups of genes related by sequence similarity across all datasets, as determined by BLAST (Altschul et al., 1990), with retrievals ranked in order of decreasing similarity to the selected

**FIGURE 9.6**   Entrez GEO profiles retrieval. The query example is for the gene Nanog using keywords 'embryonic stem cell'. A list of DataSet records is retrieved along with thumbnail images of expression profiles for the gene of interest.

sequence. This feature is useful because it can provide insights into the possible function of the original sequence if it is uncharacterized or it can identify related gene family members or facilitate cross-species comparison.

iv. *Links link* (also found in Entrez GEO DataSets retrievals) connects GEO data to related data from other NCBI Entrez resources such as Gene, OMIM, HomoloGene, SAGEMap, MapViewer, PubMed, GenBank, UniGene, etc.

Subset effect flags, which are specifiable using 'Flag' qualifiers in Entrez GEO Profiles, identify genes that show marked expression level changes between experimental variables. By default, retrievals are ranked based on the presence of these flags to sharpen the identification of potentially interesting genes. When clicking on the gene expression profile thumbnail, a more detailed image of the expression profile is shown (Figure 9.7).

As shown in Figures 9.8 and 9.9, DataSet records contain cluster heat maps, which are interactive images of precomputed unsupervised hierarchical clusters and prespecified $k$-means clusters that allow visualization, selection and download of particular clusters. The images are interactive in that the users can move a box to select a region of interest. DataSet records also contain features to compare 'query group A versus query group B' to identify expression profiles that meet user-defined statistical differences ($t$-test or fold difference) within two condition groups of specified samples within a DataSet.

**Title:** <u>GDS2294</u> / 161072_at / Nanog / Mus musculus
**Summary:** Analysis of embryonic stem cells lacking histone deacetylase 1 (HDAC1). HDACs catalyze the removal of acetyl groups from core histones resulting in chromatin compaction. Results provide insight into the role of HDAC1 in transcriptional regulation.

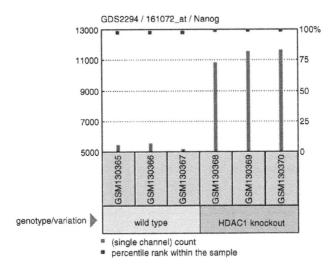

Graph caption help

( Display values )

**Samples:**
> <u>GSM130365</u>: wild_type_embryonic_stem cells_biological_rep1
> <u>GSM130366</u>: wild_type_embryonic_stem cells_biological_rep2
> <u>GSM130367</u>: wild_type_embryonic_stem cells_biological_rep3
> <u>GSM130368</u>: HDAC1_knock_out_embryonic_stem_cells_biological_rep1
> <u>GSM130369</u>: HDAC1_knock_out_embryonic_stem_cells_biological_rep2
> <u>GSM130370</u>: HDAC1_knock_out_embryonic_stem_cells_biological_rep3

**FIGURE 9.7**   An Entrez gene expression profile. The expanded chart shows the gene expression profile within a particular DataSet. The example displays the difference in expression level of the Nanog gene in wildtype compared to HDAC1 knockout mouse samples.

For the benefit of researchers who prefer to use their own analysis tools, all GEO data can be downloaded in bulk via anonymous FTP at ftp://ftp.ncbi.nih. gov/pub/geo/DATA. Available files include SOFT- and MINiML-formatted Platform and Series families, SOFT-formatted DataSets and original supplementary data types. Various software packages are available to handle GEO data formats, including the `GEOquery` package in Bioconductor. This particular package can be used to convert GDS and GSE records to data structures used by other packages in Bioconductor for further analysis: the `limma` and `Biobase` packages uses a `MAList` and `ExpressionSet` data structure, respectively.

**FIGURE 9.8**    A GEO DataSet record. The page summarizes all the available information for a particular dataset and provides entry points for data retrieval or analysis.

## 9.4 OTHER REPOSITORIES AND WAREHOUSES

Although not recommended by MGED for submitting data, there are many other useful repositories and warehouses or integrated resources that are publicly available and some of these are listed in Table 9.2. Also, a listing of microarray data and other gene expression databases is provided by Nucleic Acids Research (NAR) (http://www3.oup.co.uk/nar/database/cat/9) and the latest NAR database issue (January 2008). Among them is the Stanford Microarray Database (SMD; http://smd.stanford.edu), which was established in 1999 and serves as both a research tool and an archive. It is also MIAME-supportive and can export or import MAGE-ML (Gollub et al., 2003; Killion et al., 2003; Demeter et al., 2007). SMD provides a wide selection of web-accessible tools for processing, analyzing, visualizing and sharing microarray data in a research environment and these include: ANOVA analysis to detect spatial and print-tip bias on the array; a tool from BioConductor's `ArrayQuality` package to view diagnostic plots and assess microarray data quality; KNNImpute to estimate missing values in a data set (Troyanskaya et al., 2001); singular value decomposition (SVD)

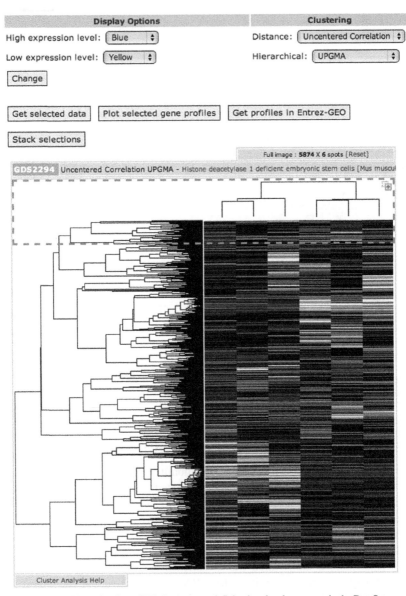

**FIGURE 9.9**   Clustering from GEO dataset record. Selecting the clustergram in the DataSet record (Figure 9.8) brings up the clustering window, which allows the application of different clustering metrics and the selection of particular clusters.

analysis to determine unique orthogonal gene and corresponding array expression patterns (Alter et al., 2000); and GO TermFinder, which uses Gene Ontology (GO) terms to determine whether a list of genes produced by any number of microarray analysis software has a significant enrichment of GO terms (Boyle et al., 2004).

Other databases are more focused on particular biological areas of interest to ask targeted biological questions of a collective transcriptome data set. For example, Oncomine is a warehouse of microarray data from a wide range of cancer related studies (Rhodes et al., 2004, 2007; Tomlins et al., 2007) and is accessible from: http://www.oncomine.org. It currently houses and displays data from 24 250 microarrays from 342 studies covering 40 different cancer types (statistics from April 2008). The microarray data is processed to allow for data comparability via a semi-automated analysis pipeline to keep up with the rapidly growing amount of published data. Several web-based data mining and visualization tools can be used to search this processed data in the database for particular gene expression profiles or to compare different condition groups such as 'cancer versus normal' or 'cancer versus cancer'. In addition, external filters can be added to rank ordered genes based on their differential expression and these filters are based on GO annotations, InterPro protein domains and families, KEGG, Biocarta pathways, chromosome localization and transcription factor binding sites. One of the main aims of Oncomine is to facilitate biomarker and therapeutic target discovery. A portfolio of analysis methods are used to identify genes with marked overexpression in subsets of cases in given datasets, these include: Molecular Concepts Analysis; Interactome Analysis or Meta-Analysis and Cancer Outlier Profile Analysis (COPA). All these methods are discussed further in Rhodes et al. (2007). As an example of the utility of Oncomine, the COPA tools was used to identify *ERG* and *ETV1* as candidate oncogenes in prostate cancer (Tomlins et al., 2005).

Many other data warehouses are purely designed to integrate resources for particular biological themes or organisms, combining data from genomics, transciptomics and proteomics, such as WormBase (Rogers et al., 2008), the Scaccharomyces database (Hong et al., 2008) or FlyMine (Lyne et al., 2007). Although there are currently relatively few gene expression or Chip–Chip datasets for *Drosophila* and *Anopheles*, FlyMine[1] has developed a powerful query tool that can be used to access integrated data via the web at a number of different levels, from simple browsing to construction of complex queries. Researchers can use existing query templates or build their own queries using the query builder tool. Accounts can be set up to store queries and results, which can be downloaded in various formats. The underlying InterMine platform is fully open source and available as a very flexible platforms for data integration supporting virtually any biological system. Another recently established database,

---

1 Flymine is built using the InterMine framework: http://www.intermine.org/.

t0020

**TABLE 9.2** A Selection of Microarray Data Repositories and Warehouses

| Database | URL | Reference |
|---|---|---|
| 4DXpress | http://ani.embl.de/4DXpress | Haudry et al. (2008) |
| ArrayExpress* | http://www.ebi.ac.uk/arrayexpress | Parkinson et al. (2007) |
| CanGEM (Cancer Genome Mine) | http://www.cangem.org | Scheinin et al. (2008) |
| CATdb (Complete Arabidopsis Transcriptome database) | http://urgv.evry.inra.fr/CATdb | Gagnot et al. (2008) |
| CIBEX (Center for Information Biology gene Expression database)* | http://cibex.nig.ac.jp | Tateno and Ikeo (2004) |
| COXPRESdb (Coexpression gene database) | http://coxpresdb.hgc.jp/ | Obayashi et al. (2008) |
| Cyclebase.org (cell-cycle database) | http://www.cyclebase.org | Gauthier et al. (2008) |
| FlyBase | http://www.flybase.org | Wilson et al. (2008) |
| FlyMine | http://www.flymine.org | Lyne et al. (2007) |
| GEO (Gene Expression Omnibus)* | http://www.ncbi.nlm.nih.gov/geo | Barrett et al. (2007) |
| $M^{3D}$ (Many Microbe Microarrays Database) | http://m3d.bu.edu | Faith et al. (2008) |
| MethyCancer | http://methycancer.genomics.org.cn | He et al. (2008) |
| Oncomine | http://www.oncomine.org | Rhodes et al. (2007), Tomlins et al. (2007) |
| PlexDB (Plant Expression Database) | http://www.plexdb.org/ | Shen et al. (2005) |
| RED (Rice Expression Database) | http://red.dna.affrc.go.jp/RED/ | Yazaki et al. (2002) |
| RefDIC (Reference genomics Database of Immune Cells) | http://refdic.rcai.riken.jp | Hijikata et al. (2007) |
| SymAtlas | http://symatlas.gnf.org/SymAtlas/ | Su et al. (2002, 2004), Walker et al. (2004) |
| TED (Tomato Expression Database) | http://ted.bti.cornell.edu/ | Fei et al. (2006) |
| TMAD (Stanford Tissue Microarray Database) | http://tma.stanford.edu | Sharma-Oates et al., 2005 |
| ToxDB (*Toxoplasma gondii* Database) | http://www.toxodb.org | Gajria et al. (2008) |
| WormBase | http://www.wormbase.org | Rogers et al. (2008) |
| YGD (Yeast Genome Database) | http://www.yeastgenome.org/ | Hong et al. (2008) |
| yMGV (yeast Microarray Global Viewer) | http://transcriptome.ens.fr/ymgv/ | Marc et al. (2001) |

* MGED for the deposition of microarray data. Not all these databases are microarray data repositories, many are data warehouses designed to integrate resources for a particular biological theme.

4Dxpress (expression database in four dimensions) is a central public repository to integrate expression patterns for zebrafish, *Drosophila*, medaka and mouse (Haudry et al., 2008). Researchers can query anatomy ontology-based expression annotations across various species and genes are cross-referenced to the orthologues in other species. Genes are also linked to public microarray data in ArrayExpress.

## 9.5 SUMMARY

The use of standards is essential for recording or describing scientific data and controlled vocabularies provide a way to capture experimental metadata so that databases can be built for storing and retrieving complex data. In the case of microarray data, several international projects have matured over the past few years that provide a framework for capturing all the relevant details of microarray experiments. The MIAME standard is the major framework underpinning the development of useful databases that enable scientists to explore deposited gene expression data at the level of single genes, particular biological systems or other specified factors. While some have grumbled about the amount of information that is captured by the full MIAME standard, it is essential if others are to use microarray data that they do so with a full understanding of the biology underpinning the experiment. Without such standards, data would be difficult to logically store and even more difficult to effectively query. The MGED society manages the development of MIAME and also supports the development of the MAGE data exchange format that facilitates the transfer of microarray data between different data systems. The major data repositories, ArrayExpress, GEO and CIBEX, house both raw microarray data and also serve as warehouses for processed experimental data, facilitating gene-based queries of multiple expression profiles.

We cannot emphasize enough the benefit to the wider community of making gene expression data available in an accessible format. In most cases, a microarray study is performed to address a particular biological question: genes are identified and studied further, hopefully gaining new insights into the biology. If the entirety of the gene expression data are made public, as they really should be when the results of the study are published, other scientists have the opportunity to mine the data, perhaps combining it with other microarray studies to uncover new facets of biological systems, often completely unrelated to the biology that inspired the initial studies. The practice of making available a summary of gene expression data, often as a list of ratios in an inflexible PDF format, on a publishers or laboratory web site, severely limits the wider utility of microarray data. While many publishers now require that data are deposited in a public repository, many do not and we believe this is an unsatisfactory situation. We point readers to the malaria and cancer studies we describe in Chapter 11 as examples of how availability of raw data enables meta-analysis and data-mining that goes far beyond the initial objectives of a study. In addition, as we have described in the preceding chapters, many analytical tools are developed with public datasets, providing computational resources for the community at large. In many ways the situation with microarray data is analogous to the position nucleic acid sequence databases were in at the start of the DNA sequencing revolution: data needed to be captured and made available in a widely accepted format. It goes without saying that the incredible benefits to the biological research community, particularly when genome sequence began to appear, of having freely available data outweigh the minor inconvenience of data submission. As with the sequence databases, microarray database providers have worked hard to streamline data submission processes are much as possible. While we recognize that more metadata need to be captured for a microarray experiment than for a sequence deposit, it is difficult to believe that scientists are performing experiments without knowing the details of the protocols they use and the provenance of the biological material. When commercial microarrays are being used, platform descriptions, including probe sequences, are usually already in the database. If homemade arrays are being used, the data regarding the array layout and probe identifiers must be available for data tracking and analysis. For an average laboratory microarray study, perhaps involving 20 or so arrays, the submission process can easily be completed in an afternoon. The curators and annotators working in the database centers are helpful and can assist with any problems, help files are extensive. There is consequently no excuse for not submitting data.

Other databases tend to be organism specific or focus on particular biological areas and tend to be populated with data taken from public sources. Some of these provide various query tools that can be both visual and informative, for example, the Oncomine warehouse of cancer-related gene expression data. Many of these databases are integrated resources for particular biology that combine data from genomics, transciptomics and proteomics to facilitate either basic or complex queries across multiple data types. Examples here include Flymine, WormBase or 4Dxpress. Sufficing to say, none of these initiatives would be possible or as powerful as they are if individuals do not make their data available.

For laboratories embarking upon a large number of microarray studies there are several stand alone small scale databases that facilitate in house data storage and, in some cases, data-mining Examples here include NOMAD (Not Another MicroArray Database; http://sourceforge.net/projects/ucsf-nomad/), which is relatively straight forward to install and manage. There are other initiatives that can help labs store relatively large datasets. For example, the BASE2 (BioArray Software Environment; http://base.thep.lu.se/) is a freely available web-based database and while it requires some bioinformatics expertise to install and maintain, it offers a well-developed route for storing microarray data. Similarly, the Stanford Microarray Database (SMD; http://genome-www5.stanford.edu/), which is MIAME supportive and serves as both an archive and a research tool, will store submitted data for a modest cost. Of course, modest experiments can be dealt with in individual laboratories with flat files or even spreadsheets. A note of caution; when using spreadsheets some software can alter names or even values and must be carefully evaluated to ensure data is not unwittingly transformed!

## REFERENCES

Alter O, Brown PO, Botstein D. (2000) Singular value decomposition for genome-wide expression data processing and modelling. *Proc Natl Acad Sci U S A* **97:** 10101–10106.

Altschul SF, Gish W, Miller W, Myers EW, Lipman DJ. (1990) Basic local alignment search tool. *J Mol Biol* **215:** 403–410.

Ashburner M, Mungall CJ, Lewis SE. (2003) Ontologies for biologists: a community model for the annotation of genomic data. *Cold Spring Harb Symp Quant Biol* **68:** 227–235.

Ashburner M, Ball CA, Blake JA, Botstein D, Butler H, Cherry JM, Davis AP, Dolinski K, Dwight SS, Eppig JT, Harris MA, Hill DP, Issel-Tarver L, Kasarskis A, Lewis S, Matese JC, Richardson JE, Ringwald M, Rubin GM, Sherlock G. (2000) Gene ontology: tool for the unification of biology. The Gene Ontology Consortium. *Nat Genet* **25:** 25–29.

Ball CA, Brazma A, Causton H, Chervitz S, Edgar R, Hingamp P, Matese JC, Parkinson H, Quacken-bush J, Ringwald M, Sansone SA, Sherlock G, Spellman P, Stoeckert C, Tateno Y, Taylor R, White J, Winegarden N. (2004) Submission of microarray data to public repositories. *PLoS Biol* **2:** E317.

Barrett T, Edgar R. (2006) Mining microarray data at NCBI's Gene Expression Omnibus (GEO). *Methods Mol Biol* **338:** 175–190.

Barrett T, Troup DB, Wilhite SE, Ledoux P, Rudnev D, Evangelista C, Kim IF, Soboleva A, Tomashevsky M, Edgar R. (2007) NCBI GEO: mining tens of millions of expression profiles – database and tools update. *Nucleic Acids Res* **35:** D760–D765.

Birney E, Andrews D, Caccamo M, Chen Y, Clarke L, Coates G, Cox T, Cunningham F, Curwen V, Cutts T, Down T, Durbin R, Fernandez-Suarez XM, Flicek P, Graf S, Hammond M, Herrero J, Howe K, Iyer V, Jekosch K, Kahari A, Kasprzyk A, Keefe D, Kokocinski F, Kulesha E, London D, Longden I, Melsopp C, Meidl P, Overduin B, Parker A, Proctor G, Prlic A, Rae M, Rios D, Redmond S, Schuster M, Sealy I, Searle S, Severin J, Slater G, Smedley D, Smith J, Stabenau A, Stalker J, Trevanion S, Ureta-Vidal A, Vogel J, White S, Woodwark C, Hubbard TJ. (2006) Ensembl 2006. *Nucleic Acids Res* **34:** D556–D561.

Boyle EI, Weng S, Gollub J, Jin H, Botstein D, Cherry JM, Sherlock G. (2004) GO:TermFinder – open source software for accessing Gene Ontology information and finding significantly enriched Gene Ontology terms associated with a list of genes. *Bioinformatics* **20:** 3710–3715.

Brazma A, Parkinson H, Sarkans U, Shojatalab M, Vilo J, Abeygunawardena N, Holloway E, Kapushesky M, Kemmeren P, Lara GG, Oezcimen A, Rocca-Serra P, Sansone SA. (2003) ArrayExpress – a public repository for microarray gene expression data at the EBI. *Nucleic Acids Res* **31:** 68–71.

Brazma A, Hingamp P, Quackenbush J, Sherlock G, Spellman P, Stoeckert C, Aach J, Ansorge W, Ball CA, Causton HC, Gaasterland T, Glenisson P, Holstege FC, Kim IF, Markowitz V, Matese JC, Parkinson H, Robinson A, Sarkans U, Schulze-Kremer S, Stewart J, Taylor R, Vilo J, Vingron M. (2001) Minimum information about a microarray experiment (MIAME)-toward standards for microarray data. *Nat Genet* **29:** 365–371.

Consortium TGO. (2008) The Gene Ontology project in 2008. *Nucleic Acids Res* **36:** D440–D444.

Demeter J, Beauheim C, Gollub J, Hernandez-Boussard T, Jin H, Maier D, Matese JC, Nitzberg M, Wymore F, Zachariah ZK, Brown PO, Sherlock G, Ball CA. (2007) The Stanford Microarray Database: implementation of new analysis tools and open source release of software. *Nucleic Acids Res* **35:** D766–D770.

Edgar R, Domrachev M, Lash AE. (2002) Gene Expression Omnibus: NCBI gene expression and hybridization array data repository. *Nucleic Acids Res* **30:** 207–210.

Faith, J. J., M. E. Driscoll, Fusaro VA, Cosgrove EJ, Hayete B, Juhn FS, Schneider SJ, Gardner TS. (2008) "Many Microbe Microarrays Database: uniformly normalized Affymetrix compendia with structured experimental metadata." *Nucleic Acids Res* **36**(Database issue): D866–70.

Fei, Z., X. Tang, Alba R, Giovannoni J. (2006) "Tomato Expression Database (TED): a suite of data presentation and analysis tools." *Nucleic Acids Res* **34**(Database issue): D766–70.

Flicek P, Aken BL, Beal K, Ballester B, Caccamo M, Chen Y, Clarke L, Coates G, Cunningham F, Cutts T, Down T, Dyer SC, Eyre T, Fitzgerald S, Fernandez-Banet J, Graf S, Haider S, Hammond M, Holland R, Howe KL, Howe K, Johnson N, Jenkinson A, Kahari A, Keefe D, Kokocinski F, Kulesha E, Lawson D, Longden I, Megy K, Meidl P, Overduin B, Parker A, Pritchard B, Prlic A, Rice S, Rios D, Schuster M, Sealy I, Slater G, Smedley D, Spudich G, Trevanion S, Vilella AJ, Vogel J, White S, Wood M, Birney E, Cox T, Curwen V, Durbin R, Fernandez-Suarez XM, Herrero J, Hubbard TJ, Kasprzyk A, Proctor G, Smith J, Ureta-Vidal A, Searle S. (2008) Ensembl 2008. *Nucleic Acids Res* **36:** D707–D714.

Gagnot, S., J. P. Tamby, Martin-Magniette ML, Bitton F, Taconnat L, Balzergue S, Aubourg S, Renou JP, Lecharny A, Brunaud V. (2008) "CATdb: a public access to Arabidopsis transcriptome data from the URGV-CATMA platform." *Nucleic Acids Res* **36**(Database issue): D986-90.

Gajria B, Bahl A, Brestelli J, Dommer J, Fischer S, Gao X, Heiges M, Iodice J, Kissinger JC, Mackey AJ, Pinney DF, Roos DS, Stoeckert CJ, Wang H, Brunk BP. (2008) ToxoDB: an integrated Toxoplasma gondii database resource. *Nucleic Acids Res* **36:** D553–D556.

Gauthier, N. P., M. E. Larsen, Wernersson R, de Lichtenberg U, Jensen LJ, Brunak S, Jensen TS. (2008) "Cyclebase.org–a comprehensive multi-organism online database of cell-cycle experiments." *Nucleic Acids Res* **36**(Database issue): D854–9.

Gollub J, Ball CA, Binkley G, Demeter J, Finkelstein DB, Hebert JM, Hernandez-Boussard T, Jin H, Kaloper M, Matese JC, Schroeder M, Brown PO, Botstein D, Sherlock G. (2003) The Stanford Microarray Database: data access and quality assessment tools. *Nucleic Acids Res* **31:** 94–96.

Harris MA, Clark J, Ireland A, Lomax J, Ashburner M, Foulger R, Eilbeck K, Lewis S, Marshall B, Mungall C, Richter J, Rubin GM, Blake JA, Bult C, Dolan M, Drabkin H, Eppig JT, Hill DP, Ni L, Ringwald M, Balakrishnan R, Cherry JM, Christie KR, Costanzo MC, Dwight SS, Engel S, Fisk DG, Hirschman JE, Hong EL, Nash RS, Sethuraman A, Theesfeld CL, Botstein D, Dolinski K, Feierbach B, Berardini T, Mundodi S, Rhee SY, Apweiler R, Barrell D, Camon E, Dimmer E, Lee V, Chisholm R, Gaudet P, Kibbe W, Kishore R, Schwarz EM, Sternberg P, Gwinn M, Hannick L, Wortman J, Berriman M, Wood V, de la Cruz N, Tonellato P, Jaiswal P, Seigfried T, White R. (2004) The Gene Ontology (GO) database and informatics resource. *Nucleic Acids Res* **32:** D258–D261.

Haudry Y, Berube H, Letunic I, Weeber PD, Gagneur J, Girardot C, Kapushesky M, Arendt D, Bork P, Brazma A, Furlong EE, Wittbrodt J, Henrich T. (2008) 4DXpress: a database for cross-species expression pattern comparisons. *Nucleic Acids Res* **36:** D847–D853.

He, X., S. Chang, Zhang J, Zhao Q, Xiang H, Kusonmano K, Yang L, Sun ZS, Yang H, Wang J. (2008) "MethyCancer: the database of human DNA methylation and cancer." *Nucleic Acids Res* **36** (Database issue): D836-41.

Hijikata A, Kitamura H, Kimura Y, Yokoyama R, Aiba Y, Bao Y, Fujita S, Hase K, Hori S, Ishii Y, Kanagawa O, Kawamoto H, Kawano K, Koseki H, Kubo M, Kurita-Miki A, Kurosaki T, Masuda K, Nakata M, Oboki K, Ohno H, Okamoto M, Okayama Y, O-Wang J, Saito H, Saito T, Sakuma M, Sato K, Sato K, Seino K, Setoguchi R, Tamura Y, Tanaka M, Taniguchi M, Taniuchi I, Teng A, Watanabe T, Watarai H, Yamasaki S, Ohara O. (2007) Construction of an open-access database that integrates cross-reference information from the transcriptome and proteome of immune cells. *Bioinformatics* **23**(21): 2934–2941.

Hong EL, Balakrishnan R, Dong Q, Christie KR, Park J, Binkley G, Costanzo MC, Dwight SS, Engel SR, Fisk DG, Hirschman JE, Hitz BC, Krieger CJ, Livstone MS, Miyasato SR, Nash RS, Oughtred R, Skrzypek MS, Weng S, Wong ED, Zhu KK, Dolinski K, Botstein D, Cherry JM. (2008) Gene Ontology annotations at SGD: new data sources and annotation methods. *Nucleic Acids Res* **36:** D577–D581.

Hull D, Wolstencroft K, Stevens R, Goble C, Pocock MR, Li P, Oinn T. (2006) Taverna: a tool for building and running workflows of services. *Nucleic Acids Res* **34:** W729–732.

Kapushesky M, Kemmeren P, Culhane AC, Durinck S, Ihmels J, Korner C, Kull M, Torrente A, Sarkans U, Vilo J, Brazma A. (2004) Expression Profiler: next generation – an online platform for analysis of microarray data. *Nucleic Acids Res* **32:** W465–470.

Killion PJ, Sherlock G, Iyer VR. (2003) The Longhorn Array Database (LAD): an open-source, MIAME compliant implementation of the Stanford Microarray Database (SMD). *BMC Bioinform* **4:** 32.

Lyne R, Smith R, Rutherford K, Wakeling M, Varley A, Guillier F, Janssens H, Ji W, McLaren P, North P, Rana D, Riley T, Sullivan J, Watkins X, Woodbridge M, Lilley K, Russell S, Ashburner M, Mizuguchi K, Micklem G. (2007) FlyMine: an integrated database for *Drosophila* and *Anopheles* genomics. *Genome Biol* **8:** R129.

Marc P, Devaux F, Jacq C. (2001) yMGV: a database for visualization and data mining of published genome-wide yeast expression data. *Nucleic Acids Res* **29:** e63.

Obayashi, T., S. Hayashi, Shibaoka M, Saeki M, Ohta H, Kinoshita K. (2008) "COXPRESdb: a database of coexpressed gene networks in mammals." *Nucleic Acids Res* **36**(Database issue): D77–82.

Parkinson H, Kapushesky M, Shojatalab M, Abeygunawardena N, Coulson R, Farne A, Holloway E, Kolesnykov N, Lilja P, Lukk M, Mani R, Rayner T, Sharma A, William E, Sarkans U, Brazma A. (2007) ArrayExpress – a public database of microarray experiments and gene expression profiles. *Nucleic Acids Res* **35:** D747–D750.

Rayner TF, Rocca-Serra P, Spellman PT, Causton HC, Farne A, Holloway E, Irizarry RA, Liu J, Maier DS, Miller M, Petersen K, Quackenbush J, Sherlock G, Stoeckert CJ, White J, Whetzel PL, Wymore F, Parkinson H, Sarkans U, Ball CA, Brazma A. (2006) A simple spreadsheet-based, MIAME-supportive format for microarray data: MAGE-TAB. *BMC Bioinform* **7:** 489.

Rhodes DR, Yu J, Shanker K, Deshpande N, Varambally R, Ghosh D, Barrette T, Pandey A, Chinnaiyan AM. (2004) ONCOMINE: a cancer microarray database and integrated data-mining platform. *Neoplasia* **6:** 1–6.

Rhodes DR, Kalyana-Sundaram S, Mahavisno V, Varambally R, Yu J, Briggs BB, Barrette TR, Anstet MJ, Kincead-Beal C, Kulkarni P, Varambally S, Ghosh D, Chinnaiyan AM. (2007)

Oncomine 3.0: genes, pathways, and networks in a collection of 18,000 cancer gene expression profiles. *Neoplasia* **9:** 166–180.

Rogers A, Antoshechkin I, Bieri T, Blasiar D, Bastiani C, Canaran P, Chan J, Chen WJ, Davis P, Fernandes J, Fiedler TJ, Han M, Harris TW, Kishore R, Lee R, McKay S, Muller HM, Nakamura C, Ozersky P, Petcherski A, Schindelman G, Schwarz EM, Spooner W, Tuli MA, Van Auken K, Wang D, Wang X, Williams G, Yook K, Durbin R, Stein LD, Spieth J, Sternberg PW. (2008) WormBase 2007. *Nucleic Acids Res* **36:** D612–D617.

Sarkans U, Parkinson H, Lara GG, Oezcimen A, Sharma A, Abeygunawardena N, Contrino S, Holloway E, Rocca-Serra P, Mukherjee G, Shojatalab M, Kapushesky M, Sansone SA, Farne A, Rayner T, Brazma A. (2005) The ArrayExpress gene expression database: a software engineering and implementation perspective. *Bioinformatics* **21:** 1495–1501.

Sharma-Oates A, Quirke P, Westhead DR. (2005) Tmadb: a repository for tissue microarray data. *BMC Bioinform* **6:** 218.

Shen, L., J. Gong, Caldo RA, Nettleton D, Cook D, Wise RP, Dickerson JA. (2005) "BarleyBase–an expression profiling database for plant genomics." *Nucleic Acids Res* **33**(Database issue): D614–8.

Su AI, Cooke MP, Ching KA, Hakak Y, Walker JR, Wiltshire T, Orth AP, Vega RG, Sapinoso LM, Moqrich A, Patapoutian A, Hampton GM, Schultz PG, Hogenesch JB. (2002) Large-scale analysis of the human and mouse transcriptomes. *Proc Natl Acad Sci U S A* **99**(7): 4465–4470.

Tateno Y, Ikeo K. (2004) International public gene expression database (CIBEX) and data submission. *Tanpakushitsu Kakusan Koso* **49:** 2678–2683.

Tomlins SA, Mehra R, Rhodes DR, Cao X, Wang L, Dhanasekaran SM, Kalyana-Sundaram S, Wei JT, Rubin MA, Pienta KJ, Shah RB, Chinnaiyan AM. (2007) Integrative molecular concept modeling of prostate cancer progression. *Nat Genet* **39:** 41–51.

Tomlins SA, Rhodes DR, Perner S, Dhanasekaran SM, Mehra R, Sun XW, Varambally S, Cao X, Tchinda J, Kuefer R, Lee C, Montie JE, Shah RB, Pienta KJ, Rubin MA, Chinnaiyan AM. (2005) Recurrent fusion of TMPRSS2 and ETS transcription factor genes in prostate cancer. *Science* **310:** 644–648.

Troyanskaya O, Cantor M, Sherlock G, Brown P, Hastie T, Tibshirani R, Botstein D, Altman RB. (2001) Missing value estimation methods for DNA microarrays. *Bioinformatics* **17:** 520–525.

Walker JR, Su AI, Self DW, Hogenesch JB, Lapp H, Maier R, Hoyer D, Bilbe G. (2004) Applications of a rat multiple tissue gene expression data set. *Genome Res* **14**(4): 742–749.

Wheeler DL, Barrett T, Benson DA, Bryant SH, Canese K, Chetvernin V, Church DM, Dicuccio M, Edgar R, Federhen S, Feolo M, Geer LY, Helmberg W, Kapustin Y, Khovayko O, Landsman D, Lipman DJ, Madden TL, Maglott DR, Miller V, Ostell J, Pruitt KD, Schuler GD, Shumway M, Sequeira E, Sherry ST, Sirotkin K, Souvorov A, Starchenko G, Tatusov RL, Tatusova TA, Wagner L, Yaschenko E. (2008) Database resources of the National Center for Biotechnology Information. *Nucleic Acids Res* **36:** D13–D21.

Wilson RJ, Goodman JL, Strelets VB, the FlyBase, Consortium. (2008) FlyBase: integration and improvements to query tools. *Nucleic Acids Res* **36:** D588–D593.

Yazaki J, Kishimoto N, Ishikawa M, Endo D, Kojima K, MicroArray Center, Kikuchi S. (2002) The Rice Expression Database (RED): gateway to rice functional genomics. *Trends in Plant Science* **12:** 563–564.

# Beyond Expression Arrays: Genome Analysis

As we described in previous chapters, microarrays for gene expression analysis are now firmly established research tools in the biologists armory. However, the incredible specificity of DNA hybridization means that, in principal, microarrays are amenable to any experiment where large-scale analysis of nucleic acid is desired. The increasing miniaturization of the technology is opening up fantastic opportunities for exploring whole genomes at an unparalleled level of resolution (Hoheisel, 2006) and the availability of very high-density microarrays is beginning to bring the analysis of complete metazoan genomes, hitherto restricted to genome centers, into the hands of any researcher. Microarrays are being applied to improving genome annotations, exploring chromatin structure and genome organization, precisely mapping the in vivo binding sites of transcription factors, scanning genomes for changes in DNA copy number, scanning for polymorphisms and even resequencing substantial amounts of DNA. In this chapter we shall explore the types of genome array available and their uses in basic biology, microarray applications with a more medical focus will be discussed more fully in the following chapter but we will introduce the relevant array platforms used in this area here.

## 10.1 GENOME ARRAYS

In Chapter 3 we focused on the design of gene expression arrays, which generally contain a single specific probe or probe set per gene. The increasing array densities now available allow the fabrication of genome tiling arrays that interrogate the transcriptome or genome in much greater detail. Here we describe the range of exon and tiling arrays that are currently available and consider how they are used to explore genome biology. Figure 10.1 illustrates, with reference to a

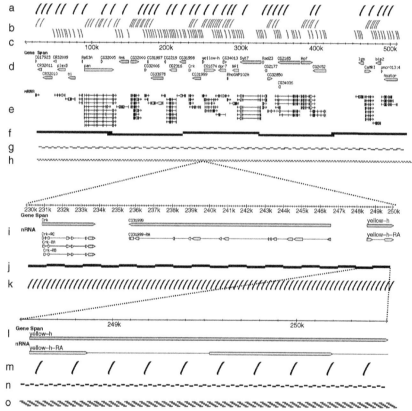

**FIGURE 10.1**   Strategies for covering the genome with different types of probes. (a) Probes for a typical expression array, a single oligonucleotide per gene *38 probes*. (b) Exon probes, a single probe is designed against each exon with the plus and minus strands shown separately *143 probes*. (c) Scale bar in 10 kb increments. (d) Gene models from FlyBase. (e) Transcript structures from FlyBase. (f) BAC tile with 100 kb average insert size *5 probes*. (g) 5 kb amplicon tile, *100 probes*. (h) 1 kb amplicon tile, *500 probes*. (i) Blow up of a 20 kb regions with gene models and transcript structures. (j) 1 kb amplicon tile, *20 probes*. (k) Partial oligo tile, a 50mer on average every 250 bp, *80 probes*. (l) Blow up of a 2 kb region with gene models and transcript structure. (m) Partial oligonucleotide tile, a 50mer probe every 200 bp, *10 probes*. (n) Full oligonucleotide tile, 25mer probes cover every base, *80 probes*. (o) Overlapping oligonucleotide tile, a 25mer probe every 10 bp, *200 probes*.

500 kb region of the *Drosophila* genome, the different types of arrays that can, in principal, be designed. These may utilize probes ranging in size from large BAC clones down to small oligonucleotides. Before proceeding we should define what is meant by tiling and the various approaches taken when generating tiling arrays. A true tiling array contains probes representing every nucleotide or an organism genome and in practice can only currently be achieved for microbial or small metazoan genomes. For example, in the case of the bakers yeast, *Saccharomyces cerevisiae*, the genome size of approximately 12 million base pairs, means that probes amounting to 24 million bases are needed to represent both strands of the genome. This requires less than one million 25mers or around 350 000 70mers, readily achievable with current in situ synthesis technologies. Indeed the smaller microbial genomes are amenable to representation as overlapping tiles, where every base is represented several times. The ultimate density being a tile with 1 base resolution: less than four million 25mers can tile both strands of the 1.8 Mb *Haemophilus influenzae* genome, effectively generating an array with 25-fold redundancy. As genome size increases two issues arise: the first is obviously a simple numerical issue, either more probes are needed to cover larger genomes or probe lengths must be increased so that each element covers more of the genome. The latter decreases resolution but also effectively limits the array technology that can be used since very high-density in situ synthesized arrays are limited to probe lengths of approximately 70 nucleotides or less. The second problem comes with the increasing quantities of repetitive DNA associated with metazoan genomes since designing probes for each instance of a repetitive region makes little sense and can confound analysis. Therefore, the genome sequence of higher eukaryotes must be segmented or repeat masked prior to tiling and, as a consequence, any array will never be a true genome tile.

An alternative to full genome tiling is to take a partial tiling approach where probes are spaced at regular intervals but do not interrogate every base. Most commercial array providers take this route: Affymetrix use 25mer probes every 35–38 bases, providing actual coverage of 65–70% of the genome, whereas Nimblegen or Agilent design longer 50–70mer probe every 90–300 bp of sequence, covering 16–70% of the genome. Unlike sequencing, where every base in a sample is examined, the resolution of any array-based approach is, to some extent, determined by the size distribution of the labeled target. Since this will be, in the vast majority of cases, a heterogeneous sample, then partial tiling arrays can be almost as effective as complete tiles since missing bases are covered by the size distribution of the targets. However, it is clear that the more closely spaced the probes are, the better the potential resolution will be.

At the other extreme, the genome can be fully tiled with much larger probes, generally ranging in size from 1 kb (amplicons) to >100 kb (BACs). Although arrays constructed with these probes are much lower resolution than very high-density arrays, they do provide an accessible route to complete coverage for even very large genomes. For example, just over 30 000 BACs have been used to cover the entire human genome (Ishkanian et al., 2004): a density within the reach of in-house printing technology.

## 10.2 EXON ARRAYS

As should be obvious from the above, tiling arrays represent a nonbiased view of a genome; probes are placed at regular intervals along the sequence without any heed to sequence annotation. An approach that is intermediate between genome tiling and gene-based expression arrays uses probes designed against specific-annotated features, generally exons and/or splice junctions. These types of arrays are of considerable utility in exploring and validating genome annotations. For example an array based analysis of the Genscan predicted genes accompanying the first draft of the human genome sequence generated probes against over 400 000 exons and found evidence suggesting that at least 35% of these exons were expressed in the two-cell types sampled (Shoemaker et al., 2001). At the time, this effort was considerable with over one million probes synthesized across 50 arrays, it is now possible to obtain complete human exon coverage on 2 or 4 arrays (e.g. Clark et al., 2007). In principal, the probe design tools described in Chapter 3 may be used for exon arrays but it should be born in mind that it can be difficult to design probes for specific exons since the target sequences are constrained. Unlike a probe for a gene-based expression array, where the best region of the target can be selected, exons may be very small and have limited probe design flexibility or show extensive homology with other exons. As a consequence, care should be taken when inferring gene expression levels using exon-arrays, and the use of focused analysis software is recommended. For example, Xing et al. (2006) use a coexpression index to select the most reliable probes for expression estimates. However, this type of analysis may miss exons utilized in a limited number of tissues or associated with minor splice variants. Splice junction probes are, by definition, fixed by the juxtaposition of two exons and consequently have almost no scope for optimal probe design other than slightly varying the contribution each exon sequence makes to the composite probe (Johnson et al., 2003). It should be obvious that longer oligonucleotides have greater utility when making splice junction probes since higher specificity can be achieved. As an illustration the design strategy for the Affymetrix Human Exon arrays is shown in Figure 10.2 along with the probes needed to detect a set of splice variants from one of the transcripts.

### 10.2.1 Using Exon Arrays

Exon arrays are now available commercially, principally from Affymetrix, and are beginning to see extensive use, primarily in mammalian systems, for the

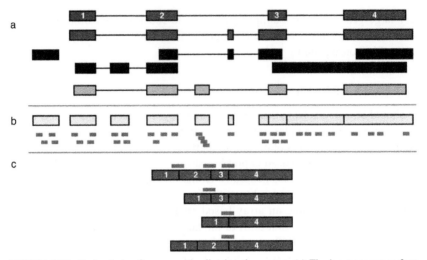

**FIGURE 10.2**  Probe design for exon and splice junction arrays. (a) The input sequences from Ensembl, RefSeq, EST clones and predictions are combined to generate a set of nonoverlapping probe selection regions (PSRs). (b) PSRs and the location of the designed probes. (c) For the transcript on the top line of (a), potential alternative splice variants are diagrammed and probes designed against each of the splice junctions (a and b are adapted from technical descriptions on the Affymetrix web site).

analysis of alternative splicing (Pan et al., 2004; Frey et al., 2005; Kwan et al., 2008), tissue specific gene expression (Clark et al., 2007) and evolutionary aspects of gene expression (Xing et al., 2007). In most respects, exon arrays are treated much in the same way as standard gene expression arrays, though obviously care must be taken when examining signals from individual exons since there is a potential for cross hybridization with related sequences. One positive aspect of the exon array design is that they can provide multiple independent expression measures for a transcript and therefore more reliable measures of gene expression (Abdueva et al., 2007; Ghosh et al., 2007). In terms of sample preparation, it should be obvious that full-length labeled targets are a prerequisite and we summaries the approaches taken in this respect below in Section 10.6.1.

## 10.3 TILING ARRAYS

While the number of independent probes required to tile even the smallest metazoan genome is very large (Table 10.1), almost 10 million to fully tile both strands of the 120 Mb *Drosophila* genome, single-array densities in this range are almost achievable. For example, the current generation of Affymetrix GeneChips are synthesized with a feature size of 5 $\mu$m providing around six million probes on a single GeneChip. At this density, a single strand of the fly or *Caenorhabditis elegans* genomes are available on a single array. A density

**TABLE 10.1** Genome Tiling Probes

| Probe | # To cover the human genome | # To cover the *Drosophila* genome |
|---|---|---|
| BAC | 20 000 | 1200 |
| 10 kb Amlicon | 200 000 | 12 000 |
| 1 kb Amplicon | 2 000 000 | 120 000 |
| Partial oligonuclotide (50mcr/250 bp) | 16 000 000 | 960 000 |
| Partial oligonucleotide (25mer/35 bp) | 116 000 000 | 7 000 000 |
| Full oligonucleotide (70mer) | 58 000 000 | 3 400 000 |
| Full oligonucleotide (25mer) | 160 000 000 | 9 600 000 |
| Overlapping oligonucleotide (10 bp) | 400 000 000 | 24 000 000 |

The number of various different probes required to tile the human or *Drosophila* genomes. The nonrepetitive fraction of the human genome is estimated here at two billion bases (a slight overestimate) and the *Drosophila* euchromatic genome is 120 million bases. For oligonucleotide probes, the number required is doubled since complete coverage requires both strands of the haploid genome be represented, this is obviously not the case for double-stranded amplicon or BAC probes.

increase of at least an order of magnitude is required before the human genome can be accommodated on a single array, however, both Affymetrix and Nimblegen have array sets (7 and 38 arrays, respectively) that partially tile the entire nonrepetitive fraction of the human genome (Table 10.2). All of the main array manufacturers are actively developing higher densities and it is only a matter of time before the entire human genome is at least partially tiled on a single array.

As will be obvious from Table 10.2, unless one has access to a very high-density in situ synthesizer, producing a high-resolution metazoan genome-tiling array in your own lab is not a realistic proposition. However, for more focused studies of a few megabases or for low-resolution studies, in-house production is certainly a cost-effective option when hundreds or thousands of arrays are required. Instances where home made genome arrays may be considered are, for example, Human Genome BAC arrays, low-resolution (5–10 kb) amplicon arrays for smallish genomes such as Yeast, *Drosophila* or *C. elegans*, higher resolution (<1 kb) amplicon or partial oligonucleotide arrays for around 20 megabase regions. We outline the basics that need to be considered when building such an array before going on to consider the types of experiments that may be of interest.

## 10.4 AMPLICON AND BAC ARRAYS

As with cDNA or amplicon-based gene expression arrays (Chapter 3), amplicon tiling arrays are constructed from PCR generated probes. Most often probe DNA is amplified from genomic DNA using a specific primer pair, although it is possible to use arrayed clone libraries such as those generated

**TABLE 10.2** Commercially Available High-density Oligonucleotide Arrays for Genomics Applications

| Company | Array | Tiling density | Coverage |
|---|---|---|---|
| Affymetrix | Human promoter 1.0R | 25mer/35 bp | 10 kb from 23 000 genes |
| Affymetrix | Human tiling 2.0R (7 arrays) | 25mer/35 bp | Genome (1 billion bases) |
| Nimblegen | Human HG18 promoter (2 arrays) | 50–70mer/100 bp | 5 kb from 23 000 genes |
| Nimblegen | Human genome (38 arrays) | 50mer/100 bp | Genome (732 million bases) |
| Agilent | Human promoter (2 arrays) | 60mer/215 bp | 8 kb from 17 000 genes |
| Affymetrix | Arabidopsis tiling 1.0R | 25mer/35 bp | Genome (80 million) |
| Nimblegen | Arabidopsis genome (3 arrays) | 50mer/90 bp | Genome (58 million) |
| Agilent | Arabidopsis genome (2 arrays) | 60mer/212 bp | Genome (24.4 million) |
| Affymetrix | C. elegans tiling 1.0R | 25mer complete | Genome (80 million) |
| Nimblegen | C. elegans genome (3 arrays) | 50mer/86 bp | Genome (58 million) |
| Agilent | C. elegans Genome (2 arrays) | 60mer/182 bp | Genome (24.4 million) |
| Affymetrix | Drosophila tiling 2.0R | 25mer/38 bp | Genome (80 Million) |
| Nimblegen | Drosophila genome (3 arrays) | 50mer/97 bp | Genome (80 million) |
| Agilent | Drosophila genome (2 arrays) | 60mer/233 bp | Genome (24.4 million) |
| Affymetrix | S. cerevisiae tiling 1.0R | 25mer 5 bp resolution | 20× genome (80 million) |
| Nimblegen | S. cerevisiae genome | 50mer 32 bp resolution | 1.6× genome (19 million) |
| Agilent | S. cerevisiae genome | 50mer complete | Genome (12 million) |

For the three main providers we indicate the arrays available for human (similar arrays are available for mouse and some other mammals) and selected model organisms. The tiling density column indicate the probe size and average spacing between probes. The coverage column indicates the approximate number of nucleotides represented on each array or array set.

during clone-based genome sequencing projects if these are available. In the latter case, the sequence of each clone will be known and mapped to the genome and all that is required is to amplify the insert from the clone using vector-specific primers (Hollich et al., 2004). For most tiling projects, however, such clones will not be available due to the wide adoption of Shotgun sequencing methods and amplicons are designed based on the genome sequence. It is wise to use the most up to date genome assembly available since draft assemblies often omit repeat regions, which can seriously affect the reliability of PCR amplification when the real genome is tackled. Primers are generally selected from a genome sequence using standard primer design software tools such as Primer 3 (Rozen and Skaletsky, 2000) and for relatively small regions (<100 kb) this can be handled manually with relatively little difficulty. However, for larger regions or for 'difficult' sequences, such as portions of vertebrate genomes that contain substantial amounts of repetitive DNA, it is wiser to use some kind of management software. We designed a simple interface, Mammot (Ryder et al., 2006), that is freely available and facilitates all stages of a tiling array project from primer design through amplicon management to data presentation (Figure 10.3). Such a tool can be very useful for keeping track of the inevitable cycles of primer redesign required when particular amplicons fail to yield clean PCR products. An online tool for designing amplicon tiling arrays is also available (Bertone et al., 2006). Such amplicon-based arrays for selected regions of eukaryotic genomes, including an entire Human chromosome, have been successfully used to explore, for example, transcription, transcription factor binding or DNA replication (Iyer et al., 2001; Horak et al., 2002; Rinn et al., 2003; MacAlpine et al., 2004; Birch-Machin et al., 2005).

The amplicon tiling approach is just about feasible for an entire small metazoan genome, for example, a group of us have tiled the entire euchromatic *Drosophila* genome with 1 kb amplicons (SR, K. White and G. Micklem). The task was computationally complex in terms of designing a tiling path that covers the genome without gaps and with minimal overlap while at the same time selecting PCR primers to reliably amplify 120 000 different products that all perform under a set of standard conditions. It was also technically demanding to prepare 120 000 PCR products, deal with specific failures through primer redesign and ensure approximately equal amounts of product were obtained. Frankly, with the recent availability of oligonucleotide-based genome tiles for *Drosophila* we would not recommend this route to anyone: primer costs alone for the project approached $1 million, a sum of money that buys a lot of commercial tiling arrays.

As we alluded to above, the use of BAC clones as array probes allows low-resolution, but almost complete coverage of very large genomes such as human: there are always, of course, 'unclonable' regions of genomes that are not present in even the largest library. In principal BAC arrays are similar to any amplicon or

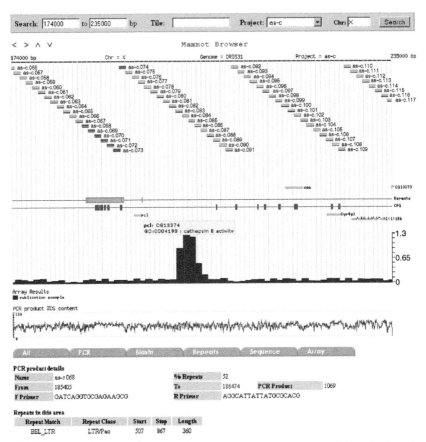

**FIGURE 10.3** The Mammot tiling array tool. The browser window shows a 61 kb region from the *Drosophila X* chromosome tiled with 1 kb amplicons. Below the tiling path, gene models, repeat regions and CpG rich regions are indicated. The histogram below this shows the results of a ChIP-array experiment with log 2 enrichment plotted. Below this, a console provides details for individual amplicons: in this window %GC over the 1069 bp product, the details of any repetitive sequences and the specific primers used to amplify the product.

cDNA array where double-stranded DNA is spotted. Rather than try to spot intact BAC DNA, which due to the large clone size will make viscose solutions at optimal spotting concentrations, each BAC DNA sample is digested with a four-cutter restriction enzyme, amplification linkers are ligated onto the ends of the digested products and the resulting pool of DNA amplified by PCR using primers recognizing the linker sequence. Such a strategy means that the library of BAC clones representing the genome is prepared once and the resulting pool of PCR products for each clone used as a reusable source of printing products (Watson et al., 2004). The utility and analysis of BAC arrays will be described in the next chapter.

## 10.5 OLIGONUCLEOTIDE TILES

Designing a full oligonucleotide tiling array for a region is straight forward since every base has to be covered. As with amplicon arrays, a repeat
masked version of the relevant genome sequence is advisable and some effort
should be made to establish the relationship between the sequences in the
tiled region compared to the rest of the genome. For example, if a pseudogene sequence is within the tiled region then gene expression-based hybridization signals due to labeled sample deriving from the *bona fide* gene
elsewhere in the genome may confound analysis. Assuming a nonoverlapping tile, the only decision to be made is the relative frame of the tiling with
respect to the genome sequence: in other words what base to begin tiling
from. This can affect the quality of the individual probes, therefore it is
prudent to try several different start points and assess the thermodynamic
properties of the probe sets (i.e. $T_m$ and $\Delta G$ distribution) to identify the frame
with the most uniform properties.

In the case of partial tiling arrays, the design is more complex and these
issues have recently been well-reviewed by Bertone et al. (2006). While there
is more latitude in positioning the individual probes, therefore more opportunity for selecting optimal sequences, probes must be positioned to ensure
even coverage. As exemplified in Figure 10.4, optimal coverage has evenly
spaced probes, in this case interrogating every 300 bp, whereas a suboptimal
path has probes spaced too close or too far apart. An example of the design
path followed for such a project is the long oligonucleotide partial tiling of the
entire human genome described by Lee et al. (2006). For this study of Polycomb complex binding in Human embryonic stem cells, a set of 115 arrays,
each containing approximately 40 000 unique 60mer probes, were synthesized by ink-jet in situ synthesis. The tiling design placed a probe approximately every 300 bp, tiling transcription units at a slightly higher density than
noncoding sequence. First the genome was divided into transcribed and intergenic regions with several annotation databases used to define transcript
structures. 10 kb of upstream sequence was added to each transcript to capture
proximal regulatory sequences, thus defining a transcript set that was then
arranged in a linear order across the human genome. The intervening
sequences between the transcript sets were separately extracted from the
repeat masked genome sequence. For both sets of sequence, unique contiguous regions less than 100 bp were removed and each remaining region between 101 and 300 bp selected. Regions between 301 and 640 bp were divided
into two evenly sized fragments and regions over 640 bp equally divided so
that no fragment was larger than 320 bp. All of these fragments were then
used as the input for the probe design software ArrayOligoSelector to select
optimal 60mers (Bozdech et al., 2003). However, to avoid the spacing problem highlighted in Figure 10.4, which OligoArraySelector cannot explicitly

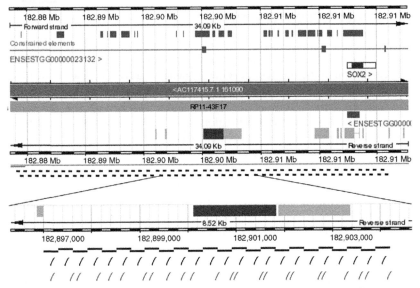

**FIGURE 10.4**   Partial tiling array design. A 34 kb region from human chromosome 3 encompassing the *SOX2* gene taken from the Ensembl browser. Below the scale bar, shaded boxes represent conserved sequence blocks. The boxes joined by a thin line represent Exons from EST sequence. Below the gene model, clone and tile-path contigs are shown followed by EST derived transcripts on the opposite strand. The light and dark boxes above the scale represent predicted or actual regulatory features. Underneath the scale we show the region segmented into 300 bp fragments for tiling design. Below this an expanded view of an 8.5 kb region focusing on the regulatory features with the 300 bp tile shown. The last two lines represent probe locations. The darker upper track represents an optimal probe layout where there is even spacing between each probe. The lighter lower track represents a poor design with probes spaced either to close or to far apart. Notice in particular the uneven coverage over the predicted regulatory blocks.

deal with, all potential oligos above an acceptable quality score within a fragment were selected and this set of oligos parsed to select the best probe within a defined distance from the preceding probe. The final design generated 4 652 484 probes covering over 90% of the repeat masked genome with 95% of the probes within 450 bp of each other. As should be obvious this array design was technically challenging, however, the principals described in the paper are generally applicable to partial tiling designs for smaller regions of the genome and provide a useful guide for anyone interested in generating their own tiling array. More generally, a fast optimal tiling algorithm has been developed by Berman and colleagues and shown to perform well with various genomes, offering a ready made solution for in-house design (Berman et al., 2004). More recently, these authors have revised their methods and provided a convenient web interface for designing tiling arrays (http://tiling.gersteinlab. org; Bertone et al., 2006) that is recommended for those with limited bioinformatics support.

## 10.6 USING TILING ARRAYS

As we describe above, tiling arrays can be used to explore any facet of genome organization where a particular DNA feature can be enriched from the genome and they are finding most use for gene expression analysis or for mapping chromatin features such as transcription factor binding sites. We will briefly introduce the experimental methods underlying these areas before considering the analytical approaches necessary for interpreting tiling array data.

### 10.6.1 Transcript Analysis

The idea behind expression profiling with tiling arrays is to understand what regions of the genome are represented as mRNA rather than the traditional approach of using gene-based probes to determine the expression of defined genes. The general picture emerging from this type of unbiased transcriptional analysis is that far more of the metazoan genome is represented as messenger RNA than is suggested by our models of gene structure and the current challenge is understanding the nature and function of the unexpected transcription (Johnson et al., 2005). At a genome wide scale, several metazoans have now been characterized at varying degrees of resolution, ranging from relatively low-coverage partial tiles to very high-density partial tiles containing over 70% of the bases in a genome. The concept of using tiling arrays to refine genome annotation was demonstrated for the human genome by Shoemaker et al. (2001) for regions of chromosome 22. As exemplified in Figure 10.5, an overlapping tiling array composed of 60mer probes spaced every 10 bases on both strands of a genomic region is capable of narrowing the potential intron–exon boundary to a short sequence window ($\sim$20 bp) that can subsequently be confirmed with other methods such as EST sequence or RT-PCR. In addition, 5$'$ and 3$'$ untranslated regions for a given transcript may also be defined. Of course tiling the human genome at 10 bp resolution is not currently feasible, however, considerable information can be obtained from high-density tiling arrays and such studies have been reported for *Arabidopsis* (Yamada et al., 2003), *Drosophila* (Stolc et al., 2004; Manak et al., 2006), *C. elegans* (He et al., 2007) Rice (Li et al., 2006) and Human (Bertone et al., 2004).

It is now clear that using tiling arrays to explore genome transcription reveals a considerable amount of unannotated transcription in all the genomes where it has been looked at. In the case of *Drosophila*, where a detailed evaluation of some of these novel sequences was possible, many signals represent new 5$'$ exons located many kilobases away from annotated genes whereas other features identify new transcription units (Figure 10.6; Manak et al., 2006). These authors estimate as much as 30% of the fly genome is represented in mature mRNA, leading them to conclude that at least 85% of the genome is transcribed to generate this processed mRNA. In other cases, validation by RT-PCR, EST

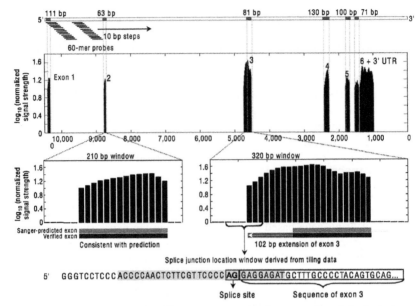

**FIGURE 10.5**  Using overlapping tiling arrays to annotate intro–exon boundaries in the human genome. A 10 kb region from chromosome 22 contains a transcript detected on predicted exon arrays. The 10 kb region was tiled on both strands with overlapping 60mers (top) and arrays hybridized with testis mRNA. The hybridization signals across the 10 kb region are shown in the boxed panel, identifying 6 exons. Zooming in on exons 2 and 3 shows the distribution of normalized hybridization signals for these exons. The hybridization signals for exon 2 agree with the prediction but for exon 3 suggest a 102 base extension, which is supported by the presence of a consensus AG splice acceptor site in the genomic sequence. The grey shading on the sequence tract indicates the extent of hybridization and typically allow the splice junction to be narrowed to a 20 base window. *Source*: From Shoemaker (2001).

sequence or cross-species conservation suggest variable levels of support for the hybridization signals representing novel genes or exons of known genes (Johnson et al., 2005). In addition to the potential exonic sequences, a considerable amount of anti-sense transcription is detected, at least some of which is believed to have regulatory functions.

There are several reasons for treating tiling array data with caution when defining gene structures, these can be divided into platform-specific effects, sample processing artefacts and problems with data extraction. As we indicate above, the probe sequences on tiling arrays are constrained by the resolution required and as a consequence some of the probes may be less than optimal. This can result in spurious hybridization signals due to cross-hybridization with samples derived from elsewhere in the genome. These may result from hybridization with expressed pseudogenes, related gene families or simply be a result of lack of probe specificity due to sequence composition. To control for these effects, the anticipated cross hybridization of every probe with sequences other

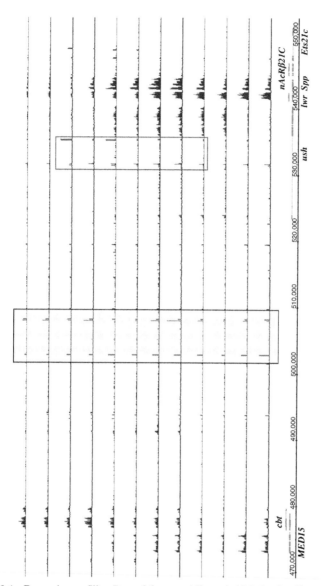

**FIGURE 10.6** Expression profiling *Drosophila* on an Affymetrix high-density tiling array. RNA extracted at 12 sequential 2 h time-points through *Drosophila* embryonic development was hybridized to an Affymetrix *Drosophila* Genome Array 1.0. The figure shows a 90 kb region of chromosome arm *2L*, containing the seven genes indicated by the gene models above the genomic coordinates, with the hybridization signals from the 12 time points from the earliest at the bottom to the latest at the top. The figure illustrates a typical finding where the boxed grey region shows reproducible hybridization signals in a region devoid of annotated transcription units. The signals in the grey-boxed region are unlikely to represent alternative 5′ ends for *ush* since their temporal profile is not in sync with *ush*. They may represent new 5′ exons for *cbt* or a novel transcript. The area

than the intended target must be determined (generally via a BLAST search) and potentially problematic probes flagged.

Sample preparation is an important issue for tiling experiments examining gene expression. Ideally, directly labeled mRNA, prepared using one of the methods described in Chapter 4, should be used (Kampa et al., 2004). However, this is not yet routine and samples have most frequently been labeled via a DNA intermediate generated by oligo-dT primed reverse transcription of total RNA or the polyA+ fraction. Since there is a considerable amount of nuclear RNA that is not processed or exported to the cytoplasm, expression profiling with tiling arrays should use RNA isolated from the cytosolic fraction, preventing contamination by unprocessed or partially processed nuclear transcripts that may lead to spurious signals from features such as introns. If an accurate view of the transcriptome is desired then reverse transcription reaction conditions that favor the generation of full-length cDNA is essential (Castle et al., 2003). This is in contrast to labeling for gene-specific expression arrays, where probes are biased towards the 3' end of the transcript and the frequently used labeling procedures generate shorter cDNAs. In some protocols, double-stranded cDNA is generated that is subsequently labeled, this procedure obviously results in a loss of strand specificity but is useful if it is necessary to keep the number of single strand oligonucleotide probes down. As a consequence of using a labeled double-stranded sample, observed intergenic or intronic hybridization signals may be due to hybridization with complementary sequences from elsewhere in the genome. Finally, some of the signals observed on tiling arrays may be due to contaminating DNA that inevitably copurifies with RNA, although some of this can be removed by DNase treatment or mitigated with (expensive) control hybridizations where samples are labeled without a reverse transcription step.

## 10.6.2 Chromatin

The large-scale analysis of genomic DNA with microarrays is revolutionizing the way that we explore genome organization via the ability to enrich for specific properties of in vivo derived chromatin. If a particular DNA-bound protein, DNA structure or modification can be separated from bulk chromatin then it is, in principal, a simple matter to identify that sequence with a specific microarray probe. As we have seen earlier, there are microarray platforms that facilitate analysis of entire genomes, opening up the possibility that the regulatory code underpinning the expression of the genome can be explored in a comprehensive

---

bounded by the light box represents signals within the large intron of *ush*, again the temporal profile of these signals suggests they are unlikely to be part of the *ush* gene and may represent a new intronic gene. The figure renders the processed data from Manak et al. (2006) in the Affymetrix Integrated Genome Browser.

*Source*: Manak et al. (2006).

way. More recently, the ability to explore polymorphisms in the human genome at a large scale with microarrays, be these single nucleotide differences or larger alterations in DNA copy number, hold out the promise of a much better understanding of how genotype relates to phenotype (Trevino et al., 2007). We deal with the genotyping applications in Chapter 11.

### 10.6.3 ChIP-Array

One of the first genome-wide applications of microarray technology outside of expression profiling was the mapping of in vivo protein binding sites. ChIP-array, ChIP-on-Chip or CHIP-Chip, as it is variably referred to, is most commonly based on the cross-linking chromatin immunopurification developed over a number of years by various groups (reviewed in Kuo and Allis, 1999). Proteins can be covalently attached to the DNA they are associated with in vivo by the addition of small cross-linking agents such as formaldehyde or, less commonly, using ultraviolet light treatment (Walter and Biggin, 1997). Chromatin is then randomly sheared, generally by sonication or controlled nuclease treatment, to an average size of 0.5–2 kb and specific protein–DNA complexes enriched from the general chromatin population by immunopurification with antibodies recognizing the protein of interest. Prior to microarray technology, enriched DNA sequences were identified by sequencing or hybridization to Southern blots (Orlando et al., 1997). The demonstration that chromatin immunopurification could be analyzed at a genome-wide scale on arrays came from two groups working with yeast (Ren et al., 2000; Iyer et al., 2001). These labs showed that microarrays containing PCR amplicon probes representing all of the intergenic sequences from *S. cerevisiae* were effective at identifying most or all of the in vivo binding sites for specific transcription factors. In yeast, the method has been used on a truly global scale with the binding sites for over 75% of the predicted transcription factors in *S. cerevisiae* mapped across the genome in a single study (Lee et al., 2002). The technique was rapidly adopted by those working with higher organisms but these studies faced two challenges: first was the development of specific in vivo immunopurification strategies for higher organisms and the second was overcoming the problem of metazoan genome size.

### 10.6.4 Purifying Chromatin

Unlike the single cell yeast, multicellular organisms are composed of hundreds, thousands or even millions of different cell types, presenting a considerable challenge if one wishes to examine tissue-specific binding of a particular transcription factor. Initial studies with metazoan systems were carried out using well-characterized antibodies in tissue culture cell systems, where large amounts of homogeneous chromatin can be relatively easily isolated: on the order of $10^7$ cells being required for standard protocols. Where factors are ubiquitously expressed and expected to occupy the same sites in different cell types, then whole tissue or organism analysis is feasible. In animals however, most tissues are composed of different cell types and a particular transcription factor may have a different repertoire of binding sites in each cell type. Thus specific cells must be purified, which may be difficult to achieve for around 10 million cells, or analysis must be restricted to an overview of potential binding sites in a range of cell types. As the molecular biology necessary for relatively unbiased DNA amplification has improved, the analysis of specific tissues (Adryan et al., 2007) or even purified cell types is now a realistic possibility. Most current ChIP protocols are based on the procedure developed by Farnham and colleagues (e.g. O'Geen et al., 2006; Sandmann et al., 2006), avoiding the inefficient density gradient ultracentrifugation steps of the original protocols. When this method is coupled with careful small sample procedures such as carrier-ChIP (O'Neill et al., 2006), it may be possible to carry out ChIP-array studies with as few as 100 cells. It is probably fair to say that, as long as specific cells can be tagged in some way, the technique is only limited by the persever-ance of the researcher in purifying the desired cell population. It is worth remembering that when formaldehyde or another small molecule is used as a cross-linking agent then chromatin can be fixed prior to purifying particular cell types if the tissue in question is sufficiently permeable to the cross-linking agent. Thus the in vivo state can be preserved before manipulations that could alter gene expression profiles or transcription factor binding, providing more oppor-tunity for selecting particular subpopulations from a sample by, for example, fluorescence-based sorting strategies.

In yeast the introduction of antigen tags into open reading frames in situ by homologous recombination is straightforward, facilitating the development of a set of uniform immunopurification conditions (Lee et al., 2002). This is less easy in higher eukaryotes and consequently many metazoan ChIP studies rely on specific antibodies that are shown to perform well in ChIP assays. There are currently over 2000 commercially available antibodies listed on the Chi-pOnChip.org website but very few have been validated in ChIP assays. Every antibody has different characteristics and individual biological samples will vary in their permeability to formaldehyde or the ease with which chromatin can be solubilized. As a consequence, it is impossible to provide a detailed 'one size fits all' protocol and careful evaluation of immunopurification conditions are re-quired for each experiment. A known in vivo binding site is useful for assaying the specificity and degree of enrichment obtained with a specific antibody. Quantitative real-time PCR can be used to measure cross-linking conditions, chromatin purification, antibody titre and postimmunopurification amplification and labeling. As an example, for an analysis of heat shock factor (Hsf) binding in *Drosophila* embryos (Birch-Machin et al., 2005), we prepared an affinity puri-fied polyclonal antisera against Hsf and assayed immunopurified DNA for known binding sites by quantitative PCR. We observed almost 100-fold enrichment comparing anti-Hsf immunopurified chromatin with a pre-immune sera control. In addition, we compared enrichment of DNA from the 5'-end of *hsp* genes (where Hsf is known to bind) with DNA from the 3'-end and again found similar levels of enrichment. Encouragingly, enrichment levels were the same with DNA before and after amplification by ligation mediated PCR, suggesting the amplification step may not be overly detrimental.

Where a known in vivo target is not available some workers recommend carrying out replicate ChIP-array assays with at least two different antibodies raised against the same protein (e.g. Sandmann et al., 2006). In metazoans it is also possible, in some circumstances, to use tagged proteins if appropriate transgenic strategies are available. This is relatively straightforward if the target tissue is a cultured cell line, since an appropriately tagged construct can be introduced by transfection or a stable cell line generated. In whole animals it may be possible to replace an endogenous coding sequence with a tagged version via homologous recombination. An alternative approach, is to use a transcription factor tagged with Green Fluorescent Protein, expressed from endogenous regulatory sequences and carried in a genetic background mutant for the endogenous factor. Immunopurification with GFP-specific antibodies can then be used (Adryan et al., 2007). Protein tagging screens, where artificial exons are introduced into endogenous genes by transposon-based strategies (Morin et al., 2001), can also generate tagged DNA-binding proteins but these must be evaluated to ensure the tagged gene has wild-type function. As with yeast, when tags are used then an untagged isogenic strain provides an excellent control. Typical experimental designs that include relevant controls are discussed below.

### 10.6.5 DamID

One ingenious method that overcomes the limitation of available antibodies is the DamID technique (van Steensel and Henikoff, 2000). Here, a fusion between the DNA-binding or chromatin protein of interest and the *E. coli* DNA adenine methyltransferase (Dam) is generated and expressed at very low levels in vivo. Dam methylates the adenosine residues in GATC motifs and, since there is very little or no adenine methylation in eukaryotes, the enzyme can be used to specifically mark eukaryotic DNA. If Dam is tethered to a particular DNA-binding protein then the methylation activity of the fusion should be enriched at the binding site of the protein of interest. This enriched DNA can be isolated by virtue of the activity of adenine methylation sensitive restriction enzyme *Dpn*I, which cuts G$^m$ATC but not GATC, thus producing a series of small restriction fragments where Dam has been active. If *Dpn*I digested DNA from a cell line or tissue expressing a transcription factor-Dam fusion is compared with a similar sample expressing only the Dam enzyme, then the binding site of the transcription factor can be inferred (Figure 10.7). This 'Dam-only' control is

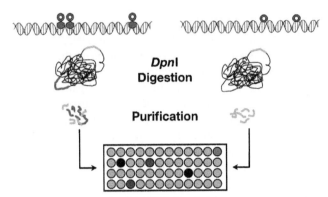

**FIGURE 10.7**   A schematic representation of the DamID technique. In the experimental strain on the left, a fusion protein between the factor of interest and Dam is expressed at low levels. Where specific binding sites are present (dark grey) the Dam will locally methylate but there will also be a degree of nonspecific methylation (light grey). On the right, a control strain expresses only the Dam, not tethered to another factor, which provides a measure of background methylation (light grey). DNA is purified from both strains and digested with *Dpn*I, which only cuts GATC when the Adenine is methylated. This releases small fragments of DNA, including sequences from the factor-binding site. After purification or methylation specific PCR amplification, fragments are differentially labeled and cohybridized to an array. Dark spots represent elements containing probes from the factor binding sites and the shades of grey represent nonspecific or background levels.

critical and cannot be replaced with other controls due to the relatively high activity of the enzyme. Specifically methylated DNA was originally isolated by sucrose gradient ultracentrifugation but this step is now more commonly replaced by a more efficient methyl fragment-specific PCR strategy (Greil et al., 2006). The technique has been used extensively in *Drosophila* but also, more recently in human (Vogel et al., 2007) and *Arabidopsis* (Germann et al., 2006). Although elegant, the method does suffer from some limitations that users need to be aware of. Dam is a highly processive enzyme, it has evolved to methylate the entire *Escherichia coli* genome in a very short period of time. As a consequence, if high levels of the Dam fusion are expressed then resolution is lost because the methylation spreads away from the tethering site, saturating considerable portions of the genome. Thus, Dam-fusions must be expressed at very low levels, effectively precluding the use of specific or inducible promoters to target expression to specific cells or tissues. Tissue-specific mapping of protein–DNA interactions with the Dam method are therefore not currently possible, although we have generated low-activity methylases via targeted mutagenesis that may circumvent this problem in the future.

### 10.6.6 Arrays

In the case of the second difficulty, metazoan genome size, as we have seen, even the largest genomes are beginning to come into the practical range of most laboratories budgets. With the human genome, various array strategies have been employed to overcome the difficulty posed by dealing with the whole genome. These include focusing on particular regions or chromosomes (Horak et al., 2002; Cawley et al., 2004), using arrays tiling only predicted promoter regions (Boyer et al., 2005) or arrays containing probes representing CpG islands (Yan et al., 2001). While an entire mammalian genome analysis is still outwith the reach of all but the wealthiest research labs (Bertone et al., 2004; Lee et al., 2006), smaller metazoan genomes are easily analyzed today (e.g. Schwartz et al., 2006) and it can only be a matter of time before whole genome ChIP-array is routinely available for mammals. It is worth noting that, depending upon the transcription factor of interest, it may be possible to use arrays containing full-length cDNA arrays in ChIP-array studies. If the factor of interest binds close to the transcriptional start site then sheared chromatin from the immuno-purification will overlap with the 5′-end of a full-length cDNA and thus can be detected with a cDNA array. While this approach has been used in *Drosophila* (e.g. van Steensel and Henikoff, 2000; Birch-Machin et al., 2005) it can be difficult to interpret the results since variable hybridization results may confound analysis.

## 10.6.7 Experimental Design and Controls

With expression profiling either a comparison between two samples (ratio measure) or an 'absolute' measure of hybridization intensity are used to infer gene expression levels and, in principal, the analysis of ChIP-array type experiments is similar: the goal is to identify elements on the array that show elevated hybridization compared to the signal level representative of the majority of the genome. In order to do this, DNA from enriched chromatin must be compared to a control sample and there are a variety of options available for assessing the hybridization or the immunopurification (Figure 10.8). To assess enrichment of particular sequences, the experimental sample should be compared with input chromatin prior to purification. On a two-color array this is achieved by cohy-bridizing input chromatin and the enriched material. If the experimental sample is amplified then a similar amount of input chromatin should be treated in the same way. This control corrects for differences in hybridization intensity between different probe elements that are due to sequence-specific probe effects, such as cross hybridization with multicopy sequences, and allows a more principled selection of enriched signals. For single channel arrays, two separate hybridizations need to be carried out. While the input chromatin control provides an assessment of probe performance and relative enrichment it does not address the reliability of the immunopurification. We have assessed some of the experimental variability associated with ChIP-array and found that replicate immunopurifications from a single chromatin sample using the same antibody generates 'IP-specific' signals (Birch-Machin et al., 2005), indicating that

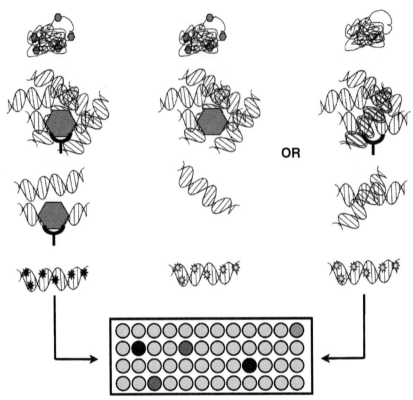

**FIGURE 10.8**   Schematic representation of the ChIP-array technique. On the left, a chromatin-binding protein distributed through the genome is represented as a grey hexgon. After cross-linking, chromatin is isolated and sheared. An antibody specific to the protein is used to enrich chromatin; the DNA associated with the protein in vivo is also enriched. The sample is usually then amplified and hybridized to an array with a control. Middle and right: potential controls. In the middle a mock immunopurification or immunopurification with a nonspecific antibody is used to generate control chromatin. On the right, the best control is to use chromatin from a strain where the gene for the factor of interest is deleted and to immunopurify with the specific antibody. A third control uses input chromatin prior to immunopurification. Experimental and control samples are differentially labeled and co-hybridized to an appropriate array; dark spots represents elements containing probes corresponding to the binding sites.

replicate immunopurifications are essential for reliable identification of binding locations. Several approaches are possible here, in yeast for example, it is sometimes possible to generate a strain carrying a deletion of the transcription factor of interest and to perform a parallel immunopurification with chromatin from the mutant strain. If an epitope tag has been used for immunopurification then a parallel immunopurification from an untagged but otherwise isogenic strain is an ideal control. In many cases however, this is not possible and alternative controls are needed. In one approach, a parallel immunopurification with preimmune sera or an unrelated antibody is used to assess nonspecific enrichment. If available, parallel immunopurifications with different antibodies recognizing the same protein can also be performed. Given the variability inherent in the immunopurification and labeling, it is sensible to use independent biological samples for the replicate immunopurifications since one can simultaneously control for technical and biological variability, selecting only those binding locations that behave consistently. Of course this approach may increase the false negative rate and if rigorous identification of binding locations is necessary then independent biological samples need to have replicated immunopurifications.

The ideal experimental design for a two-color array platform will utilize several biological samples, performing replicate imunopurifications on each and co-hybridizing the labeled DNA with input chromatin. In parallel, one of the immunopurification controls should be performed and similarly co-hybridized with input chromatin. For maximum rigor, each labeling reaction should be technically replicated with a dye-swap. For four biological samples this will require 32 arrays: an expensive proposition. Consequently many studies have taken the pragmatic decision to compare the enriched samples with the immunopurification controls rather than input chromatin, halving the number of arrays. In many cases, replicate immunopurifications are also omitted, with the independent biological samples used instead to control for immunopurification and labeling artefacts, further reducing the number of arrays required. For single channel arrays all numbers are duplicated but there is obviously no need for dye swaps. Clearly, the reliability with which signals can be identified decreases as fewer arrays are used.

## 10.7 ANALYSIS OF TILING ARRAYS

The data processing and analysis necessary for the identification of enriched chromatin with tiling arrays depends on the type of array used. Partial or full oligonucleotide tiling arrays require different treatment than amplicon or BAC arrays. The latter are typically treated in the same way that cDNA or amplicon expression arrays are, whereas the former require some additional evaluation to assess the reliability of a given probe signal.

### 10.7.1 Amplicon Arrays

A typical data processing pipeline for amplicon-based arrays will involve the normal spot finding and quantification, with or without background subtraction, and filtering for low-quality features. Within and between array normalization can follow standard approaches, some workers use a simple median normalization (e.g. Oberley et al., 2004) while others employ print-tip and intensity-dependant loess (Legube et al., 2006) or variance stabilizing normalization

(Birch-Machin et al., 2005). As Buck and Leib (Buck and Lieb, 2004) have pointed out, care with normalization is needed depending upon the type of DNA binding factor being analyzed. If, for example, a general factor with many binding sites is being studied it is possible that all the binding signals can be normalized away. It is therefore advisable to examine the intensity distributions of the raw data, prior to normalization, before deciding on an appropriate method. We note that, as was the case with array-based gene expression analysis, the increasing use of ChIP-array as an experimental tool is leading to the development of specific analytical tools that take into account the peculiarities of these data (e.g. Buck et al., 2005; Gibbons et al., 2005; Peng et al., 2007).

Ratios of experimental to control samples are calculated from the normalized data and then some threshold chosen to identify enriched array elements: here, as illustrated in Figure 10.9, difficulties can arise. In these examples of ChIP-array analysis using two different yeast factors (Iyer et al., 2001), median intensity values calculated from replicated experiments are ranked and two types of distribution are observed. The use of rank-based approaches are considered by some to be preferable for ChIP-array studies where many binding sites are identified since they are relatively insensitive to normalization methods. They are however less suitable when only a few sites are detected (Buck and Lieb, 2004). Other rank based methods have been adapted from gene expression

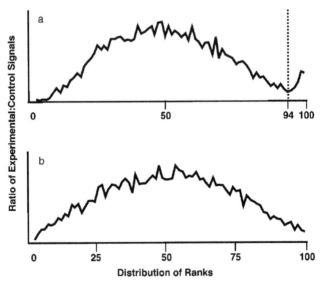

**FIGURE 10.9**   Typical profiles obtained from yeast ChIP-array studies. In both cases the distribution of rank enrichment ratios is plotted. In (a), a clearly discernable tail at the right hand side of the graph allows a cutoff to be defined. In (b), there is no clear separation of a high-ratio set from the buck distribution and thus a cutoff cannot be easily defined.
*Source*: Data from Iyer et al. (2001).

analysis to binding site data (e.g. Martin et al., 2004). The examples shown in Figure 10.9 illustrate the difficulty: in the first experiment there is a bimodal distribution of intensities, with a clear peak of high-ratio signals distinguishable from the bulk of the signals. It is therefore, in principal, easy to establish a cutoff and the top 6% of fragments can be selected as bound targets for further analysis. In the second example, however, there is a unimodal distribution and any cutoff chosen to select enriched targets from these data will be arbitrary. To overcome the problem in this particular case the ChIP-array data were combined with data from expression profiling to identify a cutoff where high-enrichment ratios were associated with genes that had particular expression characteristics: clearly this is not always possible.

An alternative method of defining binding sites is the single-array error model described by Roberts et al. (2000) for gene expression analysis and first used for ChIP-array studies in yeast (Ren et al., 2000). In this approach, the reliability of each measurement on an array is assessed according to a statistical model that takes into account the absolute intensity for each probe and the background noise for each channel to calculate a weighted average from a set of replicates. The resulting average enrichment ratio has an associated $p$-value that can be used to define a cutoff. Clearly, the lower the $p$-value threshold chosen then the fewer the number of binding events selected. The single-array error model has been successfully used in a ChIP-array analysis with human embryonic stem cells (Boyer et al., 2005). In contrast to the ranking method, the single-array error model is more accurate with small numbers of binding events in the population and less accurate with general factors identifying many binding sites (Buck and Lieb, 2004). Other appropriate statistical tests for assessing reproducibility can be used to calculate $p$-values, for example, Cyber-T (Baldi and Long, 2001) or Limma have been employed in various ChIP-array and DamID studies.

## 10.7.2 Oligo Arrays

While the methods described above may be used for the identification of peaks on oligonucleotide-based tiling arrays, since oligonucleotides can be less specific than longer amplicon probes then additional data processing is desirable (Royce et al., 2005). The most common way to deal with oligo-based tiling arrays is to include some assessment of how neighboring probes behave. At its simplest, a sliding window of specified size is used to scan along the genome and only groups of probes with signals (or ratios) above a given threshold used to define peaks. This eliminates the situation where a single probe gives a strong enrichment signal but the flanking probes are negative (e.g. Choksi et al., 2006). More rigorous approaches apply a statistical test to the probes within a window and compare this to a random sampling of intensities from a genomic region or known negative control regions. Qi et al. (2006)

developed the joint binding deconvolution probabilistic approach and Reiss et al. (2008) provide an alternative model-based approach (MeDiChI), both reported to improve the accuracy of peak-finding in ChIP-array data. With Affymetrix tiling arrays, which have short closely spaced probes, it is usual to define *maxgap* and *minrun* parameters, the maximum gap allowed between positive probe pairs and the minimum stretch of sequence containing positive probe pairs. For such arrays, Affymetrix provide tiling analysis software and others have been developed (e.g. the model-based approach, MAT; Johnson et al., 2006). Alternatively, an integrated package such as the TiMAT suite of JAVA scripts developed in Berkeley (http://bdtnp.lbl.gov/TiMAT/), may be used. Blob-like image artefacts are reported with Affymetrix tiling arrays and Song et al. (2007) have developed a software tool to compensate for these (MBR). In addition, the BioConductor suite of *R* tools has analysis software for tiling arrays and, as with standard gene expression arrays, this is an excellent source of state-of-the-art analytical approaches (Dudoit et al., 2003; Reimers and Carey, 2006; Scacheri et al., 2006). Those interested in using tiling arrays are recommended to perform literature searches or look at the BioConductor packages currently available since data analysis in this area is a rapidly developing field.

## 10.8 SUMMARY

Very high-density microarrays provide a route for the analysis of entire genomes within individual research labs. Although full coverage of large genomes, such as that of human, still requires several arrays, it is nevertheless remarkable that this level of analysis is now so easily achieved. The first burst of microarray analysis, using cDNA or oligonucleotide arrays, sparked a huge effort aimed at fully understanding the limitations of the data and developing robust analysis methods. This is the case now with exon and tiling arrays, where computational approaches that take into account the behavior of individual probes and the collective behavior of sets of probes are an active research area. It is highly likely that within a few years entire genome arrays will be routinely used for assessing gene expression, transcription factor binding or chromatin structure. While the current developments in ultra-high-throughput sequencing offer an alternative way for nucleic acid analysis, it should not be forgotten that sequence-based methods have their own biases and limitation. We believe that both platforms are complimentary rather than exclusive, each with its strengths and weaknesses. While sequencing has the potential to generate more quantitative data, the considerable ease with which array-based data can be generated and analyzed makes the microarray more suitable for high-throughput analysis. This may be advantageous, for example in diagnostic settings or where thousands of samples need to be processed in parallel.

# REFERENCES

Abdueva D, Wing MR, Schaub B, Triche TJ. (2007) Experimental comparison and evaluation of the Affymetrix exon and U133Plus2 GeneChip arrays. *PLoS ONE* **2:** e913.

Adryan B, Woerfel G, Birch-Machin I, Gao S, Quick M, Meadows L, Russell S, White R. (2007) Genomic mapping of suppressor of Hairy-wing binding sites in *Drosophila*. *Genome Biol* **8:** R167.

Baldi P, Long AD. (2001) A Bayesian framework for the analysis of microarray expression data: regularized *t*-test and statistical inferences of gene changes. *Bioinformatics* **17:** 509–519.

Berman P, Bertone P, Dasgupta B, Gerstein M, Kao MY, Snyder M. (2004) Fast optimal genome tiling with applications to microarray design and homology search. *J Comput Biol* **11:** 766–785.

Bertone P, Stolc V, Royce TE, Rozowsky JS, Urban AE, Zhu X, Rinn JL, Tongprasit W, Samanta M, Weissman S, Gerstein M, Snyder M. (2004) Global identification of human transcribed sequences with genome tiling arrays. *Science* **306:** 2242–2246.

Bertone P, Trifonov V, Rozowsky JS, Schubert F, Emanuelsson O, Karro J, Kao MY, Snyder M, Gerstein M. (2006) Design optimization methods for genomic DNA tiling arrays. *Genome Res* **16:** 271–281.

Birch-Machin I, Gao S, Huen D, McGirr R, White RA, Russell S. (2005) Genomic analysis of heat-shock factor targets in *Drosophila*. *Genome Biol* **6:** R63.

Boyer LA, Lee TI, Cole MF, Johnstone SE, Levine SS, Zucker JP, Guenther MG, Kumar RM, Murray HL, Jenner RG, Gifford DK, Melton DA, Jaenisch R, Young RA. (2005) Core transcriptional regulatory circuitry in human embryonic stem cells. *Cell* **122:** 947–956.

Bozdech Z, Zhu J, Joachimiak MP, Cohen FE, Pulliam B, DeRisi JL. (2003) Expression profiling of the schizont and trophozoite stages of *Plasmodium falciparum* with a long-oligonucleotide microarray. *Genome Biol* **4:** R9.

Buck M, Lieb J. (2004) ChIP-chip: considerations for the design, analysis and application of genome-wide chromatin immunoprecipitation experiments. *Genomics* **83:** 349–360.

Buck M, Nobel A, Lieb J. (2005) ChIPOTle: a user-friendly tool for the analysis of ChIP-chip data. *Genome Biol* **6:** R97.

Castle J, Garrett-Engele P, Armour CD, Duenwald SJ, Loerch PM, Meyer MR, Schadt EE, Stoughton R, Parrish ML, Shoemaker DD. (2003) Optimization of oligonucleotide arrays and RNA amplification protocols for analysis of transcript structure and alternative splicing. *Genome Biol* **4:** R66.

Cawley S, Bekiranov S, Ng HH, Kapranov P, Sekinger EA, Kampa D, Piccolboni A, Sementchenko V, Cheng J, Williams AJ, Wheeler R, Wong B, Drenkow J, Yamanaka M, Patel S, Brubaker S, Tammana H, Helt G, Struhl K, Gingeras TR. (2004) Unbiased mapping of transcription factor binding sites along human chromosomes 21 and 22 points to widespread regulation of noncoding RNAs. *Cell* **116:** 499–509.

Choksi SP, Southall TD, Bossing T, Edoff K, de Wit E, Fischer BE, van Steensel B, Micklem G, Brand AH. (2006) Prospero acts as a binary switch between self-renewal and differentiation in *Drosophila* neural stem cells. *Dev Cell* **11:** 775–789.

Clark TA, Schweitzer AC, Chen TX, Staples MK, Lu G, Wang H, Williams A, Blume JE. (2007) Discovery of tissue-specific exons using comprehensive human exon microarrays. *Genome Biol* **8:** R64.

Dudoit S, Gentleman RC, Quackenbush J. (2003) Open source software for the analysis of microarray data. *Biotech Suppl* , pp. 45–51.

Frey BJ, Mohammad N, Morris QD, Zhang W, Robinson MD, Mnaimneh S, Chang R, Pan Q, Sat E, Rossant J, Bruneau BG, Aubin JE, Blencowe BJ, Hughes TR. (2005) Genome-wide analysis of mouse transcripts using exon microarrays and factor graphs. *Nat Genet* **37:** 991–996.

Germann S, Juul-Jensen T, Letarnec B, Gaudin V. (2006) DamID, a new tool for studying plant chromatin profiling in vivo, and its use to identify putative LHP1 target loci. *Plant J* **48:** 153–163.

Ghosh S, Hirsch HA, Sekinger EA, Kapranov P, Struhl K, Gingeras TR. (2007) Differential analysis for high-density tiling microarray data. *BMC Bioinform* **8:** 359.

Gibbons F, Proft M, Struhl K, Roth F. (2005) Chipper: discovering transcription factor targets from chromatin immunoprecipitation microarrays using variance stabilization. *Genome Biol* **6:** R96.

Greil F, Moorman C, van Steensel B. (2006) DamID: mapping of in vivo protein–genome interactions using tethered DNA adenine methyltransferase. *Methods Enzymol* **410:** 342–359.

He H, Wang J, Liu T, Liu XS, Li T, Wang Y, Qian Z, Zheng H, Zhu X, Wu T, Shi B, Deng W, Zhou W, Skogerb½ G, Chen R. (2007) Mapping the *C. elegans* noncoding transcriptome with a whole-genome tiling microarray. *Genome Res* **17:** 1471–1477.

Hoheisel JD. (2006) Microarray technology: beyond transcript profiling and genotype analysis. *Nat Rev Genet* **7:** 200–210.

Hollich V, Johnson E, Furlong EE, Beckmann B, Carlson J, Celniker SE, Hoheisel JD. (2004) Creation of a minimal tiling path of genomic clones for *Drosophila*: provision of a common resource. *Biotechniques* **37:** 282–284.

Horak CE, Mahajan MC, Luscombe NM, Gerstein M, Weissman SM, Snyder M. (2002) GATA-1 binding sites mapped in the beta-globin locus by using mammalian chIp-chip analysis. *Proc Natl Acad Sci U S A* **99:** 2924–2929.

Ishkanian AS, Malloff CA, Watson SK, DeLeeuw RJ, Chi B, Coe BP, Snijders A, Albertson DG, Pinkel D, Marra MA, Ling V, MacAulay C, Lam WL. (2004) A tiling resolution DNA microarray with complete coverage of the human genome. *Nat Genet* **36:** 299–303.

Iyer VR, Horak CE, Scafe CS, Botstein D, Snyder M, Brown PO. (2001) Genomic binding sites of the yeast cell-cycle transcription factors SBF and MBF. *Nature* **409:** 533–538.

Johnson JM, Castle J, Garrett-Engele P, Kan Z, Loerch PM, Armour CD, Santos R, Schadt EE, Stoughton R, Shoemaker DD. (2003) Genome-wide survey of human alternative pre-mRNA splicing with exon junction microarrays. *Science* **302:** 2141–2144.

Johnson JM, Edwards S, Shoemaker D, Schadt EE. (2005) Dark matter in the genome: evidence of widespread transcription detected by microarray tiling experiments. *Trends Genet* **21:** 93–102.

Johnson WE, Li W, Meyer CA, Gottardo R, Carroll JS, Brown M, Liu XS. (2006) Model-based analysis of tiling-arrays for ChIP-chip. *Proc Natl Acad Sci U S A* **103:** 12457–12462.

Kampa D, Cheng J, Kapranov P, Yamanaka M, Brubaker S, Cawley S, Drenkow J, Piccolboni A, Bekiranov S, Helt G, Tammana H, Gingeras TR. (2004) Novel RNAs identified from an in-depth analysis of the transcriptome of human chromosomes 21 and 22. *Genome Res* **14:** 331–342.

Kuo MH, Allis CD. (1999) In vivo cross-linking and immunoprecipitation for studying dynamic Protein:DNA associations in a chromatin environment. *Methods* **19:** 425–433.

Kwan T, Benovoy D, Dias C, Gurd S, Provencher C, Beaulieu P, Hudson TJ, Sladek R, Majewski J. (2008) Genome-wide analysis of transcript isoform variation in humans. *Nat Genet* **40:** 225–231.

Lee TI, Jenner RG, Boyer LA, Guenther MG, Levine SS, Kumar RM, Chevalier B, Johnstone SE, Cole MF, Isono K, Koseki H, Fuchikami T, Abe K, Murray HL, Zucker JP, Yuan B, Bell GW, Herbolsheimer E, Hannett NM, Sun K, Odom DT, Otte AP, Volkert TL, Bartel DP, Melton DA,

Gifford DK, Jaenisch R, Young RA. (2006) Control of developmental regulators by polycomb in human embryonic stem cells. *Cell* **125:** 301–313.

Lee TI, Rinaldi NJ, Robert F, Odom DT, Bar-Joseph Z, Gerber GK, Hannett NM, Harbison CT, Thompson CM, Simon I, Zeitlinger J, Jennings EG, Murray HL, Gordon DB, Ren B, Wyrick JJ, Tagne JB, Volkert TL, Fraenkel E, Gifford DK, Young RA. (2002) Transcriptional regulatory networks in *Saccharomyces cerevisiae. Science* **298:** 799–804.

Legube G, McWeeney SK, Lercher MJ, Akhtar A. (2006) X-chromosome-wide profiling of MSL-1 distribution and dosage compensation in *Drosophila. Genes Dev* **20:** 871–883.

Li L, Wang X, Stolc V, Li X, Zhang D, Su N, Tongprasit W, Li S, Cheng Z, Wang J, Deng XW. (2006) Genome-wide transcription analyses in rice using tiling microarrays. *Nat Genet* **38:** 124–129.

MacAlpine DM, Rodriguez HK, Bell SP. (2004) Coordination of replication and transcription along a *Drosophila* chromosome. *Genes Dev* **18:** 3094–3105.

Manak JR, Dike S, Sementchenko V, Kapranov P, Biemar F, Long J, Cheng J, Bell I, Ghosh S, Piccolboni A, Gingeras TR. (2006) Biological function of unannotated transcription during the early development of *Drosophila melanogaster. Nat Genet* **38:** 1151–1158.

Martin D, Demougin P, Hall M, Bellis M. (2004) Rank difference analysis of microarrays (RDAM), a novel approach to statistical analysis of microarray expression profiling data. *BMC Bioinform* **5:** R148.

Morin X, Daneman R, Zavortink M, Chia W. (2001) A protein trap strategy to detect GFP-tagged proteins expressed from their endogenous loci in *Drosophila. Proc Natl Acad Sci U S A* **98:** 15050–15055.

O'Geen H, Nicolet CM, Blahnik K, Green R, Farnham PJ. (2006) Comparison of sample preparation methods for ChIP-chip assays. *Biotechniques* **41:** 577–580.

O'Neill LP, VerMilyea MD, Turner BM. (2006) Epigenetic characterization of the early embryo with a chromatin immunoprecipation protocol applicable to small cell populations. *Nat Genet* **38:** 835–841.

Oberley MJ, Tsao J, Yau P, Farnham PJ. (2004) High-throughput screening of chromatin immuno-precipitates using CpG-island microarrays. *Methods Enzymol* **376:** 315–334.

Orlando V, Strutt H, Paro R. (1997) Analysis of chromatin structure by in vivo formaldehyde cross-linking. *Methods* **11:** 205–214.

Pan Q, Shai O, Misquitta C, Zhang W, Saltzman AL, Mohammad N, Babak T, Siu H, Hughes TR, Morris QD, Frey BJ, Blencowe BJ. (2004) Revealing global regulatory features of mammalian alternative splicing using a quantitative microarray platform. *Mol Cell* **16:** 929–941.

Peng S, Alekseyenko A, Larschan E, Kuroda M, Park P. (2007) Normalization and experimental design for ChIP-chip data. *BMC Bioinform* **8:** 219.

Qi Y, Rolfe A, MacIsaac KD, Gerber GK, Pokholok D, Zeitlinger J, Danford T, Dowell RD, Fraenkel E, Jaakkola TS, Young RA, Gifford DK. (2006) High-resolution computational models of genome binding events. *Nat Biotechnol* **24:** 963–970.

Reimers M, Carey VJ. (2006) Bioconductor: an open source framework for bioinformatics and computational biology. *Methods Enzymol* **411:** 119–134.

Reiss DJ, Facciotti MT, Baliga NS. (2008) Model-based deconvolution of genome-wide DNA binding. *Bioinformatics* **24:** 396–403.

Ren B, Robert F, Wyrick JJ, Aparicio O, Jennings EG, Simon I, Zeitlinger J, Schreiber J, Hannett N, Kanin E, Volkert TL, Wilson CJ, Bell SP, Young RA. (2000) Genome-wide location and function of DNA binding proteins. *Science* **290:** 2306–2309.

Rinn JL, Euskirchen G, Bertone P, Martone R, Luscombe NM, Hartman S, Harrison PM, Nelson FK, Miller P, Gerstein M, Weissman S, Snyder M. (2003) The transcriptional activity of human Chromosome 22. *Genes Dev* **17**: 529–540.

Roberts CJ, Nelson B, Marton MJ, Stoughton R, Meyer MR, Bennett HA, He YD, Dai H, Walker WL, Hughes TR, Tyers M, Boone C, Friend SH. (2000) Signaling and circuitry of multiple MAPK pathways revealed by a matrix of global gene expression profiles. *Science* **287**: 873–880.

Royce T, Rozowsky J, Bertone P, Samanta M, Stolc V, Weissman S, Snyder M, Gerstein M. (2005) Issues in the analysis of oligonucleotide tiling microarrays for transcript mapping. *Trends Genet* **21**: 466–475.

Rozen S, Skaletsky H. (2000) Primer3 on the WWW for general users and for biologist programmers. *Methods Mol Biol* **132**: 365–386.

Ryder E, Jackson R, Ferguson-Smith A, Russell S. (2006) MAMMOT – a set of tools for the design, management and visualization of genomic tiling arrays. *Bioinformatics* **22**: 883–884.

Sandmann T, Jakobsen JS, Furlong EE. (2006) ChIP-on-chip protocol for genome-wide analysis of transcription factor binding in *Drosophila melanogaster* embryos. *Nat Protoc* **1**: 2839–2855.

Scacheri PC, Crawford GE, Davis S. (2006) Statistics for ChIP-chip and DNase hypersensitivity experiments on NimbleGen arrays. *Methods Enzymol* **411**: 270–282.

Schwartz YB, Kahn TG, Nix DA, Li XY, Bourgon R, Biggin M, Pirrotta V. (2006) Genome-wide analysis of Polycomb targets in *Drosophila melanogaster. Nat Genet* **38**: 700–705.

Shoemaker DD, Schadt EE, Armour CD, He YD, Garrett-Engele P, McDonagh PD, Loerch PM, Leonardson A, Lum PY, Cavet G, Wu LF, Altschuler SJ, Edwards S, King J, Tsang JS, Schimmack G, Schelter JM, Koch J, Ziman M, Marton MJ, Li B, Cundiff P, Ward T, Castle J, Krolewski M, Meyer MR, Mao M, Burchard J, Kidd MJ, Dai H, Phillips JW, Linsley PS, Stoughton R, Scherer S, Boguski MS. (2001) Experimental annotation of the human genome using microarray technology. *Nature* **409**: 922–927.

Song JS, Maghsoudi K, Li W, Fox E, Quackenbush J, Shirley Liu X. (2007) Microarray blob-defect removal improves array analysis. *Bioinformatics* **23**: 966–971.

Stolc V, Gauhar Z, Mason C, Halasz G, van Batenburg MF, Rifkin SA, Hua S, Herreman T, Tongprasit W, Barbano PE, Bussemaker HJ, White KP. (2004) A gene expression map for the euchromatic genome of *Drosophila melanogaster. Science* **306**: 655–660.

Trevino V, Falciani F, Barrera-Saldana HA. (2007) DNA microarrays: a powerful genomic tool for biomedical and clinical research. *Mol Med* **13**: 527–541.

van Steensel B, Henikoff S. (2000) Identification of in vivo DNA targets of chromatin proteins using tethered dam methyltransferase. *Nat Biotechnol* **18**: 424–428.

Vogel MJ, Peric-Hupkes D, van Steensel B. (2007) Detection of in vivo protein–DNA interactions using DamID in mammalian cells. *Nat Protoc* **2**: 1467–1478.

Walter J, Biggin MD. (1997) Measurement of in vivo DNA binding by sequence-specific transcription factors using UV cross-linking. *Methods* **11**: 215–224.

Watson SK, deLeeuw RJ, Ishkanian AS, Malloff CA, Lam WL. (2004) Methods for high throughput validation of amplified fragment pools of BAC DNA for constructing high resolution CGH arrays. *BMC Genomics* **5**: 6.

Xing Y, Kapur K, Wong WH. (2006) Probe selection and expression index computation of Affymetrix Exon Arrays. *PLoS ONE* **1**: e88.

Xing Y, Ouyang Z, Kapur K, Scott MP, Wong WH. (2007) Assessing the conservation of mammalian gene expression using high-density exon arrays. *Mol Biol Evol* **24**: 1283–1285.

Yamada K, Lim J, Dale JM, Chen H, Shinn P, Palm CJ, Southwick AM, Wu HC, Kim C, Nguyen M, Pham P, Cheuk R, Karlin-Newmann G, Liu SX, Lam B, Sakano H, Wu T, Yu G, Miranda M, Quach HL, Tripp M, Chang CH, Lee JM, Toriumi M, Chan MM, Tang CC, Onodera CS, Deng JM, Akiyama K, Ansari Y, Arakawa T, Banh J, Banno F, Bowser L, Brooks S, Carninci P, Chao Q, Choy N, Enju A, Goldsmith AD, Gurjal M, Hansen NF, Hayashizaki Y, Johnson-Hopson C, Hsuan VW, Iida K, Karnes M, Khan S, Koesema E, Ishida J, Jiang PX, Jones T, Kawai J, Kamiya A, Meyers C, Nakajima M, Narusaka M, Seki M, Sakurai T, Satou M, Tamse R, Vaysberg M, Wallender EK, Wong C, Yamamura Y, Yuan S, Shinozaki K, Davis RW, Theologis A, Ecker JR. (2003) Empirical analysis of transcriptional activity in the Arabidopsis genome. *Science* **302:** 842–846.

Yan PS, Chen CM, Shi H, Rahmatpanah F, Wei SH, Caldwell CW, Huang TH. (2001) Dissecting complex epigenetic alterations in breast cancer using CpG island microarrays. *Cancer Res* **61:** 8375–8380.

## Chapter 11

# Medical Applications of Microarray Technology

Perhaps more than any other application, the prospect of using microarray technology for exploring facets of human disease has been the focus of much attention and considerable optimism. Articles and reviews from around the turn of the century, with tantalizing titles such as 'Array of Hope' (Lander, 1999), held out the prospect that microarrays would revolutionize our exploration of the human genome and help uncover the molecular basis of disease. In many respects much of this optimism has been born out: many thousands of gene expression profiling experiments are beginning to provide insights into the molecular etiology of heterogeneous diseases such as cancer. The use of array technology to explore sequence variation in the human genome, at the level of gross genome deletions and amplifications, or by allowing truly massive SNP detection efforts via genotyping arrays, facilitates the mapping of genes implicated in simple monogenic diseases and, more importantly, offers a route for uncovering the genetic basis of much more complex multigenic diseases. In this chapter we shall look at some of the array technologies applicable to the analysis of the human genome from a biomedical perspective and highlight some of the analytical problems that may be encountered.

## 11.1 INTRODUCTION

The excitement that microarray technology generated in the field of medicine and biotechnology is perhaps best exemplified by the influential series of reviews published by Nature Genetics under the titles 'The Chipping Forecast' and 'The Chipping Forecast II', published in 1999 (http://www.nature.com/ng/journal/v21/n1s/index.html) and 2002, respectively (http://www.nature.com/ng/journal/v32/n4s/index.html). These articles provide an overview of how it was thought that microarrays could be applied and what could potentially be discovered. Of note is the much greater emphasis placed on data analysis and statistics in the second series of reviews, highlighting the sobering realization that microarray data was much more difficult to deal with than was perhaps realized in the early days of array research. As we hope we have shown in this volume, many of the issues that are of concern with respect to data analysis are being actively addressed and while the technology is by no means perfect, it is clear that significant new biology relevant to human disease can be garnered from well-designed microarray studies. In addition, the increasing efforts in the area of data standardization, well-designed controls and metadata collection, are beginning to enable data-mining studies with thousands of experiments.

As with any biology, gene expression profiling can be used to ask basic questions in disease studies. By comparing a diseased cell or tissue with the normal equivalent, the researcher may hope to uncover sets of genes that are differentially expressed, either up or down regulated, in the disease state. Such genes may then indicate what aspects of the biology or biochemistry of the cell are disrupted in the disease, suggesting avenues for therapeutic intervention. One can imagine that expression profiling may see action on a number of fronts in the war against disease and the technology is now a firmly established part of the experimental infrastructure in both academia and the pharmaceutical or biotechnology industries. The discovery pipeline is a commonly used analogy for describing the stages a research effort needs to go through when developing a new drug for a particular disease and array technology is firmly embedded in the pipeline at multiple stages (Figure 11.1; for reviews see Gerhold et al., 2002; Meloni et al., 2004; Phillips et al., 2006). Starting from a particular disease, exploratory expression profiling of a set of clinical samples can identify potential drug targets, finding, for example, specific kinases that are upregulated in the disease compared to the normal tissue. Such targets need to be validated, perhaps by examining additional clinical samples or disease models as well as carrying out genetic (loss and gain of function) studies to firmly establish that the target is indeed causative: again large scale expression analysis can be of use here. Once a particular target is shown to be robustly associated with the disease it becomes a

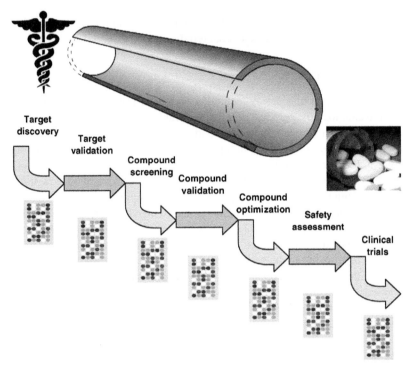

**FIGURE 11.1**   The drug discovery pipeline. Starting with a disease (symbolized by the caduceus on the left) and ending with a drug on the right, a number of stages must be traversed. Each stage becomes progressively more expensive and also results in the elimination of potential targets or candidate drugs. As described in the text, array technology may be employed as an assay at each of these stages.

target for compound screening, usually via testing very large compound libraries. Expression profiling can play a role here, if for example a specific gene expression response can be attributed to the specific target. In subsequent compound validation and optimization stages, expression analysis can help identify the most potent version of a drug and also indicate any unanticipated 'off-target' effects by highlighting unusual gene expression responses. In the area of safety assessment, animal models and/or relevant cell lines (particularly liver cells) are exposed to the drug and here gene expression profiling can be highly informative. Finally, in both early and late stage clinical trials, expression biomarkers derived from the earlier microarray studies or even whole-transcriptome arrays, may be used to monitor drug efficacy and any potential in vivo toxicity. Prior to array analysis, all of these steps were laborious and, more importantly, relied on a rather limited set of markers.

   As well as drug discovery, a huge area that utilizes expression profiling or genotyping technology is in the search for diagnostic tools. Here the objective is to identify a relevant set of biomarkers that will reliably indicate the presence of the disease. Such assays may be aimed at assessing risk, the potential or likelihood of contracting a disease; the identification of a particular genotype being an example here. In screening situations, where the presence of the disease is confirmed by a molecular profile or in prognostic testing, where variants of the disease that respond to particular treatments are identified, gene expression analysis is becoming increasingly prevalent. Similarly, monitoring gene expression may be used as a marker for the efficacy of a particular treatment, allowing an assessment of whether or not a particular drug regime is in fact working. Finally, not all diseases are genetic and the use of microarray technology for genotyping pathogenic organisms or for exploring the basic biology of pathogens to identify drugable targets or vaccine candidates is also a promising area.

Taken together, there are many potential applications for genome or expression analysis in disease studies and clinical settings. However, one should bear in mind that the data generated by genomics approaches, particularly gene expression analysis, can be complex and difficult to interpret. This can raise difficulties with regulatory authorities such as the US Food and Drug Administration (FDA) or other licensing bodies. Regulators have traditionally dealt with drug or diagnostic test approvals by examining supporting data based on a limited number of biomarkers. When supporting data potentially examines the expression of every gene in the genome then there is clearly added complexity. This is especially the case when there are no rigorously applied standards in terms of experimental design, data capture, data processing or statistical analysis: potentially every experiment is different, every analysis utilizing different thresholds and cutoffs and so on (see Frueh, 2006; Gutman and Kessler, 2006 for discussions of these issues from an FDA perspective). Obviously regulators are aware of these difficulties and have been proactive in working with the pharmaceutical industry and other interested parties to develop robust standards. In particular, the FDA have been key partners in the Microarray Quality Control project (Shi et al., 2006, 2008) and the External RNA Controls Consortium (Baker et al., 2005). Such initiatives highlight the pressing importance of developing robust standards to allow reliable assessment of genomics data (Ji and Davis, 2006). The FDA have published guidance on submitting genomics data in support of applications (Goodsaid and Frueh, 2007) but this is clearly a rapidly developing area and one can envisage that the requirements will become more stringent as quality control metrics currently being defined become de facto standards.

In this chapter we provide a series of vignettes, focused on particular aspects of disease, to highlight how array technology is currently being deployed in biomedical research. While it is clear that each disease offers a unique set of challenges in terms of sample acquisition or the translation of insights from model systems to in vivo situations, we hope that the approach we take here is instructive. We have opted to use a limited number of biological examples to allow some explanation of the background biology within a reasonable space constraint. At the same time we have tried to select the examples based on the experimental challenges or design issues they present, to give a balanced overview of the type of questions that array technologies are currently addressing. With these considerations in mind we present malaria as an example of pathogen studies and selected examples of cancer biology to represent heterogeneous disease situations.

## 11.2 INVESTIGATING PATHOGENS: MALARIA

The principal of using gene expression profiling to explore the biology of pathogens is clearly attractive: if pathogen-specific genes or pathways necessary for the life cycle of the organism can be identified then an obvious set of drug targets present themselves. However, there can be problems with the approach since it may be difficult or impossible to isolate sufficient quantities of the pathogen at relevant stages of its life cycle. This may be due to limited numbers of pathogens, restriction to tissues that are difficult to access by biopsy, or because the organism has a complex life cycle. For example, in the case of *Mycobacterium tuberculosis*, the causative agent responsible for tuberculosis, a particular stage of interest is an apparent latent stage of infection, where the bacillus is sequestered within in a host generated granuloma and enters a dormant phase. Changes in the immune status of the host can trigger the bacillus to begin replicating again and initiate an infection. The granulomas are not easily accessed in large quantities and thus the status of the bacillus with respect to the changes accompanying the switch to a replicative phase are poorly understood. Clearly the ability to characterize the transcriptome would be very informative but none of the in vitro or animal models used to study the disease recapitulate all the features of the organisms complex and heterogeneous life in humans (Murphy and Brown, 2007). Thus while aspects of the underlying biology can come from model systems it must be borne in mind that ultimately any inferences based on expression analysis need to be validated under conditions of human infection or disease.

As an example of how microarrays can be used to explore the biology of a pathogen and suggest therapeutic strategies, we focus on the principal malarial parasite *Plasmodium falciparum*. We do this for two reasons; first, malaria is an incredibly debilitating disease that is responsible for at least 1 000 000 deaths a year worldwide. It has recently been estimated that over two billion people (almost a 1/3 of the global human population) are at risk from malaria, with sub-Saharan Africa bearing the brunt of the morbidity and mortality (Guerra et al., 2008). Second, the parasite offers some interesting challenges in terms of its biology that exemplify the utility of genomics approaches. The parasite has a complex life cycle, alternating between humans and an insect vector, the mosquito *Anopheles gambiae*. When humans are bitten by a *Plasmodium* infected female mosquito (only females take blood meals, which

they require for oogenesis), sporozoites that mature in the insects salivary gland
are injected into the bloodstream where they quickly travel to the liver and
invade hepatocytes. In the liver, the sporozoites differentiate into merozoites,
which are subsequently released into the bloodstream and invade red blood cells.
Within red blood cells, merozoites replicate via intermediate stages (trophozoite
and schizont) and continue to be released infecting more red blood cells. Occasionally merozoites initiate a sexual cycle and differentiate into male and female
gametocytes, which are subsequently taken up by a mosquito during a blood
meal, where they produce more sporozoites (Figure 11.2). The decreasing
efficacy of established drugs or effective insecticides, due to rapidly evolving
parasite resistance, coupled with the lack of an effective vaccine (as well as
parasite evolution there appears to be little commercial incentive to develop
vaccines for diseases that principally affect the developing world) mean that we
urgently need to understand the biology of the parasite and find new therapeutic
targets.

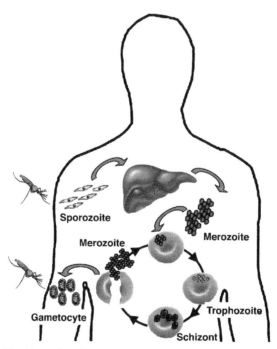

**FIGURE 11.2**   The *Plasmodium falciparum* lifecycle. A bite from an infected mosquito injects
Sporozoites into the bloodstream, which travel to the liver and differentiate into Merozoites. The
Merozoites invade red blood cells where they replicate via the Tropozoite and Schizont stages to
produce more Merozoites that, in turn, infect more red blood cells. Occasionally Gametocytes are
generated in the red blood cells, which enter the blood stream and are picked up by a mosquito taking
a blood meal. The sexual life cycle stages within the mosquito to generate new Sporozoites are not
shown.

Progress in this direction was accelerated in 2002 when the genome sequence of *P. falciparum* was published (Gardner et al., 2002) in the same month that the *A. gambiae* sequence was also released (Holt et al., 2002). The genome sequence had a few surprises in store: only 23 Mb in size, the nuclear genome is extremely AT rich (80%), a fact that complicated the sequencing and assembly. The genome is predicted to harbor a little over 5000 protein coding genes and, at the time, over 60% of the predicted genes could not be assigned any function since they showed no sequence similarity with any other sequenced genomes. Clearly this presents a challenge in terms of understanding the biology of the parasite and for identifying likely drug targets (for recent reviews of plasmodium genomics see Coppel et al., 2004; Kooij et al., 2006). While cDNA arrays had been generated prior to the genome sequence (Hayward et al., 2000; Ben Mamoun et al., 2001), these were limited and the annotated sequence allowed a more principled approach to probe design. Two groups separately designed oligonucleotide arrays based on the sequence: in one case, Le Roch et al. (2003) designed an Affymetrix GeneChip containing over 250 000 25mer probes interrogating 5159 predicted genes and an additional 100 000 probes corresponding to intergenic regions, introns and the noncoding strand of predicted genes. DeRisi and co-workers (Bozdech et al., 2003b) designed 7462 70mer probes against 4500 predicted genes with their OligoArraySelector software, using very strict specificity criteria (Bozdech et al., 2003a), and generated spotted arrays via contact printing. Both teams used their array platforms to examine gene expression during specific stages of the *Plasmodium* life cycle.

In the Le Roch et al. study, nine developmental stages were chosen: seven representing the intraerythocytic stages, isolated from a synchronously infected red blood cell culture system (two synchronization methods were independently applied to control for variation in this process). Gametocytes were also purified from similar culture systems and sporozoites were isolated from infected mosquitoes. A point to note in this experiment is that while most samples were labeled using the standard Affymetrix protocol, sporozoite stages were labeled after an RNA amplification step because of low sample yields. This raises questions regarding the comparability of the expression levels from this stage and the other stages represented by nonamplified RNA samples. Data were processed using the standard Affymetrix average intensity normalization method and a set of negative control probes used to define background hybridization levels. The authors utilize a probe-level analysis to calculate expression values, claiming a five orders of magnitude range in expression levels and suggesting that 88% of the predicted genes are expressed during at least one time point. From a technical stand-point, there is little replication in the experiment: there is a single array hybridized for each of the seven erythrocytic stages generated by each synchronization method, providing two expression measures for each stage. However, these we consider to be rather poor replicates since the samples are derived from independent methods. Having said that, to a certain extent the fact that the experiment is a time course study provides a degree of support for the temporally changing expression since it is expected that temporally regulated genes will show a degree of linearity in their expression changes through the time course. Data for genes with expression restricted to a single stage or time point should be viewed with a degree of caution in such a study. The analysis uncovers a program of coordinated expression changes during the life cycle of the parasite with approximately 40% of probes exhibiting stage-specific expression that may be informative in terms of finding intervention targets. A meta-analysis employing an unsupervised $k$-means clustering approach explored the functional relationships between sets of genes with similar expression profiles and supports the view that specific functional classes of genes are associated with specific life cycle stages. In addition, the clustering approach provides clues as to the function of the majority of *Plasmodium* genes that have no functional annotations, since their coexpression with genes of known function can be used as a guide. It should also be noted that substantial levels of antisense expression were detected using the nongenic probes on the array, a similar situation to that observed in other organisms (see Chapter 10). Altogether the study provides a valuable insight into the potential functions of many *Plasmodium* genes but aspects of the experimental design could certainly be improved, particularly the level of replication.

The second study focused on just the intraerythrocyte stages but examined these in much greater detail. Again using a red blood cell culture system, though on a much larger scale with cultures grown in a bioreactor to facilitate the isolation of relatively large populations of homogeneous samples, Bozdech et al. (2003a,b) examined gene expression in samples taken at 1 h time intervals throughout the 48 h of the intraerythrocytic life cycle. In this study, since a spotted array platform was used, each time-point sample was compared to a common reference sample by cohybridization on the same array. The reference comprised a pool of RNA samples representing all developmental stages. Again replication levels were limited, with only 8 of the 46 time-points duplicated, though in this case the very close time points provides a high degree of confidence in temporally changing expression measures. As with the Affymetrix study, standard scalar normalization was used and this was coupled with intensity-based filters, which resulted in slightly more stringent criteria for assigning expression. Just over 60% of the predicted genes were flagged as expressed in the red blood cell stages, a slightly lower figure than the ~70% found with the Affymetrix study. With such fine-grained temporal data, more sophisticated analytical methods are possible and the authors employed a fast Fourier transform to identify periodicity in the expression pattern of each gene. Rather than use a clustering method, the Fourier analysis calculates the phase and frequency for each genes expression profile throughout the 48 h time course, revealing a very striking and highly significant periodicity.

Genes may be ordered according to the phase of their expression profiles to generate a heat-map phaseogram (similar to the visualization methods used to display hierarchical clusters) and sets of genes with similar expression phases grouped. When this is done the annotations of the genes with known function suggests a highly ordered progression of biochemical processes through the intraerythrocytic development of the parasite (Figure 11.3). As the authors comment, a rigid viral-like programme of gene expression governs the progression of the plasmid through its life in the red blood cell. The tight association of particular biochemical functions with particular stages of the life cycle also suggests sets of parasite-specific functions as candidate therapeutic targets. These include gene products associated with the parasites plastid, which peak in expression during the schizont stage, a variety of proteases associated with different stages of the life cycle and a set candidate antigens for vaccine development. The overall insights from the time course study is strongly supported in a subsequent follow-up study examining similar temporal expression profiles with two other *P. falciparum* strains isolated from different geographical locations and showing different drug resistance spectra (Llinas et al., 2006). The gene expression profiles from the original study showed remarkable similarity to those generated from the two new strains, indicating that the regulatory network underpinning the intraerythrocytic life cycle is well conserved.

The unusual intraerythrocytic gene expression programme found with the *Plasmodium* strains raises questions as to how it is regulated. This is especially so when it became clear that the parasite genome is underrepresented for genes encoding likely transcriptional regulators, with only a third of the number of transcriptional regulators found in other eukaryotes (Coulson et al., 2004). This raised the possibility that aspects of the transcriptional programme could be controlled by posttranslational mechanisms, suggesting additional novel routes for drug development. The possibility of regulated RNA stability was investigated using the long-oligonucleotide arrays developed for the intraerythrocytic studies described above (Shock et al., 2007). In this study, synchronously infected red blood cell cultures were treated with the transcriptional inhibitor actinomycin D and RNA samples collected over a 4 h period. In four separate experiments, actinomycin D treatment was induced at different stages of the intraerythrocytic life cycle and derived cDNA compared with a reference pool representing all stages of the normal life cycle. In these studies array hybridizations were biologically replicated at least twice but more often three times. This particular experimental design introduces a normalization difficulty. As we described in Chapter 6, most normalization methods rely on the fact that the majority of transcripts do not show significant changes in abundance across the experiment: in this RNA decay study, many or all of the transcripts are expected to be dramatically reduced. To facilitate normalization, a set of yeast probes with no similarity to the *Plasmodium* genome sequence were printed on the array and a pool of transcripts corresponding to these probes added to each *Plasmodium*

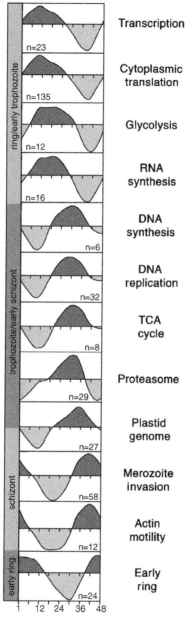

**FIGURE 11.3**   Phased gene expression during the *Plasmodium falciparum* intraerythrocytic life cycle. Each graph represent the average gene expression profile (*y*-axis, dark grey represents upregulated with respect to the average expression, light grey represents down regulation) of a set of coordinately expressed genes during the 48 h of the time course. The number of genes contributing to each class is indicated. Each graph is arranged according to its phase with respect to the time

RNA sample prior to labeling. Arrays were then normalized using the median intensities of the yeast probe signals to scale the intensities of all elements on the array. Half-life values were calculated for every gene showing significant decay using a first order decay model. The analysis revealed yet another striking feature of the *Plasmodium* gene expression programme: at early stages (ring and trophozoite) most transcripts have a relatively short half-life ($\sim$9.5 min). In contrast, by the late schizont stage, half-life measures increase dramatically to over an hour. These observations indicate that mRNA decay in *Plasmodium* is a highly regulated process that insures transcripts early in the life cycle are rapidly turned over whereas those expressed at later stages, involved in generating new merozoites for the next round of red blood cell invasion, are much more stable. Taken together, these time course expression studies demonstrate that powerful insights into the previously obscure biology of *Plasmodium* can be obtained by gene expression analysis and show how even model systems can identify potential therapeutic candidates from a genome containing many genes of unknown function.

Other expression studies continue to explore the biology of *Plasmodium*. For example, a comparison of two strains of *Plasmodium*, one normal and the other unable to produce gametocytes in red blood cell culture, using long oligonucle-otide arrays identified approximately 100 gametocyte specific genes (Silvestrini et al., 2005). Dahl et al. (2006) examined the effects of tetracycline drugs on parasite replication by expression profiling, identifying specific down-regulation of apicoplast genes, and thus suggesting potential new treatment avenues. The development of laser capture microdissection methods now open up the possibility that previously difficult to access stages of the *Plasmodium* life cycle, such as the intrahepatocytic phase, may be amenable to RNA isolation and expression profiling (Semblat et al., 2005). Closing the loop as it were, studies with the *A. gambiae* insect vector begin to shed light on gene expression responses to the blood meal in female mosquitoes (Marinotti et al., 2005), identifying sex or tissue specific transcription (Arca et al., 2005; Warr et al., 2007) and characterizing the response of the insect to insecticides (Vontas et al., 2005; Strode et al., 2006; Vontas et al., 2007). Together, these studies promise a much better understanding of the biology of the mosquito and its interactions with the parasite. The prospects for identifying better insecticides or for unco-vering weaknesses in the parasite that can be explored for drug development are good. It is important to note that the expression studies we describe here would be of limited utility were it not for the fact that, for the majority of the studies, all of the primary data are available from public data repositories such as ArrayExpress and GEO or can be accessed through the PlasmoDB and

---

course, with the major stages of the life cycle indicated to the left of the graphs. The text to the right of each graph indicates the functional classes enriched in the set of genes that make up each profile. *Source*: Bozdech et al., 2003a,b.

VectorBase databases (Bahl et al., 2003; Lawson et al., 2007). The public availability of the data is of critical importance for researchers exploring particular aspects of *Plasmodium* or *Anopheles* biology.

## 11.3 EXPRESSION PROFILING HUMAN DISEASE

As with infectious or pathogen-mediated conditions, investigations into innate human diseases such as autoimmune syndromes, neurodegenerative disorders or complex multigenic traits, including diabetes or cardiovascular diseases, are beginning to benefit from the understanding derived from global expression profiling. Perhaps more than any other disease, the heterogeneous set of conditions described with the general term cancer have been a particular focus of gene expression studies. Type the term 'cancer and microarray' into PubMed and you will be rewarded with a list of over 7000 papers, the first of which appeared at the dawn of the microarray era (DeRisi et al., 1996). The search reveals papers describing gene expression analysis of over 40 different types of cancer and highlight the potential impact that microarray technology can have on disease studies. As with any other disease, the objectives of a cancer gene expression study may be relatively straightforward, for example, identifying the basic biological processes that differentiate a cancer cell from its normal counterpart. Similarly, the possibility for monitoring the progression of a cancer or other progressive disease through temporal gene expression profiling may provide insights into the biology of the disease. The hope for such studies is that a genomic analysis will identify potential therapeutic avenues or generate reagents with diagnostic utility: so-called biomarker discovery studies. Along with these goals, and due to the incredibly heterogeneous nature of cancers – even those arising from a single cell type or tissue, the possibilities for gene expression analysis are potentially much wider than with less diverse diseases. For example, expression profiling may uncover specific molecular features that allow stratification of particular cancers that were previously undistinguishable. Here we focus on the application of gene expression profiling in cancer biology, exemplifying the type of studies carried out and analyses performed, as well as highlighting some of the pitfalls that can be encountered. Clearly this is not a comprehensive review since the field is so large and we have chosen to present a selection of studies that emphasize particular approaches.

### 11.3.1 Data Quality

Before we begin, it is pertinent to highlight some of the general problems that can be encountered in microarray-based disease studies. As we noted, cancer is a heterogeneous condition and as a consequence gene expression differences that are observed, even in simple comparative experiments, may be tainted by confounding signals. Cancer samples may be derived from a variety of sources: direct surgical biopsy of tumors, more limited needle biopsies, post mortem samples, cell lines established from particular cancers or even samples derived from fixed paraffin sections. Obviously when comparing a cancer sample with its normal counterpart, care should be taken to ensure that the samples are matched: a breast tumor sample is compared with the relevant normal breast cell type. However, this may be difficult; for example, biopsy samples may not be homogeneous, they may contain multiple cell types, mixtures of normal and malignant cells or cancer cells at different stages. In addition, it may not be clear what the normal counterpart of the cancer cell is and therefore multiple different normal samples may need to be included in the analysis. Similarly, genetic heterogeneity between different patient samples may skew results and adequate controls need to be considered to mitigate such factors. Clearly the degree of matching between normal and disease samples depends on the objectives of the experiment. If an exploratory analysis aimed at characterizing the basic biology of the system is envisaged, one could perhaps be less stringent than in the case of a biomarker discovery screen where the objective is identification of a set of reliable diagnostic tools.

From the above it should be obvious that issues of experimental design, primary data analysis and statistical method selection are paramount in disease studies. Unfortunately, there are many instances in the literature where these issues are far from clear. Recent analyses of published cancer gene expression studies reveals a worrying lack of rigor in reporting such basic statistical parameters such as sample size estimation, statistical power and multiple testing correction (Jafari and Azuaje, 2006; Dupuy and Simon, 2007). For example, the Jafari and Azuaje (2006) study of almost 300 published papers found that an incredible 60% of papers did not report the data normalization methods used and a staggering 97% of papers failed to properly describe how basic statistical tests such as ANOVA were implemented. If microarray data and the conclusions derived from its analysis are to be taken seriously it is imperative that journals take a much more stringent approach to publication and demand a more comprehensive description of analysis methods. Dupuy and Simon (2007) have published a 'checklist' of guidelines that they recommend be followed for clinical cancer studies and cancer biologists are advised to examine this carefully before embarking on large scale gene expression studies. Initiatives such as CONSORT (Consolidated Standards for Reporting Trials: Devereaux et al., 2002; Mills et al., 2005), which deal with reporting of randomized clinical trials in medical journals, could be extended to include guidelines for reporting essential statistical data from microarray studies. The concerns here are very real, it can cost millions of dollars to develop a diagnostic test and hundreds of millions to develop a drug; if such developments are founded on poor primary data, the waste of valuable R&D money is obvious. These issues have been reviewed recently (Ransohoff, 2004, 2005), highlighting the general problem of bias and reproducibility in biomarker studies.

## 11.3.2 Unsupervised Learning

The feasibility of using unsupervised clustering of gene expression data to classify and differentiate cancers is exemplified by a study on diffuse large B-cell lymphoma (DLBCL) (Alizadeh et al., 2000), and we use this, along with subsequent follow-up studies, to illustrate the basic approach. DLBCL is a clinically heterogeneous lymphoma that has highly variable treatment outcomes which cannot be differentiated on the basis of tumor cell morphology. As with many cancers, lymphomas present a challenge for diagnosis and prognosis because they derive from a normal cell lineage that is extremely complex and poorly understood (for a recent review of B-cell development see Welner et al., 2008). To untangle some of this complexity and gain better insights into the molecular basis underlying DLBCL heterogeneity, Alizadeh et al. undertook a comprehensive gene expression analysis.

The start point was a focused cDNA microarray, the Lymphochip, containing a set of approximately 18 000 cDNA clones selected from B-cell and other relevant lymphatic libraries. Approximately 25% of the genes on the array are represented by independent cDNA clones, providing a degree of intraarray replication for assessing reproducibility. The experimental design utilized a common reference approach, with a pool of cDNA derived from a set of nine different lymphoma cell lines labeled and hybridized to each array in combination with an experimental sample. A total of 96 experimental samples were examined: 48 separate DLBCL patients, 10 chronic lymphocytic leukaemia (CLL) patients and 9 follicular lymphoma (FL) patients as representatives of the common lymphoid malignancies, along with a series of lymphoma or leukaemia cell lines and various subpopulations of purified human lymphocytes. After filtering low quality spots, array data were normalized by a simple median scaling approach. In an unsupervised approach, the data are clustered on the basis of their expression similarities: genes, samples or both may be assessed in this way. When both genes and samples from the DLBCL study are clustered a remarkably clear picture emerges from the expression data: virtually all of the DLBCL samples group together and are distinct from the other cancers (Figure 11.4). The DLBCL sample expression profiles are also distinct from most of the cell lines and purified lymphocyte samples, except for germinal center B-cells. While the DLBCL samples cluster together, it is important to realize that they are not homogeneous in terms of their gene expression profiles, each sample has a unique profile. Of course much of this uniqueness will reflect heterogeneity in genetic background, differences in patient treatment regimes and technical artefacts due to sample preparation. Nevertheless, the fact that, despite these confounding effects, the DLBCL samples are grouped by the clustering algorithm clearly indicates they are related by gene expression. Extending the analysis further, the DLBC samples were reclustered, focusing on a subset of genes with expression profiles that marked germinal center B-cells

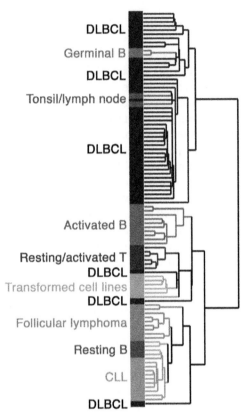

**FIGURE 11.4**   Sample clustering from the DLBCL study. The tree generated by hierarchical clustering of the various cancer and control samples in the study. Each type of sample is represented by a different shade of grey. The majority of the DLBC samples group together at the top of the tree along with germinal center samples. The major lower branch of the tree contains only three DLBCL samples along with the other cancers and the remaining lymphoid cell lines.
*Source*: Alizadeh et al., 2000.

and activated B-cells. This analysis differentiates the DLBCL samples into two groups: a class with germinal center like expression profiles and a class with activated B-cell like expression profiles (Figure 11.5a). Again we emphasize that it is not the expression of an individual gene that segregates the samples, rather the similarity in expression profiles of a relatively large set of genes. The DLBCL samples were taken from different patients over a number of years but in all cases prior to the onset of treatment (treatment regimes were reported to be comparable across the sample group). It was possible to look at the treatment success in terms of patient survival to see if the gene expression profiles that differentiate the DLBCL samples into two groups reflect different treatment outcomes. Strikingly, most of the patients that exhibited the germinal center-like

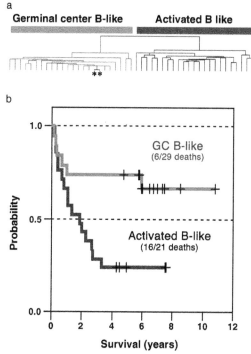

**FIGURE 11.5** Differentiating DLBCL subtypes. (a) Reclustering a subset of the genes with profiles specific to germinal center and activated B cells separates the DLBCL samples into two classes. The samples indicated with asterisk are germinal center B-cell samples. (b) Kaplan–Meier survival plots of overall patient survival; the patients with germinal center-like profiles are more likely to have successful treatment outcomes than those showing the activated B-cell like expression profiles.
*Source*: Alizadeh et al., 2000.

expression profile responded well, with over 75% still living after 5 years. In contrast, only 16% of patients showing the activated B-cell like profile were surviving after 5 years (Figure 11.5b). Importantly, the expression profiling had better prognostic value than the traditional clinical indicators.

The important feature of this study is that the prognostic expression profiles were generated in the absence of any clinical data: simply looking at the similarity in gene expression profiles across a group of samples was sufficient to allow the identification of specific sample subclasses. Of more importance, the insights from the analysis have been independently validated and extended (Rosenwald et al., 2002; Wright et al., 2003). Rosenwald et al. (2002) expression profiled 240 DLBCL biopsy samples on the Lymphochip array and, using hierarchical clustering, confirmed the classification into the two groups identified in the initial study. In addition, the much larger sample set facilitated the identification of a third expression class (type 3) unrelated to the germinal center or activated B classes. This study then used a supervised analysis based on survival to suggest underlying oncogenic events associated with each class and derive a set of 17 genes with prognostic expression profiles. However, emphasizing that this type of analysis is not necessarily straightforward, a contemporary study using a custom Affymetrix array containing probe sets representing 6817 genes (Shipp et al., 2002) suggested a different set of prognostic genes from those derived in the Alizadeh et al. (2000) study. In this case, 58 DLBCL and 19 follicular lymphomas were expression profiled and the data analyzed with a supervised approach (a weighted-voting algorithm) to discern a set of genes with expression profiles associated with DLBCL treatment outcome. Thirteen such genes were identified. Only three of the 13 prognostic genes from the Shipp et al. (2002) study were present on the lymphochip array and of these only two appeared to agree. In addition, there was no overlap between the genes identified in the Shipp and Rosenwald studies. While clearly the lack of agreement in part reflects the fact that the cDNA and oligonucleotide arrays interrogate different genes and, more significantly, almost certainly detect a different spectrum of transcripts from particular genes, it is nonetheless disconcerting that there is little agreement between the studies. An initial attempt at reconciling these data was performed by Wright et al. (2003), who combined the data from all three array studies and applied Bayesian probability methods to build a classification predictor. In another attempt at reconciling the apparently disparate microarray data, (Lossos et al., 2004) examined 36 genes reported to be predictive of survival, either from the three microarrays studies or from published single gene studies, and examined their expression in an independent set of DLBCL patients by quantitative RT-PCR. This analysis developed a model that was predictive of treatment outcome using the new patient samples as well as the microarray data from Shipp and Rosenwald.

Taken together, the prospects for developing new effective treatments based on these genomics studies are brighter than with more limited insights coming from traditional diagnostic tools. Gene expression profiling focusing on the prognostic genes identifies groups of patients that are more likely to respond well to current treatment regimes and also flags a set of patients that are unlikely to benefit from these treatments. The combined gene expression data from all the studies provide new insights into the biology of B-cell cancers and suggest avenues for developing new therapeutic strategies, particularly for the class of patients with poor prognosis.

### 11.3.3 Supervised Learning

Unsupervised analysis, as used in the DLBCL studies, is frequently a first port of call in exploratory studies aimed at understanding or differentiating cancers, as an example, see Mischel et al. (2004) for their review of DNA microarray technology in the analysis of brain cancer. There are many other examples easily found in the literature. While unsupervised analysis methods are relatively simple and clearly have utility when stratification of heterogeneous patient samples is desired, in most cases there will be clinical information associated with each sample. It would therefore seem sensible to use this clinical data to guide the analysis of the expression data. In such cases, supervised analysis methods come to the fore, and preexisting information is used to identify sets of genes that can reliably differentiate data classes. The study of Shipp et al. (2002) described above is an example of such an analysis but also highlights the difficulty of building such a classifier that translates to other datasets. The concept of using supervised methods was first reported by Golub et al. (1999) in their analysis of acute myeloid and lymphoblastic leukaemias (AML and ALL, respectively). In this pioneering study, samples from 27 ALL and 11 AML patients were expression profiled with Affymetrix arrays containing probe sets for 6817 genes. The analysis utilized a scoring system based on the $t$-statistic to assess for each gene whether its expression level was associated with the AML or ALL samples. Those genes with discriminating scores were then used to provide weighted votes for expression levels from unknown samples to assess whether they were likely to be AML or ALL samples. This concept of using a training set containing samples with known outcomes to develop a set of classifier genes that will accurately assign a test set to a class is now prevalent in disease studies (Roth, 2001; Pusztai and Hess, 2004).

There are a wide variety of supervised methods currently available, some of the most common being variants on linear discriminate analysis, support vector machines, neural networks or $k$-nearest neighbor approaches (reviewed in Dudoit et al., 2002, see Chapter 8 for details of these and other supervised methods). A recent analysis of published studies reporting success in generating good classifiers from gene expression data concluded that many of the these studies were overoptimistic in their claims and in fact did not classify patients better than chance (Michiels et al., 2005). The analysis indicates that the selection of samples used to define the training set can influence the performance of the classifier and concluded that larger cohorts are needed. A similar conclusion was reached by Ambroise and McLachlan (2002) in their study of several published experiments. In this case the authors recommend more robust cross-validation and error estimation techniques. Recent progress in the field includes the development of more sophisticated analytical methods such as the module map (Segal et al., 2004) or random forests (Cutler and Stevens, 2006) approaches and, when coupled with the increasing amounts of data available, promise more robust classification in the future (for reviews see Segal et al., 2005; Hanauer et al., 2007). More recent approaches are tending to focus less on individual genes but rather utilize functional information (such as Gene Ontology annotations) to identify classes of genes or biological processes that may provide class discrimination (Fishel et al., 2007; Huang and Chow, 2007; Maglietta et al., 2007). It is very likely that, as functional information about individual genes in the genome accumulates and improved analytical approaches are developed, the ability to reliably type cancers by their gene expression signatures will become routine. Hopefully this will provide improved diagnostics and better prognostic indicators.

## 11.3.4 Larger Scale Studies

In the approaches described above, the objectives were to uncover discriminatory signatures for a specific class of cancers and, as we note, many such studies have been performed. Rather than look at individual cancers it is possible to look at expression data from many different malignancies, originating from different tissues, and ask whether there are common signatures. For example, Chung et al. (2002) report an analysis of different public breast and lung cancer data sets, identifying several interesting gene expression signatures by simple hierarchical clustering. A much larger scale analysis (Rhodes et al., 2004b) examined 40 datasets comprising profiles from almost 4000 cancers to identify several common gene expression signatures related to neoplastic transformation. These authors have made such approaches broadly applicable via the development of the Oncomine database (Rhodes et al., 2004a, 2007), which contains expression data from over 18 000 cancer microarrays. As an example of the utility of the dataset, the authors have used Oncomine to identify expression signatures and associate specific transcription factors with each signature (Rhodes et al., 2005). The possibilities for using such very large data collections for comprehensive integrative analysis are exciting and hold out the prospect of a much better understanding of basic cancer biology (Hanash, 2004; Rhodes and Chinnaiyan, 2005).

The approaches we outline above are, in principal, extendable to any disease condition where comparisons with normal tissue can identify disease-specific changes in gene expression. The ability to explore the underlying biology of the disease state by finding pathways disrupted in diseases is a powerful tool in the early stages of drug discovery and development. In addition, the possibility of uncovering highly specific diagnostic markers in the form of gene expression signatures opens the possibility for more accurate and streamlined disease diagnosis. While the analytical methods for gene expression data are clearly still evolving, it is clear that we have come a long way in a relatively short period of time. It is a little over 10 years since the first cancer gene expression study and the changes in terms of array platforms and their analysis have been dramatic. It is to be hoped that the next decade will provide equally impressive improvements.

## 11.4 ARRAY-CGH

In the preceding section we emphasized, with a few selected examples, how gene expression profiling can contribute to exploring aspects of human disease.

However, it is not only in the area of gene expression analysis that microarray technology is contributing to our efforts at tackling disease. The genome itself is a fruitful area for exploration and two technologies in particular are having major impacts in biomedical studies: comparative genome hybridization (CGH) and single nucleotide polymorphism (SNP) genotyping. In this section we review both of these technologies with a focus on the types of array platform in use and the analytical methods being developed for reliable data processing.

It has been known for some time that genomic deletions and duplications are important in human disease (Lupski, 1998; Emanuel and Shaikh, 2001) and are often associated with various cancers (Thiagalingam et al., 2002). In addition, it is becoming increasingly apparent that variations in DNA copy number make a major contribution to genetic variation in human populations (Freeman et al., 2006). Consequently, the ability to examine genomic DNA for alterations to the normal diploid condition is highly desirable. Array-CGH, as the name implies, is a technique for comparing genomic DNA via microarray hybridization. It derives from earlier methods comparing DNA from clinical samples and controls by in situ hybridization to metaphase chromosome spreads (du Manoir et al., 1993). The method was translated to a microarray format by several groups, employing spotted arrays containing genomic DNA clones (Solinas-Toldo et al., 1997; Pinkel et al., 1998) or cDNA clones (Pollack et al., 1999) and since then has evolved into a powerful method for detecting variation in genomic DNA copy number, culminating in a genome-wide analysis of copy number variation in hundreds of human genomes (Redon et al., 2006). The method is illustrated in Figure 11.6. While simple in concept, there are a number of hurdles that must be overcome in order to make reliable estimates of DNA copy number from microarray hybridizations. First, the human genome is large and, as we describe in Chapter 10, it is not trivial to represent the entire genome on an array. Second, human DNA is highly repetitive, unique coding sequences represent less than 2% of the genome, which can complicate the analysis of any hybridization based assay. Several methods have been developed to overcome these limitations and array-CGH is now a well-established tool for comprehensive analysis of the human genome. The area has been well reviewed in recent years and interested readers are directed to these papers for a more detailed introduction (Snijders et al., 2003; Pinkel and Albertson, 2005; Vissers et al., 2005; Bejjani and Shaffer, 2006; Lockwood et al., 2006; Carter, 2007; Lee et al., 2007).

## 11.4.1 Arrays for CGH

As we note above, the early development of array-CGH employed spotted arrays containing genomic DNA or cDNA clone probes. In the former, probes derived from lambda, cosmid or P1 clones, ranging in size from 30–150 kb, were used to cover fairly restricted regions of the human genome. Obviously, the limitation of such arrays is that only a small region of the genome is represented and while

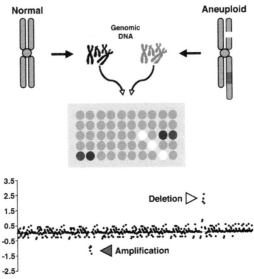

**FIGURE 11.6**  Array-CGH. DNA is prepared from a normal reference source and a sample of interest, in this case a genome containing deletions and duplications. The DNA samples are labeled with different fluorescent dyes and hybridized to an appropriate array. Increases in copy number result in higher signals in the sample channel (black elements) and deletions yield higher signals in the control channel (white elements). The log ratios for each probe ($y$-axis) are plotted in order across the genome ($x$-axis) with deletions showing a positive ratio and amplifications a negative ratio.

they are of little utility in genome wide studies they enable high-throughput analysis of selected regions. As we describe in Chapter 10, restricted spotting densities limited early attempts at covering the entire genome and in most instances relatively low resolution partial tiling arrays, placing a BAC probe every Mb or so, provided a pragmatic solution (Snijders et al., 2001). More recently, as array densities improved and BAC clones accurately mapped to the genome sequence became available, large insert BAC clones (~150 kb average size) have been used to cover virtually the entire human genome. In one case, 32 430 overlapping BACs cover most of the human entire genome (approximately 1.5-fold coverage) and provide a resolution of approximately 80 kb (Ishkanian et al., 2004). A second effort prepared probes with 26 574 BAC clones from the 'Golden Path' used to generate the reference human genome sequence and covered approximately 94% of the genome (Fiegler et al., 2006). Both of these arrays have been shown to perform well and have been validated for use in array-CGH studies by assaying DNA from cell lines containing known duplications or deletions or by spike in hybridizations with known amounts of cloned DNA or sorted chromosomes. A fairly rapid first pass assay for checking a human genome tile is to hybridize with DNA from male and female cells, one expects the signals from X and Y chromosome probes to reflect the 2:1 and 1:0 ratios of these chromosomes in the starting samples. Similar assays with known aneuploid lines can also be employed (i.e. DNA from a trisomy 21 line should show a 3:2 ratio when compared with normal DNA). Methods for generating probes for a tiling array are described in the literature and were outlined in Chapter 3 (Hollich et al., 2004; Fiegler et al., 2007a). As we shall discuss below, there are two problems that can be encountered with large insert libraries: sensitivity and specificity. Will each probe be able to detect small levels of aneuploidy, on the order of a few hundred base pairs, and will highly repetitive sequences swamp any meaningful signals?

An alternative to using genomic clones as array probes is to use cDNA clones representing some or all of the coding sequence in the human genome. While such arrays are limited in their ability to detect deletion or amplification of intergenic sequences, from a biomedical standpoint they can be much more sensitive for detecting alterations in gene copy number. This approach is exemplified by the work of Pollack et al. (1999), who prepared spotted arrays containing over 5000 mapped cDNA clones and use these to reliably detect known gene amplifications and deletions. In a follow up study the utility of the cDNA array approach was highlighted when a set of breast cancer lines were both expression profiled and used for array-CGH analysis on an array containing approximately 6600 mapped cDNA clones (Pollack et al., 2002). In this study, alterations in gene expression levels could be associated with changes in DNA copy number, suggesting that over 10% of gene expression variation in the breast cancer lines is due to gene amplification or deletion. In principal, any cDNA array should be of utility in array-CGH studies if the primary focus is detecting alterations in gene copy number. Since cDNA derived probes contain considerably lower levels of highly repetitive DNA they should suffer less from specificity issues, although caution must be exercised when interpreting ratios from such arrays since pseudogenes may contribute to the signal observed with an individual probe. It is therefore prudent to characterize the likely cross-hybridizing regions of the genome for each probe. In terms of sensitivity, there may also be issues for individual probes. Since cDNAs are derived from processed RNAs where introns have been removed, each clone in effect represents a miniature partial tile that samples fragments (exons) across a particular region of the genome. Many human genes are large, spanning tens or hundreds of kilobases (even megabases), and individual exons are relatively small (average exon size is 200 bp or 0.000006% of the human genome), therefore each probe detects a tiny fraction of the human genome. The sensitivity issue generally means that more sample DNA is required for cDNA arrays compared with large insert-based arrays.

As we describe in Chapter 10, the densities now achievable with in situ synthesis methods permit the fabrication of arrays containing millions of oligonucleotide probes and allow partial tiling of the human genome at densities approaching a 60mer probe every kb. One of the first demonstrations of the use of oligonucleotide probes in array-CGH studies utilized approximately 19 000

60mer probes designed to be specific reporters of human genes (Carvalho et al., 2004). In some respects these are similar to the cDNA array approach except well-designed probes are less likely to suffer from specificity problems since they should only detect a unique target in the genome. This work assessed the performance of the oligo array in comparison with a 1 Mb resolution BAC-based array and claimed better resolution and accuracy than achievable with the low resolution BAC array. A similar study utilized commercially available Agilent mouse and human expression arrays to demonstrate sufficient sensitivity for detecting gene duplications (Brennan et al., 2004). With in situ synthesized arrays much higher densities are achievable, in principal single base resolution with overlapping tiling arrays if sufficient funds were available to generate approaching two billion probes! To date, arrays synthesized using Agilent and Nimblegen technologies, containing 244 000 and 385 000 probes, respectively, have been utilized for array CGH studies (for a discussion see Scherer et al., 2007). The use of oligo based arrays for CGH studies has been recently reviewed (Ylstra et al., 2006) and discussed with respect to other platforms (Carter, 2007). The current view is that while oligonucleotide probes may, in principal, offer much better resolution compared with BAC-based arrays, they tend to suffer from poor signal to noise levels. However, it will be interesting to evaluate the performance of the 2.1 million probe Nimblegen arrays in array-CGH studies since they provide a dramatic increase in density, allowing a 60mer every 1.2 kb. Until these arrays have been fully evaluated, the current trend is to use BAC arrays to broadly define regions of the genome containing potential copy number variants and then employ high-resolution oligonucleotide tiles (probe every 100–200 bp) to map 2–5 Mb regions in some detail (for examples see Selzer et al., 2005; Sharp et al., 2006).

One method that is reported to overcome some of the problems associated with the relatively poor signal to noise ratio characteristic of oligonucleotide CGH-array is representational oligonucleotide microarray analysis (ROMA) (Lucito et al., 2003; Sebat et al., 2004). In this technique, the complexity of the input genomic DNA is reduced using restriction endonuclease digestion followed by PCR amplification (see below) before hybridizing to specially designed oligonucleotide arrays. Since the reduced complexity input DNA is generated by restriction enzyme digestion, the fragments generated are, in principal, predictable from the genome sequence and these fragments are used to design specific long oligonucleotide probes. The ROMA technique generates and analyses DNA fragments corresponding to around 2% of the genome spaced at approximately 17 kb intervals. The representation is more random than approaches using gene-based probes, allows for more sensitive probe design and can eliminate specificity problems by using principled probe design approaches.

Finally, there has been increasing interest in using genotyping arrays for CGH studies. As we describe below, Affymetrix GeneChip type arrays and Illumina Bead-arrays have been developed for the high-throughput genome-wide detection of single nucleotide polymorphisms. It was realized that such arrays provide intensity information as well as binary calls for SNPs and, with appropriate data analysis, information on copy number could be derived (Redon et al., 2006). Such arrays offer upwards of half a million probes with an average spacing of approximately 2 kb. The study by Redon et al. (2006) catalogues copy number variation among 270 different human samples using both a whole genome BAC tile array and an Affymetrix 500K genotyping array, providing the most comprehensive survey of genomic DNA diversity published to date. In this analysis the BAC arrays showed an approximately fivefold lower standard deviation than the genotyping array and detected approximately three times more copy number variation. On the other hand, the average resolution of the genotyping arrays was approximately 3 times better than the BAC array (81 versus 228 kb). In general, the conclusion from this comprehensive study is that large insert and genotyping array platforms are highly complementary. A variety of commonly used array-CGH platforms have recently been compared (Greshock et al., 2007) with similar conclusions regarding complimentarity and we suggest any potential user determine the goals of their experiment before deciding on a particular platform when a choice is available.

## 11.4.2 Sample Preparation for Array-CGH

As we indicate above, one of the problems associated with Array-CGH stems from the highly repetitive nature of the human genome and the fact that single copy sequences are at a relatively low concentration in a genomic DNA sample. The first of these is problematic when using large insert probes and usually dealt with by including a large quantity of human Cot-1 DNA in the hybridization solution. Cot-1 DNA is obtained by isolating the rapidly reassociating fraction from denatured genomic DNA (that fraction which reassociates with a Cot value of 1) and largely consists of highly repetitive sequences: Cot-1 DNA is available commercially from several suppliers. When included in the hybridization solution with labeled genomic DNA samples, Cot-1 DNA will hybridize with repetitive sequences in the sample rendering them unavailable for interactions with repetitive sequences contained within the probe DNA. While relatively effective, suppression with Cot-1 DNA can never entirely remove signals due to repetitive DNA and it is therefore imperative that the performance of individual array platforms is assessed in this respect by hybridization with reference samples of known DNA content and suspect probes flagged.

Initial array-CGH studies generally utilized DNA extracted from cell lines, thus large quantities of input are available for labeling by directly incorporating fluorescent dyes with standard molecular biology techniques such as nick translation or random priming. For most protocols $\sim$200 ng to 1 $\mu$g are sufficient for a hybridization using large insert clones. With cDNA clone based arrays, which are generally less sensitive, 2–3 $\mu$g of input DNA are usually required. However, in some instances, for example the analysis of primary tumors, it may be desirable to carry out hybridizations with more limited sample quantities. In such cases, whole-genome amplification techniques must be employed, of which a variety are available (for reviews of current methods see Pinkel and Albertson, 2005; Lockwood et al., 2006). Recently the analysis of single cells, though with somewhat limited resolution, have been reported using multiple displacement amplification (Le Caignec et al., 2006; Spits et al., 2006) or the GenomePlex method (Fiegler et al., 2007b; Geigl and Speicher, 2007) opening up interesting diagnostic possibilities for array-CGH.

As we described above, one method of improving sensitivity is to reduce the complexity of the input DNA, exemplified by the ROMA technique (Lucito et al., 2003). Here input DNA is digested with a restriction enzyme, chosen on the basis that it has a reasonably even distribution of sites in the genome, is not affected by DNA methylation and leaves 4 bp overhanging ends after digestion (*Bgl*II and *Xba*I have been used). Adapters are ligated to the ends of the digested DNA to provide primer sequences for PCR amplification. Since PCR selectively amplifies small fragments in a heterogeneous mixture (the authors of the technique estimate 200–1200 bp are selected), only a fraction of the genome is amplified and the resulting products are subsequently labeled for array hybridization. As we note above, the preferentially amplified regions are determined in silico to provide a set of target regions for probe design. A similar method is used to generate labeled fragments for hybridization to Affymetrix genotyping arrays since these are designed to detect SNPs harbored within 100–1000 bp amplicons (Redon et al., 2006). Whatever the amplification method employed, it is absolutely critical that experimental and reference samples are treated in exactly the same way, preferably at the same time and with the same reagents, in order to minimize misrepresentation of copy number variation that is due to differences in amplification.

One area that has received comparatively little attention in the array-CGH field, highlighted by Scherer et al. (2007), is the reference genome with which experimental samples are compared. As these authors point out, there is no commonly recognized reference genomic DNA sample that can be used across multiple experiments, indeed the reference human genome sequence is a composite derived from hundreds of different sources. Clearly the reference samples for a population-based survey of normal genome variation may differ significantly from those employed in a particular cancer study. Thus, as is the case with gene expression analysis, careful consideration regarding the development of reference samples that can increase the comparability of genome wide array CGH data is needed at the level of the whole experimental community.

### 11.4.3 Data Analysis

The reliable identification of genome duplications or deletions from an array-CGH experiment presents a number of analytical challenges that have become apparent since the initial development of the technology. In particular, the fold changes that need to be reliably detected are smaller that normally encountered with gene expression studies. For example, a heterozygous duplication is equivalent to a single copy gain when compared to a normal genome and represents a 3:2 ratio ($log_2$ ratio of 0.59) and a heterozygous deletion a 1:2 ratio ($log_2$ $-1$). Furthermore, particularly in the context of clinical samples such as cancer biopsies, the samples may be heterogeneous in terms of cell populations, further diluting the signal. Early studies used basic normalization procedures such as average intensity scaling, global median or intensity-based loess and applied fold-change thresholds to identify copy number variants. It was soon realized that these and other methods developed for the analysis of gene expression arrays were noisy and perhaps unsuitable for CGH data. More recently, there have been a number of tools developed that are specifically aimed at improving the analysis of array-CGH data as well as enabling inferences about copy number changes from hybridization to high density oligonucleotide expression arrays. Some of the CGH specific tools have been reviewed recently (Chari et al., 2006; Lockwood et al., 2006) and we refer readers to these texts for lists of currently available software. Here we highlight some of the main considerations that recent studies have highlighted as important in array-CGH analysis. Note that the particular analytical approaches are described in detail in Chapter 6, here we simply indicate the methods applied and refer readers to the detailed descriptions earlier in the book.

Several recent studies highlight the need for accurate normalization, either across arrays or between channels in two channel comparative hybridizations. The rational here is that inappropriate normalization can easily eliminate the small fold changes that are characteristic of genome changes or inadvertently increase apparent ratios when there is no underlying change in the samples. One of the first array-CGH specific normalization approaches for two-channel BAC and cDNA arrays was introduced by Khojasteh et al. (2005). Analyzing raw data from several published array-CGH experiments they identified a series of biases: an intensity dependant bias with low intensity signals; spatial bias across the array; plate bias with variations due to specific plates from the amplicon libraries and a background bias. They proposed a step-wise three stage normalization procedure to improve detection of single copy changes. The issue of spatial bias was subsequently highlighted by Neuvial et al. (2006), who proposed an improved spatial normalization approach to remove both local and global gradient bias effects. The method first estimates the overall spatial gradient and subtracts this from the log ratio data and then corrects local bias via a 2D-loess approach. The tool is available as an *R* package (MANOR). The removal of spatial bias, while clearly important, is suggested to actually remove some of the biological signal from array-CGH data, with loess approaches wrongly identifying intensity changes due to genomic DNA differences as spatial variation (Staaf et al., 2007). These authors proposed a data stratification approach ($R$ package: popLowess), partitioning the data into three populations of probes with a $k$-means clustering algorithm, and using the largest population (reflecting the unchanged genomic regions) to calculate the loess correction for the whole data. All of these methods have been applied to both BAC and long oligonucleotide-based arrays and appear to be generally applicable.

Even if array-CGH data are appropriately normalized, the problem of accurately defining copy number variants is still problematic. In the case of cancer studies, the genomic alterations are frequently large, spanning Mb regions of the genome, and are thus relatively easy to detect. However, as the focus has shifted to copy number variations characteristic of 'normal' human genomes, which are frequently much smaller (of the order of kb rather than Mb) it has become apparent that improved methods are required. Taking both smoothing and segmentation based approaches into account, Hu et al. (2007) take advantage of the noise in array-CGH data to propose a novel algorithm, CGHss (Matlab code available), to improve both spatial normalization and accuracy of copy number change boundaries. Marioni et al. (2007) have carried out an extensive analysis of their comprehensive BAC array-CGH datasets, identifying a significant spatial autocorrelation artefact, which they term wave. While unable to determine the cause of the wave artefact, they developed a loess-based correction model to reduce the effect of the wave. They also took advantage of the fact that their data were generated from a large number of individuals ($\sim$270) to develop an approach for detecting copy number variants utilizing hybridization information from all of the samples. The resulting CMVmix method (Bioconductor package *snapCGH*) is reported to increase the detection rate for copy number variation without significantly affecting the false positive rate. However, the authors note that their approach, designed for examining 'normal' copy number variation, is less suitable for detecting the large-scale genomic alterations typical of many cancers. Taking a similar idea that multiple different samples may increase reliability of detecting copy number changes, Klijn et al. (2008) propose a nonparametric statistical method (kernel regression; *KC-SMART*, Matlab code available) for analyzing multiple heterogeneous cancer biopsy samples, claiming improved detection compared to commonly employed methods such as hidden Markov Models. On the other hand, Rueda and Diaz-Uriarte (2007) claim that their Bayesian approach utilizing Markov chain Monte Carlo computations ($R$ package *RJaCGH*) is more robust approach.

Taken together, it appears that the array-CGH field is in much the same position that array-based gene expression studies were a few years ago in terms of analysis: the vagaries of the system are beginning to become apparent with the types of variability and noise starting to be characterized. It is very likely that the increasing accumulation of genome-wide experiments will provide sufficient data to drive the development of improved normalization and statistical methods in the not to distant future.

### 11.4.4 Data Visualization

Before leaving array-CGH we note that there are different considerations for data visualization than with expression based studies. The objectives of array-CGH studies are to map variation in the genome and, as a consequence, it is mapping rather than clustering tools that are most appropriate. Several groups have developed platforms for analyzing, storing and displaying array CGH data with respect to the reference genome sequence, allow zooming in and out of particular genomic regions to visualize the array data. We offer no recommendations as to particular viewers but urge some caution when 'integrated platforms' offer to process raw data before displaying it. In such cases, as we hope is clear from above, it is recommended that the normalization methods and technique for assigning copy number variation are evaluated before committing extensive experimental data to a particular software tool. A list of various viewing and analysis tools can be found in Chari et al. (2006) and Lockwood et al. (2006). For those using commercial arrays such as Affymetrix genotyping arrays, Illumina bead arrays or in situ synthesized long oligonucleotide arrays from Agilent or Nimblegen, we note that each of these companies provide copy number variation detection software for their respective platforms. Finally, we note that methods and software tools for using commonly available expression arrays to assess genomic copy number changes are beginning to appear (Auer et al., 2007; Skvortsov et al., 2007), extending the potential applications for array-CGH in both humans and other organisms.

## 11.5 SNPS AND GENOTYPING

Single nucleotide polymorphisms (SNPs) are the predominant source of variation in the human genome, it is estimated that any two individuals (other than identical twins) have at least six million individual base differences in their genome sequence. Many of these nucleotide variants occur as groups of linked SNPs in the genome known as haplotypes (Figure 11.7). Haplotypes and their associated SNPs facilitate very powerful linkage disequilibrium association studies where they are used as surrogate genomic markers for complex traits. The hope is that by identifying specific SNPs that segregate with genetic traits, genes underlying such traits can be identified. This is particularly important for complex multigenic traits, for example heart disease, psychiatric disorders or metabolic diseases such as diabetes, where it is known that variation in many genes can contribute to the final phenotype (WTCCC, 2005; Frazer et al., 2007). It is consequently unsurprising that a growing focus for medical and population

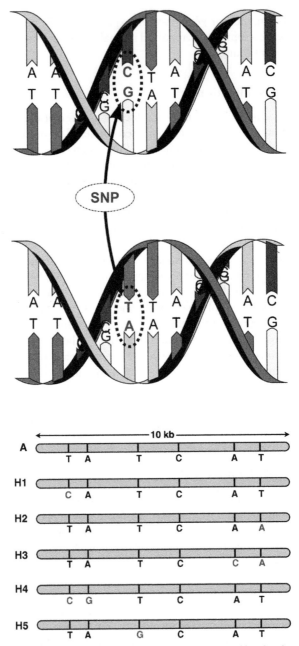

**FIGURE 11.7**  SNPs and haplotypes. A single nucleotide polymorphism is a base change in an otherwise identical stretch of sequence. Sets of particular SNPs differing from an ancestral chromosome (A in the lower panel) that are linked (generally not separated by a recombination event in the human genome) are known as haplotypes. In this case five haplotypes (H1–H5) are illustrated.

genetics studies are genome-wide association studies for identifying regions of the genome that may harbor genes responsible for complex phenotypes (Kruglyak, 2008). The ability to carry out such studies has been revolutionized by the development of microarray platforms that facilitate very high throughput and cost effective genotyping of hundreds of thousands of SNPs on a single array. There are two major microarray platforms commonly used for whole-genome association studies, Affymetrix and Illumina, which fundamentally differ in their SNP detection assays. Affymetrix arrays detect polymorphisms by a hybridization assay, whereas the Illmina platform relies upon enzymatic extension of genomic DNA captured at specific sequences on BeadArrays. Here we outline the basics behind each of these platforms and briefly discus the analytical tools used to generate SNP maps.

Whole-genome polymorphism analysis is the newest of the nucleic acid array platforms to be adopted by the research community and, as with other array technologies, there are clearly improvements to be made in terms of low level and statistical analysis methods. However, given the relatively short time that such studies have been possible, the insights are truly breathtaking. Genotyping large numbers of case-control samples with ever more SNPs have identified potential genes associated with autoimmune disease (Burton et al., 2007), type 2 diabetes (Sladek et al., 2007; Zeggini et al., 2008), Alzheimer's disease (Coon et al., 2007) and some cancers (e.g. Tomlinson et al., 2008). In one very large study, over 14 000 patient samples from seven common diseases (including coronary heart disease, bipolar disorder and rheumatoid arthritis) were examined and several candidate genes identified (WTCCC, 2007). These and many other studies hold out the very real prospect that genes involved in virtually any disease with a genetic component will be uncovered in the not to distant future (Gibbs and Singleton, 2006; Kingsmore et al., 2008).

## 11.5.1 Affymetrix

Affymetrix offer a series of genotyping array sets with increasing numbers of SNPs, including the 10K, 100K and the 500K Human Mapping GeneChips (the latter two are two array sets). The exact design of each array differs slightly but follow the basic principal outlined in Kennedy et al. (2003). A set of 25mer oligonucleotides are used to discriminate single base changes at particular genomic locations. In brief, for each bi-allelic genomic SNP a set of 5–10 probe sets covering the SNP are designed. Each probe set comprises a quartet of probes, the Perfect Match (PM) and Mis-Match (MM) pairs for each allele. Accuracy is improved by including probe quartets for both the sense and anti-sense genomic strands (Figure 11.8). The Affymetrix assay utilizes genomic DNA prepared by a similar complexity reduction approach to the one we described above for the ROMA array-CGH technique (Figure 11.9). Genomic DNA is digested with a restriction enzyme (*Hind*III, *Nsp*I, *Sty*I or *Xba*I,

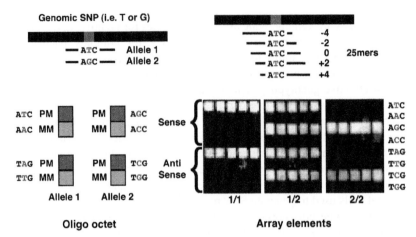

**FIGURE 11.8**   The Affymetrix genotyping platform. A particular genomic SNP (indicated by allele 1 and allele 2, top left) is detected by a set of probes offset with respect to the polymorphic nucleotide (top right). For an individual 25mer, two quartets of probes are generated (oligo octet) comprising the Perfect Match and Mis-Match oligonucleotides for both alleles on the sense and antisense strands. For the five probe sets interrogating this SNP the array hybridization results expected for each homozygote (1/1 and 2/2) and the heterozygote (1/2) are shown bottom right. Hybridization is detected at the appropriate PM oligonucleotides on both strands specifically for the polymorphic nucleotide and no signals are detected with the other probes.

depending on the particular GeneChip being interrogated) and adapters are ligated to the cohesive ends of all of the resulting genomic DNA fragments. PCR, using primers designed against the adapters, is performed under conditions that favor the amplification of 200–1100 bp fragments, resulting in an approximately 100-fold reduction in genome complexity. The amplified fraction is end labeled, fragmented and hybridized to the appropriate array. Probe sets on the array are designed against specific SNPs within the 200–1100 bp amplified fragments, hence the appropriate restriction enzyme must be used for each array. After washing and scanning genotypes are called based on the hybridization intensities for each probe cell.

## 11.5.2 Illumina

The Illumina *Infinium* platform does not use a complexity reduction step, the input is complete genomic DNA that is generally subject to a whole-genome amplification (Gunderson et al., 2006; Steemers and Gunderson, 2007). A potential area for concern with the *Infinium* platform is how reliably the whole-genome amplification step generates an accurate representation of the input DNA, since any bias at this stage could alter allele frequencies observed in the resulting genotypes. Some attempts have been made at assessing this, which suggest that the amplification is reliable (Berthier-Schaad et al., 2007).

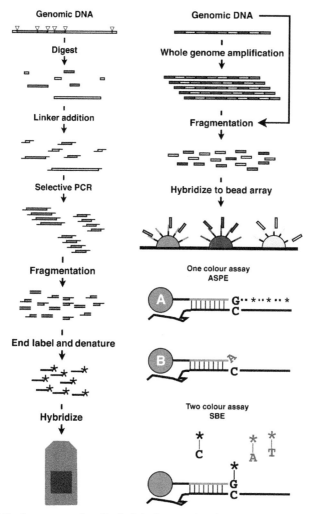

**FIGURE 11.9** Array genotyping. On the left, the procedure for gentoyping with an Affymetrix array. Genomic DNA is completely digested with a restriction enzyme liberating a range of fragment sizes. Linkers are ligated to the end of each fragment and PCR is performed under conditions favoring the amplification of fragments in the 200–1100 bp range. The amplified DNA is fragmented, end labeled and hybridized to the GeneChip. On the right, a description of the processes when using the Illumina bead array platform. Whole genomic DNA (either native if sufficient sample is available or after a whole genome amplification step) is fragmented and hybridized to a bead array where polymorphic regions are captured by unique probes tethered via their 5′ ends. Each bead-tethered oligonucleotide has a unique 50mer component (shades of grey) and a specific addressing sequence (black area). Two detection methods are available. In ASPE, two probes per SNP are used, each terminating with one of the two polymorphic bases (G or A). If the genomic DNA contains the C SNP, DNA polymerase extension can initiate from the probe, incorporating bases conjugated to a fluorescent dye. The terminal nucleotide (A) of the probe scoring the T SNP will not pair with genomic DNA containing the C SNP and hence will not prime DNA synthesis. In the case of genomic DNA

However, it is likely that successful amplification will be highly dependent upon the quality of the input DNA and this should be carefully assessed to ensure that the starting DNA is high molecular weight. Of course, if sufficient input DNA is available at the outset an amplification step is not required. The amplified DNA is captured on a genotyping BeadArray and, as with Affymetrix, a variety of formats are available for genotyping up to 500 000 SNPs in a single assay. Illumina provide smaller format arrays, allowing more samples to be genotyped against a reduced selection of SNPs, which can be useful for higher resolution scanning of selected SNPs with more samples once a particular region of interest has been identified by a genome-wide scan. A BeadChip array contains over 13 million individual beads, each with many copies of a 70mer oligonucleotide tethered to it surface by its 5′ end. The oligonucleotides comprise a unique 50mer, designed to be adjacent to the desired SNP, and a 20mer addressing sequence that is used to locate each bead on the array. After synthesis of each unique bead/oligonucleotide conjugate, the population of different beads is mixed and randomly dispersed over a substrate surface comprising etched microwells such that each bead/probe is present at around 20 copies per array. The array is addressed after manufacture by a rapid hybridization technique and a unique deconvolution file provided with each array.

There are two assays methods possible with the *Infinium* system, both relying on enzymatic extension of the bead-bound oligonucleotide (Figure 11.9). In the first, allele specific primer extension (ASPE), two oligonucleotide probes per bi-allelic SNP are utilized, with the terminal 3′ base of each probe matching the polymorphic nucleotide. If the genomic sample hybridized with the probe matches the terminal base then DNA polymerase can extend the probe sequence, copying the genomic DNA and during the extension a fluorescent reporter is incorporated. If there is a mismatch, the polymerase does not extend and there is therefore no fluorescent signal. This is a one-color assay designed to detect all possible allele classes at a given SNP by preparing specific probes for each allele. In the second approach, single base extension (SBE), the probe is again extended using the genomic target as a template. However in this case dideoxy A and T residues are labeled with one fluorescent dye, and the dideoxy C and G residues with another. Thus incorporation of an A or a T generates one signal and of a C or G a second signal. In the SBE approach only a single probe is needed for each SNP, doubling the number of independent probes that can be employed compared with the ASPE approach. Therefore, while AT or CG polymorphisms cannot be distinguished in the SBE assay it is, however, possible to carry out both

---

heterozygous for C and T SNPs, both probes incorporate label. In the two-color SBE assay, a single probe terminates with its 3′ end at the base adjacent to the SNP. DNA polymerase incorporates a fluorescently labeled dideoxy nucleotide, with one dye used for C and G residues and a second used for A and T residues. The array is scanned at two wavelengths to detect each dye thus scoring for incorporation of an A/T or C/G base.

SBE and ASPE reactions simultaneously on the same array, increasing the flexibility of the system (Steemers et al., 2006).

## 11.5.3 Analysis of Genotyping Arrays

There are many factors that confound the analysis of whole-genome genotype data, unsurprising when one considers that over a billion individual genotypes can easily be generated in a single study. It is clear that from the outset care must be taken in the selection of case and control samples. Initiatives such as the Wellcome Trust Case Control Consortium (www.wtccc.org.uk) and the Genetic Association Information Network (http://test.fnih.org/GAIN2/home_new. shtml), facilitate the controlled collection of samples and, more importantly, the uniform generation, analysis and distribution of the incredibly large data sets generated (WTCCC, 2007). The choice of controls for these studies are beyond the scope of this work but it is an important methodological question that will inevitably be more fully understood as more whole genome association studies are performed and the genetic structure of human populations better defined. Similarly, we do not discuss issues of sample size or how genotype data is assessed for association with a particular phenotypic trait. Of relevance to this discussion are the analytical methods used for genotype calling and since the Affymetrix and Illumina platforms utilize different detection technologies we shall consider them independently.

For each SNP on an array the genotype may be called as homozygous (AA or BB), heterozygous (AB) or null and, as with any array study, the aim is to minimize false positive and false negative errors. Genotyping with any platform is not perfect, genotypes calling errors may be due to poor quality data or to errors in the genotype calling algorithm. Quality control measures must be in place to ensure that low quality SNPs are rejected and as many high quality SNPs as possible retained. Miscalling errors can lead to spurious associations and when such large numbers of SNPs are being typed even small error rates can lead to unacceptable false positive levels. On the other hand, losing high quality SNPs by overconservative data filtering leads to a reduction in resolution or, at worst, failure to detect meaningful associations. In general terms, samples genotyped on either of the two array platforms are discarded or repeated if the SNP calling rate is below a predetermined rate: call rates of over 98% are frequently reported. In general, most approaches generate a plot showing the signal intensities for both SNP alleles across a set of samples and in the case of a well-called SNP, homozygotes and heterozygotes are well differentiated into clearly defined clusters on the plot. However, in some cases a particular SNP may be problematic and the genotypes are not well resolved, generating a null SNP call (Figure 11.10). Treatment of the intensity data such that reliable clusters can be generated for every SNP on the array, each including every sample is clearly the ideal for any calling algorithm. However, many difficulties arise because some array probes inaccurately report the status of the alleles being typed and calling software can flounder, especially in cases where one allele is present at a low frequency in the population.

### 11.5.4 Affymetrix

As with Affymetrix gene expression arrays, genotype calls rely on the assessment of the fluorescent intensity detected at each probe cell on the array. As we indicate above, each SNP is detected by multiple probe quartets spanning the genomic region on both strands. In brief, the original Affymetrix approach developed for the 10K arrays first employs a detection filter that assesses, for each PM/MM probe pair, if the PM signal is greater than the MM signal (Liu et al., 2003). In the feature extraction step, probes that pass the detection filter are used to calculate the relative allele signal for each SNP on both sense and antisense strands in order to determine the genotype. Unfortunately, since the hybridization behavior of the 25mer probes is strongly sequence dependent, the relative intensities for each

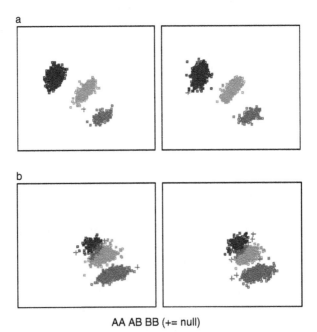

AA AB BB (+= null)

**FIGURE 11.10**   Genotyping cluster plots. In these plots the signal intensity of allele 1 (*y*-axis) is plotted against the intensity for allele 2 (*x*-axis). Normalized data from a set of samples are clustered by one of the methods described in the text. In (a), the genotypes are well resolved, with both homozygotes and the heterozygote (represented by differently shaded data points) clearly separated in the plots. Part (b) provides examples of more difficult cases, where the heterozygotes and AA homozygotes are not well resolved, correctly calling genotypes in these cases is difficult. Grey crosses represent samples that were not called by the software.

allele probe on each strand may differ dramatically. Consequently each set of probes in the quartet needs to be independently determined. As a consequence of this probe-specific behavior, individual probe pairs must be first assessed with a training set and a classification method used to predict the genotype calls when experimental samples are hybridized. The initial algorithm utilized a modification of the partitioning around medoids approach (MPAM: a clustering algorithm similar to $k$ means) to guide genotype calls. The MPAM approach requires a relatively large number of samples to build an accurate classifier and even then has difficulty with alleles present at a low frequency. In addition, the performance of the classifier is highly dependent upon the scanner settings and other experimental variables. Thus the approach can be computationally very intensive, especially if arrays are rescanned, and does not scale well for the highest density arrays. To address this issue, Affymetrix developed a dynamic model-based (DM) approach that combines information from all probe quartets interrogating a particular SNP to provide a statistical confidence metric for each genotype call (Di et al., 2005). The DM algorithm is reported to be more reliable and, in particular, handles the large 100K and 500K arrays much more efficiently. The most recent software, BRLMM (Affymetrix) removes the dependency on a training set and has grown out of the RLMM approach describe below.

Along with the analytical tools developed by Affymetrix, others have studied the problem of genotype calling and developed alternative approaches. For example, Huentelman et al. (2005) developed SNiPer, a method similar to MPAM but using alternative clustering algorithms, that is reported to improve genotype calls. However, the method has not, to our knowledge, been tested with very high-density arrays. The same authors developed another method, SNiPer-HD (Hua et al., 2007), using an expectation-maximization approach and report that this outperforms the Affymetrix DM approach with 500K arrays. All of the above methods determine the genotype calls based on the performance of all the SNPs on a single array. An alternative approach examines the performance of individual SNPs across multiple arrays. Rabbee and Speed (2006) proposed a robustly fitted linear model (*R* package *RLMM*) that combines probe-level data from multiple arrays after applying the RMA normalization method across the arrays. Using RMA they ignore the information from mismatch probes (see Chapter 6 for a description of RMA). They report a substantial improvement over DM in terms of accuracy and the overall call rate and Affymetrix have adopted this approach in their latest BRLMM software. Note that a very similar multiarray approach was proposed around the same time (Lamy et al., 2006). More recently Xiao et al. (2007) have developed a comprehensive algorithm that combines both the single array and multiple array approaches (MAMS, multiarray, multi-SNP) that requires no training. Finally, the Wellcome Case Control Consortium developed a Bayesian hierarchical mixture model using normalized signals intensities, *Chiamo*, that is reported to improve accuracy and specificity (WTCCC, 2007).

## 11.5.5 Illumina

In principle, the analysis of the nucleotide extension signals derived from the *Infinium* platform should be more straightforward since there is a unique probe for each SNP and consequently no need to assess multiple different probes per SNP. A potential advantage of the Illumina system is that multiple copies of each unique probe are present on each array thus noise or occasional nucleotide misincorporation can be average out (Gunderson et al., 2006). Assuming that the extension reaction is efficient and error free (which it is not) genotyping is achieved by collecting the fluorescent intensities for each bead, single channel in the case of the ASPE assay and two channels when the SBE format is being used. In many respects the data from the two-channel SBE assay is similar to gene expression data and suffers some of the same problems regarding, for example, differential incorporation or detection levels for each of the fluorescent dyes. Since SNPs are called based on the ratio of signals obtained with the two polymorphic probes, between channel normalization is clearly required and is a critical step. In contrast to the situation with the Affymetrix platform, as far as we are aware, there are few published reports of the development of independent analysis tools for *Infinium* arrays and most users appearing to rely on the proprietary Illumina GenCall software provided with the system (Kermani, 2004). In this approach, each array is normalized by applying an affine transformation within a section of the array (the sub-bead pool, representing $\sim$5% of the array). Genotypes are called from the normalized data using an artificial neural network approach. A model-based approach for calling genotypes, based on expectation-maximization, has recently been developed and is reported to perform better than GenCall, particularly when amplified DNA samples are genotyped (Teo et al., 2007).

The use of genotyping arrays is only going to increase in the near future and one hopes to see improvements in the reliability and efficiency of the algorithms used to call genotypes. This is particularly important when the density of genotyping platforms is increasing and will soon accommodate the simultaneous acquisition of millions of genotypes per sample. As the number of SNPs genotyped in an assay increases, the number of confounding false positives will rise. It is likely that as the amount of available data increases, particularly those derived from diverse populations, then the ability to deal with issues such as rare allele frequency will increase. It can only be a matter of time before comprehensive genotyping assays become a routine aspect of diagnostic medicine, of course assuming that cost effective resequencing does not overtake the array-based assays.

## 11.6 SUMMARY

In this overview we have tried to highlight how microarray technology is now firmly embedded in all aspects of human genetics and disease studies.

The increasingly mature gene expression platforms now available allow a comprehensive view of the transcriptome in virtually any situation where a cell population can be isolated. With several years of experience in terms of probe design, normalization methods, application of robust statistical metrics and data exploration tools, the field is beginning to stabilize. Problems of interplatform comparability are becoming issues of the past with the ongoing development of appropriate internal and external controls. While it is clear that developments will continue, particularly in the areas of data processing and analysis, we feel that improvements in these areas will be evolutionary rather than revolutionary. One area where there is certainly scope for improvement is in meta-analysis, assigning functional significance to changes in gene expression. The experiments we describe here are typical of many disease studies in as much as they generate a set of genes with altered expression profiles associated with a particular condition. Turning a gene expression profile or specific gene expression signature into biological understanding is still a difficult task involving considerable effort and detailed experimentation. In this area, our increasing functional annotation of genomes, those of model organism as well as the human genome, will provide considerable additional information to gene expression studies. In addition, the emerging area of Systems Biology, where at the transcriptome level the goal is to be able to describe the gene regulatory networks that underpin individual gene expression states, will in time provide a detailed and predictable map of the transcriptional state of any cell. Ultimately the specific biochemical pathways and biological processes perturbed in a diseased cell or tissue will be predictable from the gene expression profile.

The transcriptome of a cell is informed by the genome and, as we have seen, the structure of the genome, particularly with respect to genetic differences that can alter expression levels of genes, is still a black box to us. The incredible developments in genome analysis that facilitate the characterization of nucleotide polymorphisms and copy number alterations in individuals are opening new avenues for biomedical research. It is not unrealistic to hope that complete descriptions of all relevant genetic differences between individual humans will be achievable relatively soon. Whether this is via ultrahigh-throughput sequencing or array-based assays is difficult to predict at the moment, each technology has its strengths and weaknesses. We may soon have $1000 human genome sequencing but if array-based assays cost $10 or $100, what will be the most cost effective in terms of implementation in the clinic? Time will tell. What is clear is that in a very short space of time, the possibility of capturing the genetic fingerprint of an individual on a single array has been realized. While there clearly needs to be improvements in the platform technologies and, more importantly, the analytical tools for processing and interpreting data from array-CGH or genotyping array studies, if we use the evolution of gene expression data analysis as a guide, these improvements will be rapid.

With this explosion in human genetic data, issues regarding personal privacy, genetic discrimination and population engineering via preimplantation genetic screening raise their heads. It is clear to us that these are becoming very real issues that society needs to be informed about and debate before they are presented as *fait accompli*.

## REFERENCES

Alizadeh AA, Eisen MB, Davis RE, Ma C, Lossos IS, Rosenwald A, Boldrick JC, Sabet H, Tran T, Yu X, Powell JI, Yang L, Marti GE, Moore T, Hudson J, Lu L, Lewis DB, Tibshirani R, Sherlock G, Chan WC, Greiner TC, Weisenburger DD, Armitage JO, Warnke R, Levy R, Wilson W, Grever MR, Byrd JC, Botstein D, Brown PO, Staudt LM. (2000) Distinct types of diffuse large B-cell lymphoma identified by gene expression profiling. *Nature* **403:** 503–511.

Ambroise C, McLachlan G. (2002) Aelection bias in gene extraction on the basis of microarray gene-expression data. *Proc Natl Acad Sci U S A* **99:** 6562–6566.

Arca B, Lombardo F, Valenzuela JG, Francischetti IM, Marinotti O, Coluzzi M, Ribeiro JM. (2005) An updated catalogue of salivary gland transcripts in the adult female mosquito *Anopheles gambiae. J Exp Biol* **208:** 3971–3986.

Auer H, Newsom DL, Nowak NJ, McHugh KM, Singh S, Yu CY, Yang Y, Wenger GD, Gastier-Foster JM, Kornacker K. (2007) Gene-resolution analysis of DNA copy number variation using oligo-nucleotide expression microarrays. *BMC Genomics* **8:** 111.

Bahl A, Brunk B, Crabtree J, Fraunholz MJ, Gajria B, Grant GR, Ginsburg H, Gupta D, Kissinger JC, Labo P, Li L, Mailman MD, Milgram AJ, Pearson DS, Roos DS, Schug J, Stoeckert CJ, Whetzel P. (2003) PlasmoDB: the Plasmodium genome resource. A database integrating experimental and computational data. *Nucleic Acids Res* **31:** 212–215.

Baker SC, Bauer SR, Beyer RP, Brenton JD, Bromley B, Burrill J, Causton H, Conley MP, Elespuru R, Fero M, Foy C, Fuscoe J, Gao X, Gerhold DL, Gilles P, Goodsaid F, Guo X, Hackett J, Hockett RD, Ikonomi P, Irizarry RA, Kawasaki ES, Kaysser-Kranich T, Kerr K, Kiser G, Koch WH, Lee KY, Liu C, Liu ZL, Lucas A, Manohar CF, Miyada G, Modrusan Z, Parkes H, Puri RK, Reid L, Ryder TB, Salit M, Samaha RR, Scherf U, Sendera TJ, Setterquist RA, Shi L, Shippy R, Soriano JV, Wagar EA, Warrington JA, Williams M, Wilmer F, Wilson M, Wolber PK, Wu X, Zadro R. (2005) The External RNA Controls Consortium: a progress report. *Nat Methods* **2:** 731–734.

Bejjani BA, Shaffer LG. (2006) Application of array-based comparative genomic hybridization to clinical diagnostics. *J Mol Diagn* **8:** 528–533.

Ben Mamoun C, Gluzman IY, Hott C, MacMillan SK, Amarakone AS, Anderson DL, Carlton JM, Dame JB, Chakrabarti D, Martin RK, Brownstein BH, Goldberg DE. (2001) Co-ordinated programme of gene expression during asexual intraerythrocytic development of the human malaria parasite *Plasmodium falciparum* revealed by microarray analysis. *Mol Microbiol* **39:** 26–36.

Berthier-Schaad Y, Kao WH, Coresh J, Zhang L, Ingersoll RG, Stephens R, Smith MW. (2007) Reliability of high-throughput genotyping of whole genome amplified DNA in SNP genotyping studies. *Electrophoresis* **28:** 2812–2817.

Bozdech Z, Zhu J, Joachimiak MP, Cohen FE, Pulliam B, DeRisi JL. (2003) Expression profiling of the schizont and trophozoite stages of *Plasmodium falciparum* with a long-oligonucleotide microarray. *Genome Biol* **4:** R9.

Bozdech Z, Llinas M, Pulliam BL, Wong ED, Zhu J, DeRisi JL. (2003) The transcriptome of the intraerythrocytic developmental cycle of *Plasmodium falciparum*. *PLoS Biol* **1**: E5.

Brennan C, Zhang Y, Leo C, Feng B, Cauwels C, Aguirre AJ, Kim M, Protopopov A, Chin L. (2004) High-resolution global profiling of genomic alterations with long oligonucleotide microarray. *Cancer Res* **64**: 4744–4748.

Burton PR, Clayton DG, Cardon LR, Craddock N, Deloukas P, Duncanson A, Kwiatkowski DP, McCarthy MI, Ouwehand WH, Samani NJ, Todd JA, Donnelly P, Barrett JC, Davison D, Easton D, Evans DM, Leung HT, Marchini JL, Morris AP, Spencer CC, Tobin MD, Attwood AP, Boorman JP, Cant B, Everson U, Hussey JM, Jolley JD, Knight AS, Koch K, Meech E, Nutland S, Prowse CV, Stevens HE, Taylor NC, Walters GR, Walker NM, Watkins NA, Winzer T, Jones RW, McArdle WL, Ring SM, Strachan DP, Pembrey M, Breen G, St Clair D, Caesar S, Gordon-Smith K, Jones L, Fraser C, Green EK, Grozeva D, Hamshere ML, Holmans PA, Jones IR, Kirov G, Moskivina V, Nikolov I, O'Donovan MC, Owen MJ, Collier DA, Elkin A, Farmer A, Williamson R, McGuffin P, Young AH, Ferrier IN, Ball SG, Balmforth AJ, Barrett JH, Bishop TD, Iles MM, Maqbool A, Yuldasheva N, Hall AS, Braund PS, Dixon RJ, Mangino M, Stevens S, Thompson JR, Bredin F, Tremelling M, Parkes M, Drummond H, Lees CW, Nimmo ER, Satsangi J, Fisher SA, Forbes A, Lewis CM, Onnie CM, Prescott NJ, Sanderson J, Matthew CG, Barbour J, Mohiuddin MK, Todhunter CE, Mansfield JC, Ahmad T, Cummings FR, Jewell DP, Webster J, Brown MJ, Lathrop MG, Connell J, Dominiczak A, Marcano CA, Burke B, Dobson R, Gungadoo J, Lee KL, Munroe PB, Newhouse SJ, Onipinla A, Wallace C, Xue M, Caulfield M, Farrall M, Barton A; Biologics in RA Genetics and Genomics Study Syndicate (BRAGGS) Steering Committee, Bruce IN, Donovan H, Eyre S, Gilbert PD, Hilder SL, Hinks AM, John SL, Potter C, Silman AJ, Symmons DP, Thomson W, Worthington J, Dunger DB, Widmer B, Frayling TM, Freathy RM, Lango H, Perry JR, Shields BM, Weedon MN, Hattersley AT, Hitman GA, Walker M, Elliott KS, Groves CJ, Lindgren CM, Rayner NW, Timpson NJ, Zeggini E, Newport M, Sirugo G, Lyons E, Vannberg F, Hill AV, Bradbury LA, Farrar C, Pointon JJ, Wordsworth P, Brown MA, Franklyn JA, Heward JM, Simmonds MJ, Gough SC, Seal S; Breast Cancer Susceptibility Collaboration (UK), Stratton MR, Rahman N, Ban M, Goris A, Sawcer SJ, Compston A, Conway D, Jallow M, Newport M, Sirugo G, Rockett KA, Bumpstead SJ, Chaney A, Downes K, Ghori MJ, Gwilliam R, Hunt SE, Inouye M, Keniry A, King E, McGinnis R, Potter S, Ravindrarajah R, Whittaker P, Widden C, Withers D, Cardin NJ, Davison D, Ferreira T, Pereira-Gale J, Hallgrimsdo'ttir IB, Howie BN, Su Z, Teo YY, Vukcevic D, Bentley D, Brown MA, Compston A, Farrall M, Hall AS, Hattersley AT, Hill AV, Parkes M, Pembrey M, Stratton MR, Mitchell SL, Newby PR, Brand OJ, Carr-Smith J, Pearce SH, McGinnis R, Keniry A, Deloukas P, Reveille JD, Zhou X, Sims AM, Dowling A, Taylor J, Doan T, Davis JC, Savage L, Ward MM, Learch TL, Weisman MH, Brown M. (2007) Association scan of 14,500 nonsynonymous SNPs in four diseases identifies autoimmunity variants. *Nat Genet* **39**: 1329–1337.

Carter NP. (2007) Methods and strategies for analyzing copy number variation using DNA microarrays. *Nat Genet* **39**: S16–21.

Carvalho B, Ouwerkerk E, Meijer GA, Ylstra B. (2004) High resolution microarray comparative genomic hybridization analysis using spotted oligonucleotides. *J Clin Pathol* **57**: 644–646.

Chari R, Lockwood WW, Lam WL. (2006) Computational methods for the analysis of array comparative genomic hybridization. *Cancer Inform* **2**: 48–58.

Chung CH, Bernard PS, Perou CM. (2002) Molecular portraits and the family tree of cancer. *Nat Genet* **32**: S533–S540.

Coon KD, Myers AJ, Craig DW, Webster JA, Pearson JV, Lince DH, Zismann VL, Beach TG, Leung D, Bryden L, Halperin RF, Marlowe L, Kaleem M, Walker DG, Ravid R, Heward CB, Rogers J, Papassotiropoulos A, Reiman EM, Hardy J, Stephan DA. (2007) A high-density whole-genome association study reveals that APOE is the major susceptibility gene for sporadic late-onset Alzheimer's disease. *J Clin Psychiatry* **68:** 613–618.

Coppel RL, Roos DS, Bozdech Z. (2004) The genomics of malaria infection. *Trends Parasitol* **20:** 553–557.

Coulson RM, Hall N, Ouzounis CA. (2004) Comparative genomics of transcriptional control in the human malaria parasite *Plasmodium falciparum. Genome Res* **14:** 1548–1554.

Cutler A, Stevens JR. (2006) Random forests for microarrays. *Methods Enzymol* **411:** 422–432.

Dahl EL, Shock JL, Shenai BR, Gut J, DeRisi JL, Rosenthal PJ. (2006) Tetracyclines specifically target the apicoplast of the malaria parasite *Plasmodium falciparum. Antimicrob Agents Chemother* **50:** 3124–3131.

DeRisi J, Penland L, Brown PO, Bittner ML, Meltzer PS, Ray M, Chen Y, Su YA, Trent JM. (1996) Use of a cDNA microarray to analyse gene expression patterns in human cancer. *Nat Genet* **14:** 457–460.

Devereaux PJ, Manns BJ, Ghali WA, Quan H, Guyatt GH. (2002) The reporting of methodological factors in randomized controlled trials and the association with a journal policy to promote adherence to the Consolidated Standards of Reporting Trials (CONSORT) checklist. *Contr Clin Trials* **23:** 380–388.

Di X, Matsuzaki H, Webster TA, Hubbell E, Liu G, Dong S, Bartell D, Huang J, Chiles R, Yang G, Shen MM, Kulp D, Kennedy GC, Mei R, Jones KW, Cawley S. (2005) Dynamic model based algorithms for screening and genotyping over 100K SNPs on oligonucleotide microarrays. *Bioinformatics* **21:** 1958–1963.

du Manoir S, Speicher MR, Joos S, Schrock E, Popp S, Dohner H, Kovacs G, Robert-Nicoud M, Lichter P, Cremer T. (1993) Detection of complete and partial chromosome gains and losses by comparative genomic in situ hybridization. *Hum Genet* **90:** 590–610.

Dudoit S, Fridlyand J, Speed T. (2002) Comparison of discrimination methods for classification of tumors using gene expression. *J Am Stat Assoc* **97:** 77–87.

Dupuy A, Simon RM. (2007) Critical review of published microarray studies for cancer outcome and guidelines on statistical analysis and reporting. *J Natl Cancer Inst* **99:** 147–157.

Emanuel BS, Shaikh TH. (2001) Segmental duplications: an 'expanding' role in genomic instability and disease. *Nat Rev Genet* **2:** 791–800.

Fiegler H, Redon R, Carter NP. (2007) Construction and use of spotted large-insert clone DNA microarrays for the detection of genomic copy number changes. *Nat Protoc* **2:** 577–587.

Fiegler H, Geigl JB, Langer S, Rigler D, Porter K, Unger K, Carter NP, Speicher MR. (2007) High resolution array-CGH analysis of single cells. *Nucleic Acids Res* **35:** e15.

Fiegler H, Redon R, Andrews D, Scott C, Andrews R, Carder C, Clark R, Dovey O, Ellis P, Feuk L, French L, Hunt P, Kalaitzopoulos D, Larkin J, Montgomery L, Perry GH, Plumb BW, Porter K, Rigby RE, Rigler D, Valsesia A, Langford C, Humphray SJ, Scherer SW, Lee C, Hurles ME, Carter NP. (2006) Accurate and reliable high-throughput detection of copy number variation in the human genome. *Genome Res* **16:** 1566–1574.

Fishel I, Kaufman A, Ruppin E. (2007) Meta-analysis of gene expression data: a predictor-based approach. *Bioinformatics* **23:** 1599–1606.

Frazer KA, Ballinger DG, Cox DR, Hinds DA, Stuve LL, Gibbs RA, Belmont JW. (2007) A second generation human haplotype map of over 3.1 million SNPs. *Nature* **449:** 851–861.

Freeman JL, Perry GH, Feuk L, Redon R, McCarroll SA, Altshuler DM, Aburatani H, Jones KW, Tyler-Smith C, Hurles ME, Carter NP, Scherer SW, Lee C. (2006) Copy number variation: new insights in genome diversity. *Genome Res* **16**: 949–961.

Frueh FW. (2006) Impact of microarray data quality on genomic data submissions to the FDA. *Nat Biotechnol* **24**: 1105–1107.

Gardner MJ, Hall N, Fung E, White O, Berriman M, Hyman RW, Carlton JM, Pain A, Nelson KE, Bowman S, Paulsen IT, James K, Eisen JA, Rutherford K, Salzberg SL, Craig A, Kyes S, Chan MS, Nene V, Shallom SJ, Suh B, Peterson J, Angiuoli S, Pertea M, Allen J, Selengut J, Haft D, Mather MW, Vaidya AB, Martin DM, Fairlamb AH, Fraunholz MJ, Roos DS, Ralph SA, McFadden GI, Cummings LM, Subramanian GM, Mungall C, Venter JC, Carucci DJ, Hoffman SL, Newbold C, Davis RW, Fraser CM, Barrell B. (2002) Genome sequence of the human malaria parasite *Plasmodium falciparum*. *Nature* **419**: 498–511.

Geigl JB, Speicher MR. (2007) Single-cell isolation from cell suspensions and whole genome amplification from single cells to provide templates for CGH analysis. *Nat Protoc* **2**: 3173–3184.

Gerhold DL, Jensen RV, Gullans SR. (2002) Better therapeutics through microarrays. *Nat Genet* **32**: S547–551.

Gibbs JR, Singleton A. (2006) Application of genome-wide single nucleotide polymorphism typing: simple association and beyond. *PLoS Genet* **2**: e150.

Golub TR, Slonim DK, Tamayo P, Huard C, Gaasenbeek M, Mesirov JP, Coller H, Loh ML, Downing JR, Caligiuri MA, Bloomfield CD, Lander ES. (1999) Molecular classification of cancer: class discovery and class prediction by gene expression monitoring. *Science* **286**: 531–537.

Goodsaid F, Frueh FW. (2007) Implementing the U.S. FDA guidance on pharmacogenomic data submissions. *Environ Mol Mutagen* **48**: 354–358.

Greshock J, Feng B, Nogueira C, Ivanova E, Perna I, Nathanson K, Protopopov A, Weber BL, Chin L. (2007) A comparison of DNA copy number profiling platforms. *Cancer Res* **67**: 10173–10180.

Guerra CA, Gikandi PW, Tatem AJ, Noor AM, Smith DL, Hay SI, Snow RW. (2008) The limits and intensity of *Plasmodium falciparum* transmission: implications for malaria control and elimination worldwide. *PLoS Med* **5**: e38.

Gunderson KL, Steemers FJ, Ren H, Ng P, Zhou L, Tsan C, Chang W, Bullis D, Musmacker J, King C, Lebruska LL, Barker D, Oliphant A, Kuhn KM, Shen R. (2006) Whole-genome genotyping. *Methods Enzymol* **410**: 359–376.

Gutman S, Kessler LG. (2006) The US Food and Drug Administration perspective on cancer biomarker development. *Nat Rev Cancer* **6**: 565–571.

Hanash S. (2004) Integrated global profiling of cancer. *Nat Rev Cancer* **4**: 638–643.

Hanauer DA, Rhodes DR, Sinha-Kumar C, Chinnaiyan AM. (2007) Bioinformatics approaches in the study of cancer. *Curr Mol Med* **7**: 133–141.

Hayward RE, Derisi JL, Alfadhli S, Kaslow DC, Brown PO, Rathod PK. (2000) Shotgun DNA microarrays and stage-specific gene expression in *Plasmodium falciparum* malaria. *Mol Microbiol* **35**: 6–14.

Hollich V, Johnson E, Furlong EE, Beckmann B, Carlson J, Celniker SE, Hoheisel JD. (2004) Creation of a minimal tiling path of genomic clones for *Drosophila*: provision of a common resource. *Biotechniques* **37**: 282–284.

Holt RA, Subramanian GM, Halpern A, Sutton GG, Charlab R, Nusskern DR, Wincker P, Clark AG, Ribeiro JM, Wides R, Salzberg SL, Loftus B, Yandell M, Majoros WH, Rusch DB, Lai Z, Kraft CL, Abril JF, Anthouard V, Arensburger P, Atkinson PW, Baden H, de Berardinis V, Baldwin D, Benes V, Biedler J, Blass C, Bolanos R, Boscus D, Barnstead M, Cai S, Center A, Chaturverdi K,

Christophides GK, Chrystal MA, Clamp M, Cravchik A, Curwen V, Dana A, Delcher A, Dew I, Evans CA, Flanigan M, Grundschober-Freimoser A, Friedli L, Gu Z, Guan P, Guigo R, Hillenmeyer ME, Hladun SL, Hogan JR, Hong YS, Hoover J, Jaillon O, Ke Z, Kodira C, Kokoza E, Koutsos A, Letunic I, Levitsky A, Liang Y, Lin JJ, Lobo NF, Lopez JR, Malek JA, McIntosh TC, Meister S, Miller J, Mobarry C, Mongin E, Murphy SD, O'Brochta DA, Pfannkoch C, Qi R, Regier MA, Remington K, Shao H, Sharakhova MV, Sitter CD, Shetty J, Smith TJ, Strong R, Sun J, Thomasova D, Ton LQ, Topalis P, Tu Z, Unger MF, Walenz B, Wang A, Wang J, Wang M, Wang X, Woodford KJ, Wortman JR, Wu M, Yao A, Zdobnov EM, Zhang H, Zhao Q, Zhao S, Zhu SC, Zhimulev I, Coluzzi M, della Torre A, Roth CW, Louis C, Kalush F, Mural RJ, Myers EW, Adams MD, Smith HO, Broder S, Gardner MJ, Fraser CM, Birney E, Bork P, Brey PT, Venter JC, Weissenbach J, Kafatos FC, Collins FH, Hoffman SL. (2002) The genome sequence of the malaria mosquito *Anopheles gambiae*. *Science* **298:** 129–149.

Hu J, Gao JB, Cao Y, Bottinger E, Zhang W. (2007) Exploiting noise in array CGH data to improve detection of DNA copy number change. *Nucleic Acids Res* **35:** e35.

Hua J, Craig DW, Brun M, Webster J, Zismann V, Tembe W, Joshipura K, Huentelman MJ, Dougherty ER, Stephan DA. (2007) SNiPer-HD: improved genotype calling accuracy by an expectation-maximization algorithm for high-density SNP arrays. *Bioinformatics* **23:** 57–63.

Huang D, Chow TW. (2007) Identifying the biologically relevant gene categories based on gene expression and biological data: an example on prostate cancer. *Bioinformatics* **23:** 1503–1510.

Huentelman MJ, Craig DW, Shieh AD, Corneveaux JJ, Hu-Lince D, Pearson JV, Stephan DA. (2005) SNiPer: improved SNP genotype calling for Affymetrix 10K GeneChip microarray data. *BMC Genomics* **6:** 149.

Ishkanian AS, Malloff CA, Watson SK, DeLeeuw RJ, Chi B, Coe BP, Snijders A, Albertson DG, Pinkel D, Marra MA, Ling V, MacAulay C, Lam WL. (2004) A tiling resolution DNA microarray with complete coverage of the human genome. *Nat Genet* **36:** 299–303.

Jafari P, Azuaje F. (2006) An assessment of recently published gene expression data analyses: reporting experimental design and statistical factors. *BMC Med Inform Decis Mak* **6:** 27.

Ji H, Davis RW. (2006) Data quality in genomics and microarrays. *Nat Biotechnol* **24:** 1112–1113.

Kennedy GC, Matsuzaki H, Dong S, Liu WM, Huang J, Liu G, Su X, Cao M, Chen W, Zhang J, Liu W, Yang G, Di X, Ryder T, He Z, Surti U, Phillips MS, Boyce-Jacino MT, Fodor SP, Jones KW. (2003) Large-scale genotyping of complex DNA. *Nat Biotechnol* **21:** 1233–1237.

Kermani B. (2004) Artificial intelligence and global normalization methods for genotyping. *USPTO Patent Application* 20060224529, Illumina Inc.

Khojasteh M, Lam WL, Ward RK, MacAulay C. (2005) A stepwise framework for the normalization of array CGH data. *BMC Bioinform* **6:** 274.

Kingsmore SF, Lindquist IE, Mudge J, Gessler DD, Beavis WD. (2008) Genome-wide association studies: progress and potential for drug discovery and development. *Nat Rev Drug Discov* **7:** 221–230.

Klijn C, Holstege H, de Ridder J, Liu X, Reinders M, Jonkers J, Wessels L. (2008) Identification of cancer genes using a statistical framework for multiexperiment analysis of nondiscretized array CGH data. *Nucleic Acids Res* **36:** e13.

Kooij TW, Janse CJ, Waters AP. (2006) Plasmodium post-genomics: better the bug you know? *Nat Rev Microbiol* **4:** 344–357.

Kruglyak L. (2008) The road to genome-wide association studies. *Nat Rev Genet* **9:** 314–318.

Lamy P, Andersen CL, Wikman FP, Wiuf C. (2006) Genotyping and annotation of Affymetrix SNP arrays. *Nucleic Acids Res* **34:** e100.

Lander ES. (1999) Array of hope. *Nat Genet* **21:** S3–4.

Lawson D, Arensburger P, Atkinson P, Besansky NJ, Bruggner RV, Butler R, Campbell KS, Christophides GK, Christley S, Dialynas E, Emmert D, Hammond M, Hill CA, Kennedy RC, Lobo NF, MacCallum MR, Madey G, Megy K, Redmond S, Russo S, Severson DW, Stinson EO, Topalis P, Zdobnov EM, Birney E, Gelbart WM, Kafatos FC, Louis C, Collins FH. (2007) VectorBase: a home for invertebrate vectors of human pathogens. *Nucleic Acids Res* **35:** D503–505.

Le Caignec C, Spits C, Sermon K, De Rycke M, Thienpont B, Debrock S, Staessen C, Moreau Y, Fryns JP, Van Steirteghem A, Liebaers I, Vermeesch JR. (2006) Single-cell chromosomal imbalances detection by array CGH. *Nucleic Acids Res* **34:** e68.

Le Roch KG, Zhou Y, Blair PL, Grainger M, Moch JK, Haynes JD, De La Vega P, Holder AA, Batalov S, Carucci DJ, Winzeler EA. (2003) Discovery of gene function by expression profiling of the malaria parasite life cycle. *Science* **301:** 1503–1508.

Lee C, Iafrate AJ, Brothman AR. (2007) Copy number variations and clinical cytogenetic diagnosis of constitutional disorders. *Nat Genet* **39:** S48–54.

Liu WM, Di X, Yang G, Matsuzaki H, Huang J, Mei R, Ryder TB, Webster TA, Dong S, Liu G, Jones KW, Kennedy GC, Kulp D. (2003) Algorithms for large-scale genotyping microarrays. *Bioinformatics* **19:** 2397–2403.

Llinas M, Bozdech Z, Wong ED, Adai AT, DeRisi JL. (2006) Comparative whole genome transcriptome analysis of three *Plasmodium falciparum* strains. *Nucleic Acids Res* **34:** 1166–1173.

Lockwood WW, Chari R, Chi B, Lam WL. (2006) Recent advances in array comparative genomic hybridization technologies and their applications in human genetics. *Eur J Hum Genet* **14:** 139–148.

Lossos IS, Czerwinski DK, Alizadeh AA, Wechser MA, Tibshirani R, Botstein D, Levy R. (2004) Prediction of survival in diffuse large-B-cell lymphoma based on the expression of six genes. *N Engl J Med* **350:** 1828–1837.

Lucito R, Healy J, Alexander J, Reiner A, Esposito D, Chi M, Rodgers L, Brady A, Sebat J, Troge J, West JA, Rostan S, Nguyen KC, Powers S, Ye KQ, Olshen A, Venkatraman E, Norton L, Wigler M. (2003) Representational oligonucleotide microarray analysis: a high-resolution method to detect genome copy number variation. *Genome Res* **13:** 2291–2305.

Lupski JR. (1998) Genomic disorders: structural features of the genome can lead to DNA rearrangements and human disease traits. *Trends Genet* **14:** 417–422.

Maglietta R, Piepoli A, Catalano D, Licciulli F, Carella M, Liuni S, Pesole G, Perri F, Ancona N. (2007) Statistical assessment of functional categories of genes deregulated in pathological conditions by using microarray data. *Bioinformatics* **23:** 2063–2072.

Marinotti O, Nguyen QK, Calvo E, James AA, Ribeiro JM. (2005) Microarray analysis of genes showing variable expression following a blood meal in *Anopheles gambiae*. *Insect Mol Biol* **14:** 365–373.

Marioni JC, Thorne NP, Valsesia A, Fitzgerald T, Redon R, Fiegler H, Andrews TD, Stranger BE, Lynch AG, Dermitzakis ET, Carter NP, Tavare S, Hurles ME. (2007) Breaking the waves: improved detection of copy number variation from microarray-based comparative genomic hybridization. *Genome Biol* **8:** R228.

Meloni R, Khalfallah O, Biguet NF. (2004) DNA microarrays and pharmacogenomics. *Pharmacol Res* **49:** 303–308.

Michiels S, Koscielny S, Hill C. (2005) Prediction of cancer outcome with microarrays: a multiple random validation strategy. *Lancet* **365:** 488–492.

Mills EJ, Wu P, Gagnier J, Devereaux PJ. (2005) The quality of randomized trial reporting in leading medical journals since the revised CONSORT statement. *Contemp Clin Trials* **26:** 480–487.

Mischel P, Cloughesy T, Nelson S. (2004) DNA microarray analysis of broan cancer: molecular classification for therapy. *Nat Rev Neurosci* **5:** 782–792.

Murphy DJ, Brown JR. (2007) Identification of gene targets against dormant phase *Mycobacterium tuberculosis* infections. *BMC Infect Dis* **7:** 84.

Neuvial P, Hupe P, Brito I, Liva S, Manie E, Brennetot C, Radvanyi F, Aurias A, Barillot E. (2006) Spatial normalization of array-CGH data. *BMC Bioinform* **7:** 264.

Phillips KA, Van Bebber S, Issa AM. (2006) Diagnostics and biomarker development: priming the pipeline. *Nat Rev Drug Discov* **5:** 463–469.

Pinkel D, Albertson DG. (2005) Comparative genomic hybridization. *Annu Rev Genomics Hum Genet* **6:** 331–354.

Pinkel D, Segraves R, Sudar D, Clark S, Poole I, Kowbel D, Collins C, Kuo WL, Chen C, Zhai Y, Dairkee SH, Ljung BM, Gray JW, Albertson DG. (1998) High resolution analysis of DNA copy number variation using comparative genomic hybridization to microarrays. *Nat Genet* **20:** 207–211.

Pollack JR, Perou CM, Alizadeh AA, Eisen MB, Pergamenschikov A, Williams CF, Jeffrey SS, Botstein D, Brown PO. (1999) Genome-wide analysis of DNA copy-number changes using cDNA microarrays. *Nat Genet* **23:** 41–46.

Pollack JR, Sorlie T, Perou CM, Rees CA, Jeffrey SS, Lonning PE, Tibshirani R, Botstein D, Borresen-Dale AL, Brown PO. (2002) Microarray analysis reveals a major direct role of DNA copy number alteration in the transcriptional program of human breast tumors. *Proc Natl Acad Sci U S A* **99:** 12963–12968.

Pusztai L, Hess K. (2004) Clinical trial design for microarray predictive marker discovery and assessment. *Ann Oncol* **15:** 1731–1737.

Rabbee N, Speed TP. (2006) A genotype calling algorithm for affymetrix SNP arrays. *Bioinformatics* **22:** 7–12.

Ransohoff DF. (2004) Rules of evidence for cancer molecular-marker discovery and validation. *Nat Rev Cancer* **4:** 309–314.

Ransohoff DF. (2005) Bias as a threat to the validity of cancer molecular-marker research. *Nat Rev Cancer* **5:** 142–149.

Redon R, Ishikawa S, Fitch KR, Feuk L, Perry GH, Andrews TD, Fiegler H, Shapero MH, Carson AR, Chen W, Cho EK, Dallaire S, Freeman JL, Gonzalez JR, Gratacos M, Huang J, Kalaitzo-poulos D, Komura D, MacDonald JR, Marshall CR, Mei R, Montgomery L, Nishimura K, Okamura K, Shen F, Somerville MJ, Tchinda J, Valsesia A, Woodwark C, Yang F, Zhang J, Zerjal T, Zhang J, Armengol L, Conrad DF, Estivill X, Tyler-Smith C, Carter NP, Aburatani H, Lee C, Jones KW, Scherer SW, Hurles ME. (2006) Global variation in copy number in the human genome. *Nature* **444:** 444–454.

Rhodes D, Chinnaiyan A. (2005) Integrative analysis of the cancer transcriptome. *Nat Genet* **37:** S31–S37.

Rhodes DR, Kalyana-Sundaram S, Mahavisno V, Barrette TR, Ghosh D, Chinnaiyan AM. (2005) Mining for regulatory programs in the cancer transcriptome. *Nat Genet* **37:** 579–583.

Rhodes DR, Yu J, Shanker K, Deshpande N, Varambally R, Ghosh D, Barrette T, Pandey A, Chinnaiyan AM. (2004) ONCOMINE: a cancer microarray database and integrated data-mining platform. *Neoplasia* **6:** 1–6.

Rhodes DR, Yu J, Shanker K, Deshpande N, Varambally R, Ghosh D, Barrette T, Pandey A, Chinnaiyan AM. (2004) Large-scale meta-analysis of cancer microarray data identifies common transcriptional profiles of neoplastic transformation and progression. *Proc Natl Acad Sci U S A* **101:** 9309–9314.

Rhodes DR, Kalyana-Sundaram S, Mahavisno V, Varambally R, Yu J, Briggs BB, Barrette TR, Anstet MJ, Kincead-Beal C, Kulkarni P, Varambally S, Ghosh D, Chinnaiyan AM. (2007) Oncomine 3.0: genes, pathways, and networks in a collection of 18,000 cancer gene expression profiles. *Neoplasia* **9:** 166–180.

Rosenwald A, Wright G, Chan WC, Connors JM, Campo E, Fisher RI, Gascoyne RD, Muller-Hermelink HK, Smeland EB, Giltnane JM, Hurt EM, Zhao H, Averett L, Yang L, Wilson WH, Jaffe ES, Simon R, Klausner RD, Powell J, Duffey PL, Longo DL, Greiner TC, Weisenburger DD, Sanger WG, Dave BJ, Lynch JC, Vose J, Armitage JO, Montserrat E, Lopez-Guillermo A, Grogan TM, Miller TP, LeBlanc M, Ott G, Kvaloy S, Delabie J, Holte H, Krajci P, Stokke T, Staudt LM. (2002) The use of molecular profiling to predict survival after chemotherapy for diffuse large-B-cell lymphoma. *N Engl J Med* **346:** 1937–1947.

Roth F. (2001) Bringing out the best features of expression data. *Genome Res* **11:** 1801–1802.

Rueda OM, Diaz-Uriarte R. (2007) Flexible and accurate detection of genomic copy-number changes from aCGH. *PLoS Comput Biol* **3:** e122.

Scherer SW, Lee C, Birney E, Altshuler DM, Eichler EE, Carter NP, Hurles ME, Feuk L. (2007) Challenges and standards in integrating surveys of structural variation. *Nat Genet* **39:** S7–15.

Sebat J, Lakshmi B, Troge J, Alexander J, Young J, Lundin P, Maner S, Massa H, Walker M, Chi M, Navin N, Lucito R, Healy J, Hicks J, Ye K, Reiner A, Gilliam TC, Trask B, Patterson N, Zetterberg A, Wigler M. (2004) Large-scale copy number polymorphism in the human genome. *Science* **305:** 525–528.

Segal E, Friedman N, Koller D, Regev A. (2004) A module map showing conditional activity of expression modules in cancer. *Nat Genet* **36:** 1090–1098.

Segal E, Friedman N, Kaminski N, Regev A, Koller D. (2005) From signatures to models: understanding cancer using microarrays. *Nat Genet* **37:** S38–45.

Selzer RR, Richmond TA, Pofahl NJ, Green RD, Eis PS, Nair P, Brothman AR, Stallings RL. (2005) Analysis of chromosome breakpoints in neuroblastoma at sub-kilobase resolution using fine-tiling oligonucleotide array CGH. *Genes Chromosomes Cancer* **44:** 305–319.

Semblat JP, Silvie O, Franetich JF, Mazier D. (2005) Laser capture microdissection of hepatic stages of the human parasite *Plasmodium falciparum* for molecular analysis. *Methods Mol Biol* **293:** 301–307.

Sharp AJ, Hansen S, Selzer RR, Cheng Z, Regan R, Hurst JA, Stewart H, Price SM, Blair E, Hennekam RC, Fitzpatrick CA, Segraves R, Richmond TA, Guiver C, Albertson DG, Pinkel D, Eis PS, Schwartz S, Knight SJ, Eichler EE. (2006) Discovery of previously unidentified genomic disorders from the duplication architecture of the human genome. *Nat Genet* **38:** 1038–1042.

Shi L, Perkins RG, Fang H, Tong W. (2008) Reproducible and reliable microarray results through quality control: good laboratory proficiency and appropriate data analysis practices are essential. *Curr Opin Biotechnol* **19:** 10–18.

Shi L, Reid LH, Jones WD, Shippy R, Warrington JA, Baker SC, Collins PJ, de Longueville F. (2006) The MicroArray Quality Control: (MAQC) project shows inter- and intraplatform reproducibility of gene expression measurements. *Nat Biotechnol* **24:** 1151–1161.

Shipp MA, Ross KN, Tamayo P, Weng AP, Kutok JL, Aguiar RC, Gaasenbeek M, Angelo M, Reich M, Pinkus GS, Ray TS, Koval MA, Last KW, Norton A, Lister TA, Mesirov J, Neuberg DS, Lander ES, Aster JC, Golub TR. (2002) Diffuse large B-cell lymphoma outcome prediction by gene-expression profiling and supervised machine learning. *Nat Med* **8:** 68–74.

Shock JL, Fischer KF, DeRisi JL. (2007) Whole-genome analysis of mRNA decay in *Plasmodium falciparum* reveals a global lengthening of mRNA half-life during the intra-erythrocytic development cycle. *Genome Biol* **8:** R34.

Silvestrini F, Bozdech Z, Lanfrancotti A, Di Giulio E, Bultrini E, Picci L, Derisi JL, Pizzi E, Alano P. (2005) Genome-wide identification of genes upregulated at the onset of gametocytogenesis in *Plasmodium falciparum*. *Mol Biochem Parasitol* **143:** 100–110.

Skvortsov D, Abdueva D, Stitzer ME, Finkel SE, Tavare S. (2007) Using expression arrays for copy number detection: an example from *E. coli*. *BMC Bioinform* **8:** 203.

Sladek R, Rocheleau G, Rung J, Dina C, Shen L, Serre D, Boutin P, Vincent D, Belisle A, Hadjadj S, Balkau B, Heude B, Charpentier G, Hudson TJ, Montpetit A, Pshezhetsky AV, Prentki M, Posner BI, Balding DJ, Meyre D, Polychronakos C, Froguel P. (2007) A genome-wide association study identifies novel risk loci for type 2 diabetes. *Nature* **445:** 881–885.

Snijders AM, Pinkel D, Albertson DG. (2003) Current status and future prospects of array-based comparative genomic hybridization. *Brief Funct Genomic Proteomic* **2:** 37–45.

Snijders AM, Nowak N, Segraves R, Blackwood S, Brown N, Conroy J, Hamilton G, Hindle AK, Huey B, Kimura K, Law S, Myambo K, Palmer J, Ylstra B, Yue JP, Gray JW, Jain AN, Pinkel D, Albertson DG. (2001) Assembly of microarrays for genome-wide measurement of DNA copy number. *Nat Genet* **29:** 263–264.

Solinas-Toldo S, Lampel S, Stilgenbauer S, Nickolenko J, Benner A, Dohner H, Cremer T, Lichter P. (1997) Matrix-based comparative genomic hybridization: biochips to screen for genomic imbalances. *Genes Chromosom Cancer* **20:** 399–407.

Spits C, Le Caignec C, De Rycke M, Van Haute L, Van Steirteghem A, Liebaers I, Sermon K. (2006) Whole-genome multiple displacement amplification from single cells. *Nat Protoc* **1:** 1965–1970.

Staaf J, Jonsson G, Ringner M, Vallon-Christersson J. (2007) Normalization of array-CGH data: influence of copy number imbalances. *BMC Genomics* **8:** 382.

Steemers FJ, Gunderson KL. (2007) Whole genome genotyping technologies on the BeadArray platform. *Biotechnol J* **2:** 41–49.

Steemers FJ, Chang W, Lee G, Barker DL, Shen R, Gunderson KL. (2006) Whole-genome genotyping with the single-base extension assay. *Nat Methods* **3:** 31–33.

Strode C, Steen K, Ortelli F, Ranson H. (2006) Differential expression of the detoxification genes in the different life stages of the malaria vector *Anopheles gambiae*. *Insect Mol Biol* **15:** 523–530.

Teo YY, Inouye M, Small KS, Gwilliam R, Deloukas P, Kwiatkowski DP, Clark TG. (2007) A genotype calling algorithm for the Illumina BeadArray platform. *Bioinformatics* **23:** 2741–2746.

Thiagalingam S, Foy RL, Cheng KH, Lee HJ, Thiagalingam A, Ponte JF. (2002) Loss of heterozygosity as a predictor to map tumor suppressor genes in cancer: molecular basis of its occurrence. *Curr Opin Oncol* **14:** 65–72.

Tomlinson IP, Webb E, Carvajal-Carmona L, Broderick P, Howarth K, Pittman AM, Spain S, Lubbe S, Walther A, Sullivan K, Jaeger E, Fielding S, Rowan A, Vijayakrishnan J, Domingo E, Chandler I, Kemp Z, Qureshi M, Farrington SM, Tenesa A, Prendergast JG, Barnetson RA, Penegar S, Barclay E, Wood W, Martin L, Gorman M, Thomas H, Peto J, Bishop DT, Gray R, Maher ER, Lucassen A, Kerr D, Evans DG; CORGI Consortium, Schafmayer C, Buch S, Völzke

H, Hampe J, Schreiber S, John U, Koessler T, Pharoah P, van Wezel T, Morreau H, Wijnen JT, Hopper JL, Southey MC, Giles GG, Severi G, Castellví-Bel S, Ruiz-Ponte C, Carracedo A, Castells A; EPICOLON Consortium, Försti A, Hemminki K, Vodicka P, Naccarati A, Lipton L, Ho JW, Cheng KK, Sham PC, Luk J, Agundez JA, Ladero JM, de la Hoya M, Caldés T, Niittymäki I, Tuupanen S, Karhu A, Aaltonen L, Cazier JB, Campbell H, Dunlop MG, Houlston RS. (2008) A genome-wide association study identifies colorectal cancer susceptibility loci on chromosomes 10p14 and 8q23.3. *Nat Genet* **40:** 623–630.

Vissers LE, Veltman JA, van Kessel AG, Brunner HG. (2005) Identification of disease genes by whole genome CGH arrays. *Hum Mol Genet* **14:** R215–223.

Vontas J, David JP, Nikou D, Hemingway J, Christophides GK, Louis C, Ranson H. (2007) Transcriptional analysis of insecticide resistance in *Anopheles stephensi* using cross-species microarray hybridization. *Insect Mol Biol* **16:** 315–324.

Vontas J, Blass C, Koutsos AC, David JP, Kafatos FC, Louis C, Hemingway J, Christophides GK, Ranson H. (2005) Gene expression in insecticide resistant and susceptible *Anopheles gambiae* strains constitutively or after insecticide exposure. *Insect Mol Biol* **14:** 509–521.

Warr E, Aguilar R, Dong Y, Mahairaki V, Dimopoulos G. (2007) Spatial and sex-specific dissection of the *Anopheles gambiae* midgut transcriptome. *BMC Genomics* **8:** 37.

WTCCC (Wellcome Trust Case Control Consortium). (2005) A haplotype map of the human genome. *Nature* **437:** 1299–1320.

WTCCC (Wellcome Trust Case Control Consortium). (2007) Genome-wide association study of 14,000 cases of seven common diseases and 3,000 shared controls. *Nature* **447:** 661–678.

Welner RS, Pelayo R, Kincade PW. (2008) Evolving views on the genealogy of B cells. *Nat Rev Immunol* **8:** 95–106.

Wright G, Tan B, Rosenwald A, Hurt EH, Wiestner A, Staudt LM. (2003) A gene expression-based method to diagnose clinically distinct subgroups of diffuse large B cell lymphoma. *Proc Natl Acad Sci U S A* **100:** 9991–9996.

Xiao Y, Segal MR, Yang YH, Yeh RF. (2007) A multi-array multi-SNP genotyping algorithm for Affymetrix SNP microarrays. *Bioinformatics* **23:** 1459–1467.

Ylstra B, van den Ijssel P, Carvalho B, Brakenhoff RH, Meijer GA. (2006) BAC to the future! Or oligonucleotides: a perspective for micro array comparative genomic hybridization (array CGH). *Nucleic Acids Res* **34:** 445–450.

Zeggini E, Scott LJ, Saxena R, Voight BF, Marchini JL, Hu T, de Bakker PI, Abecasis GR, Almgren P, Andersen G, Ardlie K, Boström KB, Bergman RN, Bonnycastle LL, Borch-Johnsen K, Burtt NP, Chen H, Chines PS, Daly MJ, Deodhar P, Ding CJ, Doney AS, Duren WL, Elliott KS, Erdos MR, Frayling TM, Freathy RM, Gianniny L, Grallert H, Grarup N, Groves CJ, Guiducci C, Hansen T, Herder C, Hitman GA, Hughes TE, Isomaa B, Jackson AU, J½rgensen T, Kong A, Kubalanza K, Kuruvilla FG, Kuusisto J, Langenberg C, Lango H, Lauritzen T, Li Y, Lindgren CM, Lyssenko V, Marvelle AF, Meisinger C, Midthjell K, Mohlke KL, Morken MA, Morris AD, Narisu N, Nilsson P, Owen KR, Palmer CN, Payne F, Perry JR, Pettersen E, Platou C, Prokopenko I, Qi L, Qin L, Rayner NW, Rees M, Roix JJ, Sandbaek A, Shields B, Sjögren M, Steinthorsdottir V, Stringham HM, Swift AJ, Thorleifsson G, Thorsteinsdottir U, Timpson NJ, Tuomi T, Tuomilehto J, Walker M, Watanabe RM, Weedon MN, Willer CJ, Wellcome Trust Case Control ConsortiumIllig T, Hveem K, Hu FB, Laakso M, Stefansson K, Pedersen O, Wareham NJ, Barroso I, Hattersley AT, Collins FS, Groop L, McCarthy MI, Boehnke M, Altshuler D. (2008) Meta-analysis of genome-wide association data and large-scale replication identifies additional susceptibility loci for type 2 diabetes. *Nat Genet* **40:** 638–645.

# Other Array Technologies

Throughout this work we have focused on the use of DNA microarrays as sensors for detecting nucleic acids, be that reverse transcribed RNA or genomic DNA. However, the technology is versatile and, in principal, may be used for other assays; for example, directly monitoring DNA–protein interactions or as high-throughput systems for phenotypic screening of cells. In addition, the principal of generating dense arrays of biomolecules is emerging as a powerful tool in proteomics where a variety of functional array-based assays have been developed. In this chapter we briefly review some of these technologies with a particular focus on protein-binding DNA arrays, cell-based arrays and protein arrays. While relatively new, it is hoped that these emerging applications of microarray-based technologies will also have a large impact on biomedical research just as gene expression arrays.

## 12.1 PROTEIN-BINDING ARRAYS

The use of nucleic acid microarrays is not limited to the detection of other nucleic acids by hybridization, in principal any biomolecule that interacts with DNA can be assayed by a microarray-based approach. Perhaps the most powerful implementation of this type of technology has been the development of protein binding microarrays (PBMs) for assaying protein–DNA interactions (Carlson and Brent, 1999; Berger and Bulyk, 2006; Bulyk, 2006a). PBMs contain double-stranded DNA that is interrogated with proteins that bind DNA and the interaction quantified by detecting directly or indirectly labeled proteins. The initial implementation of the method utilized photolithograpically synthesized arrays of oligonucleotides up to 40 bases in length. The

Microarray Technology in Practice

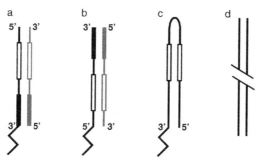

**FIGURE 12.1**  Protein binding array elements. (a) Photolithographically generated probes are synthesized in the 3′–5′ direction with the 3′-end tethered to the array. Each probe contains a priming sequence (filled box), a spacer and the recognition sequence of interest (open box). After array synthesis, each probe is made double-stranded by DNA polymerase synthesis initiated from the priming sequence (grey elements). (b) Conventionally synthesized probes are generated with 5′ amino linkers with the same general structure as those described in (a). The probes may be made double-stranded in solution prior to printing or by on array DNA synthesis. In the latter case, probes should be tethered to the array by 3′ linkers as shown in (a) since priming from the array surface out is probably more efficient because the DNA polymerase may be obstructed from reaching the terminal bases close to the array surface when synthesis proceeds towards the array. (c) Hairpin probes may be synthesized conventionally or by in situ synthesis, in the case illustrated the probes are synthesized in situ with the 3′-end tethered. Each probe contains a region of self-complementarily separated by a loop such that the probe folds back on itself to form a double-stranded sequence. (d) Large PCR derived amplicons representing portions of the genome may also be used for PBMs.

oligonucleotides, synthesized in the 3′–5′ direction, included the recognition sequences for restriction enzymes. After the photolithography each probe was converted into a double-stranded form on the array by priming DNA-polymerase driven synthesis from a primer complimentary to the 3′-end of the tethered probes (Figure 12.1a). The first experiments utilized a fluorescent label incorporated during the on-array DNA polymerase step and the protein-binding assay, the activity of a restriction enzyme at its target site but not a related site, was easily visualized by specific loss of fluorescence when probes contain exact matches to the target site are cut (Bulyk et al., 1999).

While the mask-based photolithography method used to make the first PBMs is capable of generating very high-density arrays, there are some limitations due to the relatively low yield of full-length oligonucleotides that can be achieved (Chapter 3). A second approach, exploring the specificity of zinc-finger DNA-binding domains, synthesized a set of 37mer oligonucleotides, each containing a 'universal' primer sequence and individually one of all possible 3-base combinations making up the central core of a 9-base zinc finger recognition sequence. The test 37mers were made double-stranded by DNA synthesis primed from the priming sequence and arrayed on derivatized slides (Figure 12.1b). Incubating these arrays with different zinc finger domains, and detecting binding via a fluorescently labeled antibody recognizing the zinc finger proteins, allowed an evaluation of the relative strength of the interaction between each Zn-finger protein and all of the DNA sequences on the array by comparing the fluorescence signals. In addition, by comparing the microarray signals for one of the proteins with independently generated quantitative binding data, the authors were able to construct a calibration curve relating microarray fluorescence to the dissociation constant for the protein–DNA interaction (Bulyk et al., 2001).

Bulyk and colleagues have gone on to further develop the PBM technique into a powerful method for the analysis of DNA-binding specificity. In one series of studies they demonstrate the use of *Saccharomyces cerevisiae* amplicon arrays, similar to those used for ChIP-array studies (Ren et al., 2000), for detecting potential binding sites for several yeast transcription factors (Mukherjee et al., 2004). As with the Zinc-finger experiments described above, transcription factors bound to DNA fragments on the array are detected by virtue of a fluorophore-tagged antibody recognizing an epitope introduced into the cloned transcription factor. To assess relative binding strength, protein-binding array signals are normalized with respect to a set of control arrays (from the same print run) stained with SyberGreen to quantify DNA concentration. Plotting the distribution of log ratios of protein fluorescence–DNA concentration from a set of normalized replicates allows the identification of array fragments with significant levels of binding. Encouragingly, the data from the PBM study is in good agreement with ChIP-array studies, indicating that this technique, at least for some factors, can provide a very rapid in vitro screen for potential genomic binding locations. The relevant methods for carrying out this type of PBM have recently been described in some detail (Bulyk, 2006b).

While elegant, the use of relatively large amplicons as probes restricts the analysis to a binary bound or unbound, assay. The amplicons are relatively long and may contain multiple binding sites for the factor being analyzed, consequently the signal cannot be easily related to binding strength. To address this Bulyk and colleagues (Berger et al., 2006) developed a 'universal' microarray. In this approach, 60mer olignucleotides are synthesized by in situ ink-jet synthesis. Each oligonucleotide is synthesized with a 3' linker, attaching the probes to the array, followed by a 24 base primer sequence. Each element then contains one of a set of permutated 35 base sequences representing all possible 10mers. The authors used a combinatorial mathematics technique known as De Bruijn sequencing to ensure that the variable 35 bases of the 42 000 probes on the array contained all possible 10 base binding site permutations. This ingenious technique allows coverage of a complex sequence space ($4^{10}$ possible 10mers = 1 048 576) in a highly compact way. As previously, arrays were rendered double-stranded by primer extension from the common priming sequence. Trace amounts of Cy3-conjugated dUTP allow an assessment of synthesis efficiency and occasional incorporation of a Cy5 labeled primer can be used to monitor primer-annealing efficiency. As before, the arrays are interrogated with tagged transcription factors synthesized in *Escherichia coli* and detected by a fluorescently labeled antibody recognizing the tag. Since most transcription factors are believed to recognize sequences shorter than 10 bases and many bind to palindromic sequences, the use of the de Bruijn sequence approach means that shorter sequences are represented multiple times on the array in slightly different contexts, i.e. each 8mer occurs at least 16 times. Thus the approach not only provides independent sequences on the array for assessing binding but also, by virtue of accommodating every possible mismatch, allows an exploration of binding site preferences and the generation of position weight matrices for each factor. In addition, the analysis can reveal nucleotides critical for binding and those that are more flexible. It is clear that this approach has the capacity to support a very detailed and highly parallel analysis of in vitro DNA-binding specificity, certainly more rapidly than standard SELEX approaches. Of particular importance, the authors make the sequences of all their probes freely available to academic or not-for-profit researchers, facilitating the wider adoption of the technology.

A related method termed cognate site identification (CSI: Warren et al., 2006) synthesizes 34 nucleotide oligonucleotides by maskless photolithography. Each oligonucleotide is designed as a self-complimentary palindrome containing a core of eight permuted bases with an unpaired loop allowing hairpin formation (Figure 12.1c). The authors estimate that under the conditions they employ, approximately 95% of the probes formed duplexes on the array. Thus all possible double-stranded 8mer sequences are interrogated on the array. Warren et al. (2006) demonstrate the effectiveness of the approach by assaying the binding of an engineered polyamide as well as the cooperative binding of a polyamide and the *Drosophila* transcription factor Extradenticle (Exd). In the published reports both the polyamides and Exd were directly labeled with a fluorescent reporter, however, in principal the secondary antibody detection method used by Bulyk and colleagues described above should also be applicable. More recently, the authors have developed a computational approach (CSI-Tree) for modeling the nucleotide dependencies of DNA-binding proteins from CSI or PBM data (Keleş et al., 2008). As is noted by Keleş et al., the ability to use such methods for defining DNA–protein interactions opens up a variety of possibilities for designing and assaying specific therapeutic agents that target defined DNA sequence. While these technologies are in their infancy, the ability to reliably synthesize over two million long oligonucleotides offers new dimensions in the systematic combinatorial analysis of DNA-binding. When data from PBM techniques are combined with in vivo data generated by ChIP-array studies, we anticipate a much better understanding of the protein–DNA interactions that underpin the regulation of gene expression.

## 12.2 CELL AND TISSUE ARRAYS

For diagnostic or functional analysis of cell phenotypes it may be desirable to analyze cells directly rather than extracted nucleic acids or proteins. This is particularly the case when sample quantities are limited and sensitive in situ detection methods such as immuno-histochemistry are available. In addition, in clinical diagnostic settings it is frequently the case that a particular bio-marker is robust and known to perform well as a diagnostic test. In such a circumstance, expensive whole genome expression analysis may be considered overkill. It is therefore useful to be able to interrogate many tissue or cell samples simultaneously in high throughput and the tissue microarray technique can facilitate this.

## 12.2.1 Tissue Arrays

The concept of arraying cells or tissue samples for high-throughput analysis was demonstrated by Kononen and colleagues (Kononen et al., 1998; Nocito et al., 2001) when they showed that samples from paraffin-embedded tumor samples could be arrayed on a microscope slide and interrogated by parallel immuno-histochemistry or in situ hybridization. In the tissue microarray (TMA) technique, small samples from up to 1000 different tumors are collected by a small coring tool and arranged in a single new paraffin tissue block (Figure 12.2). Sections from the composite tissue block are cut and deposited on a microscope slide where they can be used for a variety of prognostic or diagnostic analyses. The TMA slides can be interrogated with antibodies to detect particular proteins or by in situ hybridization to detect transcripts or assay for changes in DNA copy number. Several studies have utilized TMAs for exploring aspects of cancer biology, including breast cancer (Torhorst et al., 2001; Ruiz et al., 2006; Zerkowski et al., 2007) and renal cell cancer (Mertz et al., 2007) to name but two. Rimm and colleagues have developed an automated image analysis tool (AQUA), which facilitates the evaluation of TMA data (Camp et al., 2002; Rubin et al., 2004; McCabe et al., 2005).

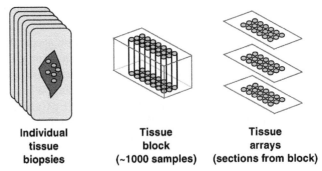

| Individual | Tissue | Tissue |
| tissue | block | arrays |
| biopsies | (~1000 samples) | (sections from block) |

**FIGURE 12.2**  Tissue arrays. Hundreds or thousands of paraffin embedded tissue biopsy samples (dark grey area) are individually sampled multiple times (light grey circles). Single samples from individual biopsies are arranges in a new paraffin block to form a tissue block. The tissue block is sectioned and each section placed on a suitable substrate to generate the tissue array.

It is clear that the ability to interrogate many clinical samples in parallel offers tremendous opportunities for diagnostics or for biomarker discovery. It is particularly powerful when limited quantities of biopsy samples are available since a single biopsy can be used to generate many tissue blocks and each tissue block can yield many arrays, in effect providing a route for amplifying limited tissue resources (Rimm et al., 2001). The method appears to perform well compared with other histological methods that utilize much more tissue. (Kyndi et al., 2008) and has recently been well-reviewed (Hewitt, 2006; Pick et al., 2008; Voduc et al., 2008). As the range of TMA experiments increases, there is a concomitant need for computational tools for storage, integration and querying of the derived data. A number of groups have developed freely available infrastructure (e.g. Coombes et al., 2002; Liu et al., 2002; Sharma-Oates et al., 2005; Della Mea et al., 2006), with Viti and colleagues recently proposing a web-based solution facilitating the annotation of TMA data with tissue and sequence ontologies (Viti et al., 2008). While clearly powerful for diagnostic applications, TMA technology is restricted to a certain extent since it deals with fixed biopsy samples. In principal, more powerful analysis could be performed if live cells were assayed in a high-throughput way.

## 12.2.2 Cell Arrays

Functional genomics seeks to provide a description of the biological function of every gene in the genome and while gene expression assays can provide clues as to gene function, ultimately the phenotypic consequences of removing a gene need to be determined. In the case of yeast, where systematic knockout efforts have generated mutations in virtually all 6000 genes encoded in the genome, the results generated are clearly very powerful (Winzeler et al., 1999; Giaever et al., 2002). However, such genetic studies are currently outwith the reach of most metazoan systems with the important exception of *Caenorhabditis elegans*, where systematic RNA-interference at the whole organism level provides powerful insights into gene function (Kamath et al., 2003). An alternative method for high-throughput analysis in metazoans is to interrogate cell lines rather than the whole organism. Cell-based assays can be used to assess the consequences of over or ectopic expression of genes. A particularly powerful technique is the use of high-throughput RNA-interference, which is proving to be a highly tractable method for exploring cellular phenotypes (Kuttenkeuler and Boutros, 2004; Silva et al., 2004a). Large-scale RNAi screens are generally carried out in microtitre plates using well-established cell lines, which are available in large quantities, and the analysis carried out with automated microscopy systems or via fluorescence-based assays in plate readers. However, if the goal is to assay primary cells, which may only be available in limited quantities, the relatively large-scale microtitre-based assays may not be feasible and an increasingly attractive alternative is to use cell-based microarrays. This technology has been recently reviewed (Castel et al., 2006) and a guide to current methodologies published (Castel et al., 2007), here we touch on some of the basics of the technique.

The reverse transfection method developed by Ziauddin and Sabatini (2001) first demonstrated the concept of using cells grown on microarray slides containing defined DNA sequences to carry out high throughput functional assays. In their method, DNA solutions, for example sets of cDNAs cloned in expression vectors, are printed at defined locations on a glass slide. The DNA can either be prepared in an aqueous gelatin solution and, after drying the array, treating the slide with a lipid-based transfection solution. Alternatively, the DNA may be mixed directly with the transfection reagent prior to printing. With either method, after printing, the array is placed in a culture dish containing cells in an appropriate medium and after a few days the cells on the microarray grow. Those cells growing over a particular DNA spot take up the expression clone by transfection and subsequently express the cDNA. Phenotypic scoring can then be carried out by a microscopic examination of cell morphology, by scanning to determine the expression of a fluorescent reporter gene, by autoradiography for radiometric assays or by antibody staining. In the original method, DNA spots of approximately 150 $\mu$m allow up to 80 cells to be assayed at each DNA element. The study also showed that transfection with multiple constructs was possible, increasing the flexibility of the assay. Assays of up to 5000 mammalian or *Drosophila* cells per array have been reported (Wu et al., 2002; Wheeler et al., 2005) and a variety of different functional assays are emerging: for example, scoring for apoptosis, (Palmer et al., 2006) or calcium flux (Mishina et al., 2004). Clearly the densities achievable with cell microarrays are limited by the number of cells transfected by a given array element and while it is possible to array tens or hundreds of thousands of elements on a single array, such densities are not feasible for cell arrays since they would result in interrogation of less than a single cell. Xu (2002) has reported the use of permeable membranes, onto which cells are 'printed' using solid microarray printing pins. In this method, the printing technique results in the formation of so-called nanocraters, which hold the printed cells. Due to the nature of the membranes used, the arrays may be floated on the surface of culture media or incubated on solid media, facilitating long-term storage and assay of high-density microorganism (e.g. yeast or *E. coli*) cell arrays. More recently, the reverse transfection method has been adapted to facilitate high-throughput RNAi-based assays by a number of groups (see Wheeler et al., 2005 for a review). Wheeler et al. (2004) describe screening *Drosophila* tissue culture cells with relatively large dsRNAs (up to 800 bp), whereas Silva et al. (2004b) and colleagues have screened mammalian cells directly with short hairpin RNAs (shRNAs) or by reverse transfection with constructs expressing shRNAs.

Up to 5600 different constructs have been assayed on a single array and, depending upon the nature of the assays, can present an analysis challenge. If the assay involves a quantitative fluorescence readout from a reporter gene or an antibody staining, the array may be treated much like a regular nucleic acid microarray and images acquired with a standard scanner. However, such instruments are generally not suitable for analysis that requires single-cell resolution or nonfluorescence assays such as cell morphology. In these cases, microscopes fitted with motorized stages and focusing controls along with automated image acquisition are required. The problem with the high-resolution microscopy is that image analysis becomes increasingly challenging. A simple problem like stitching together the images of thousands of individual spots to facilitate analysis of the entire array is not trivial (see Wheeler et al., 2005 for a discussion of the imaging problems). Sabatini and colleagues (Carpenter et al., 2006; Lamprecht et al., 2007) have developed a freely available software package, CellProfiler, to facilitate this type of image analysis as have Strömberg et al. (Stromberg et al., 2007) with their TMAx package. It is clear that this is an emerging field where software and microscopy development will go hand-in-hand to facilitate more detailed and accurate analysis of cell arrays with subcellular resolution.

## 12.3 PROTEIN ARRAYS

While analysis of the genome or transcriptome provides insights into the regulatory programmes driving cellular processes and provides clues about the specific molecular pathways responsible for biological systems under investigation, it is the proteome and its interactions that ultimately generate biological function. Unlike nucleic acids, which are readily amplified and analyzed due to their relative simple chemical structure, proteins are incredibly heterogeneous and their analysis is restricted to what can be isolated from biological samples. An emerging an increasingly powerful technology for the high-throughput analysis of proteins is the protein microarray. While similar in concept to nucleic acid arrays, there are several problems that need to be overcome, not least the isolation of protein 'probes' and the generation of labeled samples. Over the past few years several proof of concept studies have tackled some of the problems associated with arraying proteins and provide an optimistic outlook for wider adoption of the technology. There have been several reviews of protein array technologies over the past few years that summaries the development of the field (MacBeath, 2002; Glökler and Angencndt, 2003; Zhu and Snyder, 2003; Bertone and Snyder, 2005; Hall et al., 2007) and interested readers are directed to these as the start point for a more detailed discussion of relevant platforms and technologies.

### 12.3.1 Array Types

There are several different types of protein arrays currently in use that differ in terms of the type of immobilized protein they use and in the types of molecules they are designed to detect (Figure 12.3).

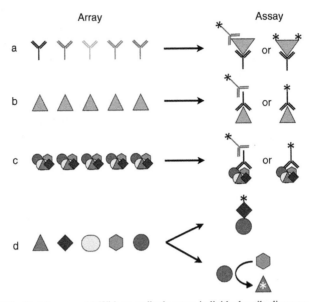

**FIGURE 12.3**  Protein arrays (a) With an antibody array, individual antibodies are spotted on the surface (shaded Y shapes) and these are interrogated with a mixture of proteins or analyte. The interactions between analyte proteins and antibody may be detected by a second antibody recognizing another epitope in the target protein (left) or by directly labeling the analyte before interrogating the array (right). (b) With epitope arrays, a few specific epitopes are used to detect the presence of a specific antibody in a sample or to evaluate the specificity of a particular antibody. Detection may be indirect, using a second antibody recognizing the common part of the interrogated antibody (left) or by directly labeling the antibody of interest (right). (c) Protein extract arrays interrogate a mixture of proteins from a cell or tissue with a specific marker (in this case an antibody), again detection may be direct or indirect. (d) Proteome arrays are generated from individual protein or peptide molecules and can be used to assay for interactions with specific ligands (top arrow: protein, DNA, carbohydrate, lipid, small molecule, etc.) which may be directly or indirectly detected. They may also be used to assay enzymatic activity where the substrate conversion is marked by a fluorescence or radioactive output.

## 12.3.1.1 *Antibody Arrays*

This type of array consists of a set of immobilized antibodies designed against antigens of interest and is used to detect specific components in complex mixtures. First demonstrated in a microarray format by two groups (MacBeath and Schreiber, 2000; Haab et al., 2001), antibody arrays grew out of previous work utilizing macroarrays or antibodies as capture agents in analytical flow-sensors (e.g. Rowe et al., 1999). While monoclonal antibodies produced via standard hybridoma technology can provide a diverse source of different antibodies, they are expensive to produce in large quantities and most current approaches utilize more high-throughput recombinant methods (see Glökler and Angenendt, 2003 for a discussion of antibody sources). Potential sources of diverse antibody array reagents include antibodies derived from phage display technology (Steinhauer et al., 2002; Angenendt et al., 2004), affinity-tagged single chain Fv antibodies (Wingren et al., 2005; Dexlin et al., 2008) or antibodies derived from wholly in vitro technologies such as ribosomal display (He and Taussig, 2002; He and Khan, 2005).

### 12.3.1.2 Antigen Arrays

In this approach one or a few specific antigens are immobilized and the arrays are used to detect the presence of specific antibodies in a serum sample (e.g. Joos et al., 2000). The objective here is not so much generating diversity on the array but rather providing a high-throughput platform for rapidly screening samples with very well characterized antigens. Since only tens of different antigens are used in this approach, recombinant proteins can be purified in large quantities. An alternative, and readily parallelisable, approach is to use light-directed in situ synthesis approaches, similar to the mask-based or maskless methods employed in the construction of oligonucleotide arrays, to directly synthesize peptides on the array (Fodor et al., 1991; Gao et al., 2003; Li et al., 2004; Bhushan, 2006). A variety of chemistries have been reported, with peptides up to 10 amino acids synthesized (Hilpert et al., 2007). A halfway house between this type of antigen arrays and tissue arrays deposits protein extracts from cells of interest on the array. The arrays are then used to assay for the presence of particular antigens by antibody staining (Figure 12.3c; Speer et al., 2005).

### 12.3.1.3 Protein Arrays

This type of platform is differentiated from antigen arrays by the diversity of different proteins represented on the array, in some cases virtually an entire proteome. Protein array technology is used to screen many different proteins in a variety of different assays, which at their most basic can be considered as one of two types: interaction screening seeks to identify specific interactions between the arrayed proteins and some component of the analyte (protein, carbohydrate, lipid, small molecule, etc). In contrast, functional assays seek to identify proteins with specific enzymatic activities such as phosphorylation. Developed in a high-density microarray format by several groups (MacBeath and Schreiber, 2000; Zhu et al., 2000; Haab et al., 2001) and first demonstrated at a whole-proteome scale in yeast by the Snyder group (Zhu et al., 2001), protein microarrays show considerable promise for global high-throughput functional proteomics.

The major problem associated with protein arrays is the source of proteins to array and a variety of technologies have been adopted (reviewed by Braun and LaBaer, 2003). Expression of cloned eukaryotic open reading frames (ORFs) in *E. coli* is a popular method for expressing recombinant proteins and has been widely used to generate proteins for arraying (e.g. Leuking et al., 1999; Braun et al., 2002). While convenient and supported by well-developed methods, eukaryotic post-translational modifications that may be necessary for protein activity do not occur in prokaryotic expression systems and, in some cases, eukaryotic proteins are not soluble or difficult to purify. To overcome such limitations a variety of strategies have been adopted: these include expression of tagged proteins in yeast, exemplified by the whole proteome array developed by Zhu et al. (2001) or the human protein arrays generated by Holz et al. (2003). Other approaches include expressing recombinant proteins with a baculovirus system (Albala et al., 2000) or using in vitro synthesis methods such as mRNA display (Keefe and Szostak 2001; Wilson et al., 2001) or cell free expression systems, to synthesize proteins directly on the array (Ramachandran et al., 2006; He et al., 2008a,b).

Protein arrays have seen in action in a variety of functional screens, best exemplified by work with the whole yeast proteome array, a platform containing protein probes representing over 80% of the predicted yeast ORFs (Zhu et al., 2001). These studies have included an analysis of antibody specificity (Michaud et al., 2003), identifying calmodulin-binding proteins, assaying proteins with specificity for particular phosphoinositides (Figure 12.4; Zhu et al., 2001), or identifying novel DNA-binding proteins (Hall et al., 2004). In a particularly impressive functional screen Snyder and colleagues determined the specificity and potential targets of 82 different protein kinases encoded in the yeast genome via protein microarrays, generating a phosphorylation map of the yeast proteome (Ptacek et al., 2005). Such studies highlight the enormous potential of proteome arrays for assaying protein function. Other assays include the identification of potential ligands for G-protein coupled receptors (Fang et al., 2002; Hong et al., 2005; Fang et al., 2006) or assessing the activity of cytochrome P450s on different prodrugs by a combined protein and cell-based microarray approach (Lee et al., 2005; Lee et al., 2008). Both of these areas are clearly of particular importance to the pharmaceutical industry.

## 12.3.2 Array Substrates

While the interactions between relatively simple nucleic acid molecules and the array substrate surface is fairly well-understood (Chapter 3), the heterogeneous nature of protein molecules presents more of a challenge when thinking about the chemical nature of the array surface. Even more challenging, protein function and the formation of particular epitopes recognized by antibodies relies upon correct protein structure, thus any protein microarray platform must ensure that the deposited proteins are properly folded and adopt their correct three-dimensional structure. This is not a trivial issue and a variety of substrates have been used: a recent review of the literature suggests that, by and large, the technologies are working and many protein arrays can be considered functional (Merkel et al., 2005). Attachment of proteins to the substrate depends on the technology used to generate the protein probes. For example, many in vivo

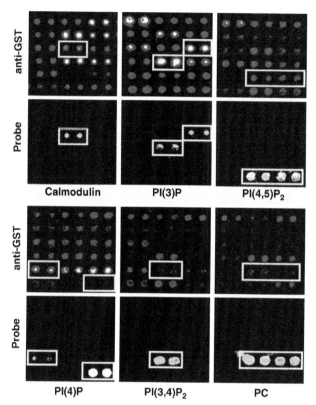

**FIGURE 12.4**   Protein microarray assays. Examples showing interrogation of a yeast whole proteome array for different specific binding interactions. The anti-GST panels detect the amount of protein
present in each array element, each protein is spotted in duplicate. Results for six different probes are
shown. Labeled calmodulin detects specific calmodulin binding proteins, the next five sets detect
specific interactions with different labeled phosphoinositides (PI) and with phosphatidylcholine (PC).
*Source*: Data from Zhu et al. (2001).

expression systems generate recombinant proteins containing a small purification tag. Both biotin and 6XHis tagged proteins have been specifically attached
via affinity interactions, with streptavidin or nickel-coated slides respectively.
The benefit of the affinity attachment approach is that all of the protein molecules are presented in a uniform orientation on the slide surface. For maximum
utility proteins independently tagged at both N- and C-termini may be considered, potentially ameliorating any attachment effects on protein function. Other
surfaces allow less defined covalent attachment (e.g. epoxy- or aldehyde-based
substrates) or nonspecific absorption (e.g. poly-lysine or nitrocellulose). In
general, the nonaffinity tag methods result in random attachment of proteins
to the substrate, which may offer a variety of surfaces for antibody detection but
potentially reduce functional enzymatic assays. Along with binding of the

protein probes, the potential for interactions between the analyte and the substrate must also be considered since this can result in high levels of nonspecific background. The area of slide surface chemistries has recently been reviewed in some detail (Zhu and Snyder, 2003; Rusmini et al., 2007), with the latter being the first port of call for those considering generating protein arrays.

With protein and appropriate substrate to hand, a variety of arraying instruments can be used to print arrays. These have included contact spotters using blunt or quill printing tips, loop-and-pin type arrayers or noncontact delivery systems such as ink-jet or electrospray devices. As we indicate in Chapter 3, the range of commercially available printing instruments is declining and options may be limited for those new to the microarray printing field. Whatever the instrument employed, it is critical to maintain good environmental control during the arraying process since proteins can be denatured by dehydration. Consequently, at least for robotic contact printing, proteins are usually arrayed in glycerol solutions with relative humidity levels between 25 and 30%. Clearly any postprinting treatment depends on the particular slide surface chemistry being employed. The reviews listed above provide discussions of these and other relevant issues.

### 12.3.3 Assays, Detection and Analysis

For compatibility with commonly used microarray scanners, fluorescent labels are generally the detection method of choice, however, some studies have used radiolabeled samples and quantified arrays by autoradiography (see Ptacek et al., 2005 for an example of phosphorylation assays using $^{33}$P-$\gamma$-ATP). Some of the issues relating to imaging of protein microarray data, including discussion of some potential new label-free approaches have been recently reviewed (Schaferling and Nagl, 2006; Yu et al., 2006) and here we briefly touch on the approaches currently prevalent in the literature.

The most common approach for antigen detection when using antibody arrays is to employ a classical sandwich assay (Nielsen and Geierstanger, 2004) where the antibody–antigen interactions detected on the array are revealed by the use of a second antibody recognizing a different epitope on the antigen of interest. Obviously this technique requires that at least two different specific antibodies are available for each assayed protein, which may be difficult for some proteins, but the method is capable of very high sensitivity and quantitative antigen detection. The alternative approach, directly labeling the protein extract of interest with fluorescent dyes such as Cy3 or Cy5 is more widely applicable but can only really generate comparative measures of antigen differences between samples. In addition, and in contrast to the sandwich assay where signal amplification is possible via the secondary labeling step, sensitivity is limited by the amount of fluorescence incorporated during the analyte labeling step. This type of antigen capture assay is exemplified by the proof of concept study by

Haabet al. (2001) or a study identifying radiation induced proteins in colon carcinoma cells (Sreekumar et al., 2001). Since fluorescent dyes are bulky and may interfere with the interaction between the analyte and the antibody, smaller labels such as biotin may be used and these can subsequently be detected by streptavidin conjugated to a fluorescent dye. Biotin labeling is particularly attractive when small molecules are used to interrogate the protein array since the small label is less likely to interfere with binding (Huang et al., 2004).

Scanning protein arrays interrogated with fluorescently labeled samples generates output images similar to those obtained with conventional DNA microarrays (see Chapter 5). Haab et al. (2001), in their comparative analysis of differentially labeled samples, used standard nucleic acid microarray data extraction approaches, background subtracting spot values and calculating a scaling factor for between channel normalization, to determine relative binding ratios between two samples. With their yeast proteome arrays, Snyder and colleagues utilize the fact that their protein probes are all GST-tagged to normalize signals with respect to the amount of protein in each spot by interrogating the array with Cy5-labeled anti-GST and using Cy3 labeled analyte in another channel (Zhu et al., 2001). Caution must be exercised here since it is possible that the antitag antibody may interfere with the activity or binding availability of the arrayed protein and it may be prudent to carry out pilot studies where the antitag and labeled analyte are assayed on independent arrays. Since the field is relatively new and there are many undefined variables in protein microarray data (protein concentration, specific affinities of interactions, etc.) the analytical methods remain somewhat rudimentary. Royce et al. (2006) have recently reviewed the relevant nucleic acid microarray analytical methods and how they may be applied to protein arrays but it remains early days in this emerging field. As far as we are aware there have been relatively few efforts focused at specifically developing protein array analysis tools. Lubomirski et al. (2007) have described a statistical framework for dealing with protein arrays but this has yet to be independently validated. Similarly, Zhu et al. (2006) have developed an analytical pipeline, ProCAT, to deal with all stages of the protein array analytical pipeline (background correction, normalization, filtering and annotation) that is available as a web-based tool (http://purelight.biology.yale.edu:8080/servlets-examples/procat.html). It is clear that, as with nucleic acid microarray data, analytical methods will improve as data accumulates. Some of the issues surrounding data analysis and quality control from the perspective of implementing protein arrays in diagnostic settings have recently been discussed (Kricka and Master, 2008).

## 12.4 SUMMARY

The technologies we have described in this chapter extend the possibilities for microarray-based assays to many fields of biological research. In principal, any biological molecule may be adaptable to a high-throughput assay if appropriate substrates and detection methods can be defined. Of course new assays generate new challenges in terms of analysis and quantification and it is clear that, as with nucleic acid microarrays, a better understanding of the molecular interactions being assayed is required before many of the new array technologies can move from a qualitative to a quantitative footing. While tackling this problem with molecules and interactions that are considerably more complex than nucleic acid hybridization is not trivial, the rewards are potentially enormous. The prospect of developing diagnostic tools that will be able to assay hundreds or even thousands of biomarkers with a relatively small patient sample is very attractive. Similarly, the potential for assaying gene function by missexpression or targeted gene knockdown in high throughput with difficulty to obtain primary cells opens up areas of functional genomics that were unimaginable a few years ago. Along with the challenges associated with technology development, issues of data analysis and data storage will also need to be tackled. Ontologies and controlled vocabularies for describing protein interactions or cellular phenotypes that transcend species boundaries need to be developed if we are to capitalize on high-throughput data.

## REFERENCES

Albala JS, Franke K, McConnell IR, Pak KL, Folta PA, Rubinfeld B, Davies AH, Lennon GG, Clark R. (2000) From genes to proteins: high-throughput expression and purification of the human proteome. *J Cell Biochem* **80:** 187–191.

Angenendt P, Wilde J, Kijanka G, Baars S, Cahill DJ, Kreutzberger J, Lehrach H, Konthur Z, Glokler J. (2004) Seeing better through a MIST: evaluation of monoclonal recombinant antibody fragments on microarrays. *Anal Chem* **76:** 2916–2921.

Berger M, Bulyk M. (2006) Protein binding microarrays (PBMs) for rapid, high-throughput characterization of the sequence specificities of DNA binding proteins. *Methods Mol Biol* **338:** 245–260.

Berger MF, Philippakis AA, Qureshi AM, He FS, Estep PW, Bulyk ML. (2006) Compact, universal DNA microarrays to comprehensively determine transcription-factor binding site specificities. *Nat Biotechnol* **24:** 1429–1435.

Bertone P, Snyder M. (2005) Advances in functional protein microarray technology. *FEBS J* **272:** 5400–5411.

Bhushan KR. (2006) Light-directed maskless synthesis of peptide arrays using photolabile amino acid monomers. *Org Biomol Chem* **4:** 1857–1859.

Braun P, LaBaer J. (2003) High throughput protein production for functional proteomics. *Trends Biotechnol* **21:** 383–388.

Braun P, Hu Y, Shen B, Halleck A, Koundinya M, Harlow E, LaBaer J. (2002) Proteome-scale purification of human proteins from bacteria. *Proc Natl Acad Sci U S A* **99:** 2654–2659.

Bulyk M. (2006) DNA microarray technologies for measuring protein–DNA interactions. *Curr Opin Biotechnol* **17:** 422–430.

Bulyk M. (2006) Analysis of sequence specificities of DNA proteins with protein binding microarrays. *Methods Enzymol* **410:** 279–299.

Bulyk M, Gentalen E, Lockhart D, Church G. (1999) Quantifying DNA–protein interactions by double-stranded DNA arrays. *Nat Biotechnol* **17:** 573–577.

Bulyk M, Huang X, Choo Y, Church G. (2001) Exploring the DNA-binding specificities of zinc-finges with DNA microarrays. *Proc Natl Acad Sci U S A* **98:** 7158–7163.

Camp RL, Chung GG, Rimm DL. (2002) Automated subcellular localization and quantification of protein expression in tissue microarrays. *Nat Med* **8:** 1323–1327.

Carlson R, Brent R. (1999) Double-stranded DNA arrays: next steps in the surface campaign. *Nat Biotechnol* **17:** 536–537.

Carpenter A, Jones T, Lamprecht M, Clarke C, Kang I, Friman O, Guertin D, Chang J, Lindquist R, Moffat J, Golland P, Sabatini D. (2006) CellProfiler: image analysis software for identifying and quantifying cell phenotypes. *Genome Biol* **7:** R100.

Castel D, Pitaval A, Debily MA, Gidrol X. (2006) Cell microarrays in drug discovery. *Drug Discov Today* **11:** 616–622.

Castel D, Debily MA, Pitaval A, Gidrol X. (2007) Cell microarray for functional exploration of genomes. *Methods Mol Biol* **381:** 375–384.

Coombes KR, Zhang L, Bueso-Ramos C, Brisbay S, Logothetis C, Roth J, Keating MJ, McDonnell TJ. (2002) TAD: a web interface and database for tissue microarrays. *Appl Bioinform* **1:** 155–158.

Della Mea V, Bin I, Pandolfi M, Di Loreto C. (2006) A web-based system for tissue microarray data management. *Diagn Pathol* **1:** 36.

Dexlin L, Ingvarsson J, Frendeus B, Borrebaeck C, Wingren C. (2008) Design of recombinat antibody microarrays for cell surface membrane proteomics. *J Proteome Res* **7:** 319–327.

Fang Y, Frutos AG, Lahiri J. (2002) Membrane protein microarrays. *J Am Chem Soc* **124:** 2394–2395.

Fang Y, Peng J, Ferrie AM, Burkhalter RS. (2006) Air-stable G protein-coupled receptor microarrays and ligand binding characteristics. *Anal Chem* **78:** 149–155.

Fodor SP, Read JL, Pirrung MC, Stryer L, Lu AT, Solas D. (1991) Light-directed, spatially address-able parallel chemical synthesis. *Science* **251:** 767–773.

Gao X, Zhou X, Gulari E. (2003) Light directed massively parallel on-chip synthesis of peptide arrays with t-Boc chemistry. *Proteomics* **3:** 2135–2141.

Giaever G, Chu AM, Ni L, Connelly C, Riles L, Veronneau S, Dow S, Lucau-Danila A, Anderson K, Andre B, Arkin AP, Astromoff A, El-Bakkoury M, Bangham R, Benito R, Brachat S, Campanaro S, Curtiss M, Davis K, Deutschbauer A, Entian KD, Flaherty P, Foury F, Garfinkel DJ, Gerstein M, Gotte D, Guldener U, Hegemann JH, Hempel S, Herman Z, Jaramillo DF, Kelly DE, Kelly SL, Kotter P, LaBonte D, Lamb DC, Lan N, Liang H, Liao H, Liu L, Luo C, Lussier M, Mao R, Menard P, Ooi SL, Revuelta JL, Roberts CJ, Rose M, Ross-Macdonald P, Scherens B, Schim-mack G, Shafer B, Shoemaker DD, Sookhai-Mahadeo S, Storms RK, Strathern JN, Valle G, Voet M, Volckaert G, Wang CY, Ward TR, Wilhelmy J, Winzeler EA, Yang Y, Yen G, Youngman E, Yu K, Bussey H, Boeke JD, Snyder M, Philippsen P, Davis RW, Johnston M. (2002) Functional profiling of the *Saccharomyces cerevisiae* genome. *Nature* **418:** 387–391.

Glökler J, Angenendt P. (2003) Protein and antibody microarray technology. *J Chromatogr B* **797:** 229–240.

Haab B, Dunham M, Brown P. (2001) Protein microarrays for highly parallel detection and quan-titation of specific proteins and antibodies in complex solutions. *Genome Biol* **2:** R0004.0001.

Hall D, Ptacek J, Snyder M. (2007) Protein microarray technology. *Mech Ageing Dev* **128:** 161–167.

Hall DA, Zhu H, Zhu X, Royce T, Gerstein M, Snyder M. (2004) Regulation of gene expression by a metabolic enzyme. *Science* **306:** 482–484.

He M, Taussig MJ. (2002) Ribosome display: cell-free protein display technology. *Brief Funct Genomic Proteomic* **1:** 204–212.

He M, Khan F. (2005) Ribosome display: next-generation display technologies for production of antibodies in vitro. *Expert Rev Proteomics* **2:** 421–430.

He M, Stoevesandt O, Taussig MJ. (2008) In situ synthesis of protein arrays. *Curr Opin Biotechnol* **19:** 4–9.

He M, Stoevesandt O, Palmer EA, Khan F, Ericsson O, Taussig MJ. (2008) Printing protein arrays from DNA arrays. *Nat Methods* **5:** 175–177.

Hewitt SM. (2006) The application of tissue microarrays in the validation of microarray results. *Methods Enzymol* **410:** 400–415.

Hilpert K, Winkler DF, Hancock RE. (2007) Peptide arrays on cellulose support: SPOT synthesis, a time and cost efficient method for synthesis of large numbers of peptides in a parallel and addressable fashion. *Nat Protoc* **2:** 1333–1349.

Holz C, Prinz B, Bolotina N, Sievert V, Bussow K, Simon B, Stahl U, Lang C. (2003) Establishing the yeast *Saccharomyces cerevisiae* as a system for expression of human proteins on a proteome-scale. *J Struct Funct Genomics* **4:** 97–108.

Hong Y, Webb BL, Su H, Mozdy EJ, Fang Y, Wu Q, Liu L, Beck J, Ferrie AM, Raghavan S, Mauro J, Carre A, Mueller D, Lai F, Rasnow B, Johnson M, Min H, Salon J, Lahiri J. (2005) Functional GPCR microarrays. *J Am Chem Soc* **127:** 15350–15351.

Huang J, Zhu H, Haggarty SJ, Spring DR, Hwang H, Jin F, Snyder M, Schreiber SL. (2004) Finding new components of the target of rapamycin (TOR) signaling network through chemical genetics and proteome chips. *Proc Natl Acad Sci U S A* **101:** 16594–16599.

Joos TO, Schrenk M, Hopfl P, Kroger K, Chowdhury U, Stoll D, Schorner D, Durr M, Herick K, Rupp S, Sohn K, Hammerle H. (2000) A microarray enzyme-linked immunosorbent assay for auto-immune diagnostics. *Electrophoresis* **21:** 2641–2650.

Kamath RS, Fraser AG, Dong Y, Poulin G, Durbin R, Gotta M, Kanapin A, Le Bot N, Moreno S, Sohrmann M, Welchman DP, Zipperlen P, Ahringer J. (2003) Systematic functional analysis of the *Caenorhabditis elegans* genome using RNAi. *Nature* **421:** 231–237.

Keefe A, Szostak J. (2001) Functional proteins from a random-sequence library. *Nature* **410:** 715–718.

Keleş S, Warren CL, Carlson CD, Ansari AZ. (2008) CSI-Tree: a regression tree approach for modeling binding properties of DNA-binding molecules based on cognate site identification (CSI) data. *Nucleic Acids Res* **36:** 3171–3184.

Kononen J, Bubendorf L, Kallioniemi A, Barlund M, Schraml P, Leighton S, Torhorst J, Mihatsch MJ, Sauter G, Kallioniemi OP. (1998) Tissue microarrays for high-throughput molecular profiling of tumor specimens. *Nat Med* **4:** 844–847.

Kricka LJ, Master SR. (2008) Validation and quality control of protein microarray-based analytical methods. *Mol Biotechnol* **38:** 19–31.

Kuttenkeuler D, Boutros M. (2004) Genome-wide RNAi as a route to gene function in *Drosophila*. *Brief Funct Genomic Proteomic* **3:** 168–176.

Kyndi M, Sorensen FB, Knudsen H, Overgaard M, Nielsen HM, Andersen J, Overgaard J. (2008) Tissue microarrays compared with whole sections and biochemical analyses. A subgroup analysis of DBCG 82 b&c. *Acta Oncol* **47:** 591–599.

Lamprecht MR, Sabatini DM, Carpenter AE. (2007) CellProfiler: free, versatile software for automated biological image analysis. *Biotechniques* **42:** 71–75.

Lee MY, Park CB, Dordick JS, Clark DS. (2005) Metabolizing enzyme toxicology assay chip (MetaChip) for high-throughput microscale toxicity analyses. *Proc Natl Acad Sci U S A* **102:** 983–987.

Lee MY, Kumar RA, Sukumaran SM, Hogg MG, Clark DS, Dordick JS. (2008) Three-dimensional cellular microarray for high-throughput toxicology assays. *Proc Natl Acad Sci U S A* **105:** 59–63.

Leuking A, Horn M, Eickhoff H, Buessow K, Lehrach H, Walter G. (1999) Protein microarrays for gene expression and antibody screening. *Anal Biochem* **270:** 103–111.

Li S, Bowerman D, Marthandan N, Klyza S, Luebke KJ, Garner HR, Kodadek T. (2004) Photolithographic synthesis of peptoids. *J Am Chem Soc* **126:** 4088–4089.

Liu CL, Prapong W, Natkunam Y, Alizadeh A, Montgomery K, Gilks CB, van de Rijn M. (2002) Software tools for high-throughput analysis and archiving of immunohistochemistry staining data obtained with tissue microarrays. *Am J Pathol* **161:** 1557–1565.

Lubomirski M, D'Andrea MR, Belkowski SM, Cabrera J, Dixon JM, Amaratunga D. (2007) A consolidated approach to analyzing data from high-throughput protein microarrays with an application to immune response profiling in humans. *J Comput Biol* **14:** 350–359.

MacBeath G. (2002) Protein microarrays and proteomics. *Nat Genet* **32:** S526–S532.

MacBeath G, Schreiber S. (2000) Printing proteins as microarrays for high-throughput function determination. *Science* **289:** 1760–1763.

McCabe A, Dolled-Filhart M, Camp RL, Rimm DL. (2005) Automated quantitative analysis (AQUA) of in situ protein expression, antibody concentration, and prognosis. *J Natl Cancer Inst* **97:** 1808–1815.

Merkel JS, Michaud GA, Salcius M, Schweitzer B, Predki PF. (2005) Functional protein microarrays: just how functional are they?. *Curr Opin Biotechnol* **16:** 447–452.

Mertz KD, Demichelis F, Kim R, Schraml P, Storz M, Diener PA, Moch H, Rubin MA. (2007) Automated immunofluorescence analysis defines microvessel area as a prognostic parameter in clear cell renal cell cancer. *Hum Pathol* **38:** 1454–1462.

Michaud G, Salcius M, Zhou F, Bangham R, Bonin J, Guo H, Snyder M, Predki P, Schweitzer B. (2003) Analysing antibody specificity with whole proteome microarrays. *Nat Biotechnol* **21:** 1509–1512.

Mishina Y, Wilson C, Bruett L, Smith J, Stoop-Myer C, Jong S, Amaral L, Pedersen R, Lyman S, Myer V, Kreider B, Thompson C. (2004) Multiplex GPCR assay in reverse transfection cell microarrays. *J Biomol Screen* **9:** 196–207.

Mukherjee S, Berger MF, Jona G, Wang XS, Muzzey D, Snyder M, Young RA, Bulyk ML. (2004) Rapid analysis of the DNA-binding specificities of transcription factors with DNA microarrays. *Nat Genet* **36:** 1331–1339.

Nielsen UB, Geierstanger BH. (2004) Multiplexed sandwich assays in microarray format. *J Immunol Methods* **290:** 107–120.

Nocito A, Kononen J, Kallioniemi A, Sauter G. (2001) Tissue microarrays (TMAS) for high-throughput molecular pathology research. *Int J Cancer* **94:** 1–5.

Palmer E, Miller A, Freeman T. (2006) Identification and characterisation of human apoptosis inducing proteins using cell-based transfection microarrays and expression analysis. *BMC Genomics* **7:** 145.

Pick E, McCarthy MM, Kluger HM. (2008) Assessing expression of apoptotic markers using large cohort tissue microarrays. *Methods Mol Biol* **414:** 83–93.

Ptacek J, Devgan G, Michaud G, Zhu H, Zhu X, Fasolo J, Guo H, Jona G, Breitkreutz A, Sopko R, McCartney RR, Schmidt MC, Rachidi N, Lee SJ, Mah AS, Meng L, Stark MJ, Stern DF, De Virgilio C, Tyers M, Andrews B, Gerstein M, Schweitzer B, Predki PF, Snyder M. (2005) Global analysis of protein phosphorylation in yeast. *Nature* **438:** 679–684.

Ramachandran N, Hainsworth E, Demirkan G, LaBaer J. (2006) On-chip protein synthesis for making microarrays. *Methods Mol Biol* **328:** 1–14.

Ren B, Robert F, Wyrick JJ, Aparicio O, Jennings EG, Simon I, Zeitlinger J, Schreiber J, Hannett N, Kanin E, Volkert TL, Wilson CJ, Bell SP, Young RA. (2000) Genome-wide location and function of DNA binding proteins. *Science* **290:** 2306–2309.

Rimm DL, Camp RL, Charette LA, Olsen DA, Provost E. (2001) Amplification of tissue by construction of tissue microarrays. *Exp Mol Pathol* **70:** 255–264.

Rowe C, Scruggs S, Feldstein M, Golden J, Ligler F. (1999) An array immunosensor for simultaneous detection of clinical analytes. *Anal Chem* **71:** 433–439.

Royce TE, Rozowsky JS, Luscombe NM, Emanuelsson O, Yu H, Zhu X, Snyder M, Gerstein MB. (2006) Extrapolating traditional DNA microarray statistics to tiling and protein microarray technologies. *Methods Enzymol* **411:** 282–311.

Rubin MA, Zerkowski MP, Camp RL, Kuefer R, Hofer MD, Chinnaiyan AM, Rimm DL. (2004) Quantitative determination of expression of the prostate cancer protein alpha-methylacyl-CoA racemase using automated quantitative analysis (AQUA): a novel paradigm for automated and continuous biomarker measurements. *Am J Pathol* **164:** 831–840.

Ruiz C, Seibt S, Al Kuraya K, Siraj AK, Mirlacher M, Schraml P, Maurer R, Spichtin H, Torhorst J, Popovska S, Simon R, Sauter G. (2006) Tissue microarrays for comparing molecular features with proliferation activity in breast cancer. *Int J Cancer* **118:** 2190–2194.

Rusmini F, Zhong Z, Feijen J. (2007) Protein immobilization strategies for protein biochips. *Biomacromolecules* **8:** 1775–1789.

Schaferling M, Nagl S. (2006) Optical technologies for the read out and quality control of DNA and protein microarrays. *Anal Bioanal Chem* **385:** 500–517.

Sharma-Oates A, Quirke P, Westhead DR. (2005) TmaDB: a repository for tissue microarray data. *BMC Bioinform* **6:** 218.

Silva J, Chang K, Hannon GJ, Rivas FV. (2004) RNA-interference-based functional genomics in mammalian cells: reverse genetics coming of age. *Oncogene* **23:** 8401–8409.

Silva JM, Mizuno H, Brady A, Lucito R, Hannon GJ. (2004) RNA interference microarrays: high-throughput loss-of-function genetics in mammalian cells. *Proc Natl Acad Sci U S A* **101:** 6548–6552.

Speer R, Wulfkuhle JD, Liotta LA, Pertricoin EF. (2005) Reverse-phase protein microarrays for tissue-based analysis. *Curr Opin Mol Ther* **7:** 240–245.

Sreekumar A, Nyati MK, Varambally S, Barrette TR, Ghosh D, Lawrence TS, Chinnaiyan AM. (2001) Profiling of cancer cells using protein microarrays: discovery of novel radiation-regulated proteins. *Cancer Res* **61:** 7585–7593.

Steinhauer C, Wingren C, Hager AC, Borrebaeck CA. (2002) Single framework recombinant antibody fragments designed for protein chip applications. *Biotechniques* **33:** S38–45.

Stromberg S, Bjorklund MG, Asplund C, Skollermo A, Persson A, Wester K, Kampf C, Nilsson P, Andersson AC, Uhlen M, Kononen J, Ponten F, Asplund A. (2007) A high-throughput strategy for protein profiling in cell microarrays using automated image analysis. *Proteomics* **7:** 2142–2150.

Torhorst J, Bucher C, Kononen J, Haas P, Zuber M, Kochli OR, Mross F, Dieterich H, Moch H, Mihatsch M, Kallioniemi OP, Sauter G. (2001) Tissue microarrays for rapid linking of molecular changes to clinical endpoints. *Am J Pathol* **159**: 2249–2256.

Viti F, Merelli I, Caprera A, Lazzari B, Stella A, Milanesi L. (2008) Ontology-based, tissue microarray oriented, image centered tissue bank. *BMC Bioinform* **9**: S4.

Voduc D, Kenney C, Nielsen TO. (2008) Tissue microarrays in clinical oncology. *Semin Radiat Oncol* **18**: 89–97.

Warren CL, Kratochvil NC, Hauschild KE, Foister S, Brezinski ML, Dervan PB, Phillips GN, Ansari AZ. (2006) Defining the sequence-recognition profile of DNA-binding molecules. *Proc Natl Acad Sci U S A* **103**: 867–872.

Wheeler D, Bailey S, Guertin D, Carpenter A, Higgins C, Sabatini D. (2004) RNAi living-cell microarrays for loss-of-function screens in *Drosophila melanogaster* cells. *Nat Methods* **1**: 1–6.

Wheeler DB, Carpenter AE, Sabatini DM. (2005) Cell microarrays and RNA interference chip away at gene function. *Nat Genet* **37**: S25–30.

Wilson DS, Keefe AD, Szostak JW. (2001) The use of mRNA display to select high-affinity protein-binding peptides. *Proc Natl Acad Sci U S A* **98**: 3750–3755.

Wingren C, Steinhauer C, Ingvarsson J, Persson E, LArsson K, Borrebaeck C. (2005) Microarrays based on affinity-tagged single chain antibodies: sensitive detection of analyte in complex proteomes. *Proteomics* **5**: 1281–1291.

Winzeler EA, Shoemaker DD, Astromoff A, Liang H, Anderson K, Andre B, Bangham R, Benito R, Boeke JD, Bussey H, Chu AM, Connelly C, Davis K, Dietrich F, Dow SW, El Bakkoury M, Foury F, Friend SH, Gentalen E, Giaever G, Hegemann JH, Jones T, Laub M, Liao H, Liebundguth N, Lockhart DJ, Lucau-Danila A, Lussier M, M'Rabet N, Menard P, Mittmann M, Pai C, Rebischung C, Revuelta JL, Riles L, Roberts CJ, Ross-MacDonald P, Scherens B, Snyder M, Sookhai-Mahadeo S, Storms RK, Veronneau S, Voet M, Volckaert G, Ward TR, Wysocki R, Yen GS, Yu K, Zimmermann K, Philippsen P, Johnston M, Davis RW. (1999) Functional characterization of the *S. cerevisiae* genome by gene deletion and parallel analysis. *Science* **285**: 901–906.

Wu R, Bailey S, Sabatini D. (2002) Cell-biological applications of transfected-cell microarrays. *Tends Cell Bio* **12**: 485–488.

Xu C. (2002) High-density cell microarrays for parallel functional determinations. *Genome Res* **12**: 482–486.

Yu X, Xu D, Cheng Q. (2006) Label-free detection methods for protein microarrays. *Proteomics* **6**: 5493–5503.

Zerkowski MP, Camp RL, Burtness BA, Rimm DL, Chung GG. (2007) Quantitative analysis of breast cancer tissue microarrays shows high cox-2 expression is associated with poor outcome. *Cancer Invest* **25**: 19–26.

Zhu H, Snyder M. (2003) Protein chip technology. *Curr Opin Chem Biol* **7**: 55–63.

Zhu H, Klemic JF, Chang S, Bertone P, Casamayor A, Klemic KG, Smith D, Gerstein M, Reed MA, Snyder M. (2000) Analysis of yeast protein kinases using protein chips. *Nat Genet* **26**: 283–289.

Zhu H, Bilgin M, Bangham R, Hall D, Casamayor A, Bertone P, Lan N, Jansen R, Bidlingmaier S, Houfek T, Mitchell T, Miller P, Dean RA, Gerstein M, Snyder M. (2001) Global analysis of protein activities using proteome chips. *Science* **293**: 2101–2105.

Zhu X, Gerstein M, Snyder M. (2006) ProCAT: a data analysis approach for protein microarrays. *Genome Biol* **7**: R110.

Ziauddin J, Sabatini DM. (2001) Microarrays of cells expressing defined cDNAs. *Nature* **411**: 107–110.

# Future Prospects

We hope this book has shown the incredible strides that have been made in the application of microarray-based analysis to a wide variety of biological systems. When we consider that the first broadly applicable report of using microarrays to explore gene expression was published in 1995 (Schena et al., 1995), it is little short of remarkable that in less than 15 years we are now in a position where whole genomes can be reliably analyzed by this relatively simple technology. Of course, nothing is perfect and it is clear that we still have some way to go if we wish to use microarray-based analysis as a quantitative platform for monitoring gene expression. While there are hopes that the new ultrahigh-throughput sequencing technologies will provide a cost effective route for generating digital (counting molecules) gene expression data (Morin et al., 2008; Torres et al., 2008), it is, in our view, premature to dismiss array-based technologies out of hand at the moment. Sequencing methods present their own problems in terms of library construction and as yet undiscovered sources of bias and variability. Some of these issues are well-recognized in the microarray field and efforts to ameliorate them ongoing. In addition, as we hope to show below, methods for array-based assays continue to evolve and offer the prospect of harnessing microarray platforms for digital nucleic acid assays. Here we briefly overview some of the improved methods and approaches that may be implemented in the not too distant future.

## 13.1 ARRAYS

In situ oligonucleotide synthesis methods now allow the generation of very high density arrays: Affymetrix GeneChips, containing over six million probes per array at a feature size of 5 $\mu$m, lead the field in this respect. At this density seven arrays are required to interrogate the whole repeat-masked human genome. Thus a relatively modest decrease in feature size will enable the entire human genome on a single array, a goal within the reach of current photoresist technology (McGall et al., 1996). It is therefore likely that within the next few years virtually all of the nonrepetitive fraction of the human genome will be widely available on a single chip. It may be argued that, for the analysis of gene expression, the currently achievable array densities, especially the light-directed maskless technology that allows the synthesis of over two million relatively long oligonucleotides, are sufficient to allow the design of arrays where multiple specific probes interrogate each of the 48 000 transcripts or $\sim$280 000 exons predicted in the human genome (Ensembl release 49). It is therefore apparent that whole genome and whole transcriptome analysis for most complex metazoan genomes is now within reach using current technologies. What remains for the development of microarray technology? We believe the future focus will be very much on improvements to the reliability of the platform in terms of probe design and data analysis.

As we describe in Chapter 3, our ability to design microarray probes that are both sensitive and specific is hampered by our relatively poor understanding of the kinetics and dynamics of solid phase hybridization reactions. Much of the current state-of-the-art in terms of probe design and data analysis are guided by Langmuir isotherms, a physical–chemical model describing the behavior of idealized solution hybridizations. It is comparatively recently that efforts are turning towards developing a better theoretical understanding of the kinetics underlying the solid-phase hybridization relevant to DNA microarray behavior (Levicky and Horgan, 2005). Halperin et al. (2006) provide an excellent review of the relevant issues. They describe how the equilibrium behavior of solid-phase hybridization differs significantly from the idealized Langmuir model and suggest a number of extensions to the model to account for competitive hybridization with tethered probes. They indicate a need for collecting additional experimental data to account for surface effects and sample interactions. Additional theoretical analysis of hybridization kinetics provided by Binder (2006) suggests corrections for nonlinear effects and nonspecific hybridization, especially with Affymetrix data, and provide methods for calculating modifications to the nearest-neighbor model (SantaLucia, 1998) to account for sequence-specific effects (Binder and Preibisch, 2006). Other models, based on extensions to the Langmuir model, have also been proposed (e.g. Heim et al., 2006; Ono et al., 2008) and quantitative data relating solution and solid-phase hybridization kinetics is beginning to appear (Fish et al., 2007). The continued development of models that accurately describe the physiochemical properties of microarray hybridizations for both long and short probes will inevitably lead to far more reliable microarray data.

One area that needs clarification is the effect of the density of probe molecules within each array element. Some studies suggest that densities above $\sim$10$^{12}$ molecules per cm$^2$ result in reduced hybridization efficiencies (Peterson et al., 2000, 2001) because of interactions between crowded probe molecules in individual array elements. More recent work suggests that the density effects are more complex and may be less important, even at densities above $10^{13}$ molecules per $cm^2$, when high ionic strength conditions typical of most hybridization reactions are used (Gong and Levicky, 2008). A second factor that needs to be addressed is the effect of probe molecule heterogeneity that is an inevitable result of all in situ synthesis methods. As we note in Chapter 3, the typical step-wise coupling efficiencies achieved with light-directed oligonucleotide synthe-sis results in a substantial proportion of less than full-length probe molecules within each array element. The effects of such probe polydispersity need to be characterized for both short and long oligonucleotides (Halperin et al., 2006; Gong and Levicky, 2008). The heterogeneity of labeled sample molecules is also an area that needs some attention since different target sizes, as well as different propensities for forming secondary structures, complicates the conversion of intensity to concentration. Thus, a better understanding of the effects of probe and target heterogeneity as well as the hybridization kinetics at different probe densities will be required if more quantitative data are to be generated from in situ synthesized microarrays.

One possibility that has been discussed for improving the quantitative aspects of gene expression data is to monitor hybridization kinetics continuously rather than assaying the end point after a prolonged period of hybridization to reach equilibrium (Bishop et al., 2008). While Halperin et al. (2006) correctly argue that nonequilibrium measures may be unreliable compared to end-point measures, primarily due to non-specific probe and target interactions compli-cating the interpretation of hybridization dynamics, it is not clear that equilib-rium is actually reached in all hybridization assays. This is particularly important since it is clear that under nonequilibrium conditions, substantial cross-hybridization can occur (e.g. Bhanot et al., 2003). Interestingly, Horne et al. (2006) suggest that perfect match and mismatch hybrids exhibit different hybridization kinetics and that this may be used as a diagnostic tool for discriminating specific hybridization, a view supported by the analysis of Wick et al. (2006). Taken together, the theoretical approaches described above combined with the collection of critical experimental data that can accurately parameterize hybridization models will result in a technology that can accurate-ly convert the observed hybridization signal into a defined number of molecules.

## 13.2 LABELING AND DETECTION

With progress being made in terms of understanding hybridization kinetics, the methods for detecting hybrids on microarrays also need attention. In an ideal world, especially in the case of gene expression analysis, each molecule in the sample mixture would receive an identical amount of label and thus hybrids detected on the array would equate to the concentration of each molecule in the original analyte. We are, unfortunately, some way from this ideal at present.

Sample labeling methods currently employed generally incorporate a moiety directly conjugated to a nucleotide trisphosphate (NTP) during cDNA or subsequent nucleic acid synthesis step (see Chapter 4). While the problems associated with directly incorporating organic fluorophores such as Cy dyes can be overcome by postsynthesis incorporation or detection, so-called indirect labeling, other drawbacks with standard fluorescent dyes are less easily addressed, including susceptibility to photobleaching, fluorescence quenching and chemical lability. In addition, accurately quantifying the amount of label each molecule incorporates is difficult and thus signal is decoupled from concentration.

Efforts to ameliorate some of these problems are underway with a variety of approaches being pursued. In terms of labeling, the use of fluorescent nanoparticles including nanobeads and semiconductor nanoparticles (or quantum dots) are a promising avenue (reviewed by Schaferling and Nagl, 2006). Quantum dots have been shown to perform well in assays detecting single nucleic acid molecules (Crut et al., 2005; Zhang and Johnson, 2005; Zhang et al., 2005) and in principal could be generally applied to microarray-based assays. Such labels will certainly generate brighter signals and have the additional benefit that multiple colors are possible allowing multiplexing. Quantum dot-based detection in a microarray format has been demonstrated (Gerion et al., 2003; Robelek et al., 2004) and the potential for using the technology with genotyping arrays has been recently described (Karlin-Neumann et al., 2007). Clearly the ability to more readily equate hybridization signal with target concentration in complex analytes is critical if we want truly quantitative microarrays. One of the ways the currently employed labeling strategies cloud the interpretation of any hybridization data is a poor understanding of the effects that the labels employed to detect hybridization have on the hybridization reaction itself. It may therefore be more beneficial to adopt posthybridization detection strategies such as the binding of biotin by streptavidin or antibody binding to a small hapten. If relatively small indirect labels are used, and they are incorporated by some type of quantitative end-labeling strategy, it may be possible to use nanoparticle-based detection methods.

## 13.3 IMAGING

Whether microarrays are interrogated with targets labeled with traditional organic dyes or quantum dots, they need to be imaged to collect the primary data: currently the methods used are a compromise between efficiency and sensitivity. As we discuss in Chapter 5, arrays may be scanned using laser/PMT-based or with CCD-based systems and these provide a dynamic range between 3.5 and 4.5 orders of magnitude. Unfortunately, transcript abundances in biological systems can span at least six orders of magnitude (Holland, 2002) and although it is possible to perform multiple scans at different scanner settings to increase the dynamic range, it would be preferable if a larger dynamic range could be captured in a single scan. Current scanning techniques and more promising methods on the horizon have recently been reviewed (Schaferling and Nagl, 2006; Timlin, 2006), here we mention a few possibilities.

Using standard fluorescent dyes, Timlin et al. (2005) describe how alternative scanning technologies, in this case hyperspectral scanning, can substantially improve data quality. In this technique, rather than collecting fluorescence emissions from narrow wavelengths as is the case with standard scanning, the entire emission spectrum for each pixel is collected and analyzed with multivariate data extraction techniques. In particular they report elimination of common scanning artefacts and a reduction in fluorescence artefacts as well as offering the prospect of more reliable analysis with multiple fluorofors. A similar hyperspectral imager using a more sophisticated illumination system has been recently reported (Glasenapp et al., 2007). Other imaging technologies such as time-resolved fluorescence or Förster resonance energy transfer (FRET, a particularly attractive imaging method when combined with fluorescent nanoparticles) may benefit from improvements in single-photon counting devices (Schaferling and Nagl, 2006). Other detection methods that do not utilize fluorescence based assays, such as surface plasmon resonance (Smith and Corn, 2003; Chen et al., 2005) and resonance light scattering (Bao et al., 2002; Francois et al., 2003), are capable of very sensitive and highly specific data capture but require alternative microarray manufacturing technologies that have not been shown yet to be applicable in high-density settings. These non-fluorescence-based assays are particularly attractive as sensitive imaging methods for protein arrays (e.g. Singh and Hillier, 2007). There is considerable potential for using such imaging methods in diagnostic tools, where far fewer probes need to be interrogated and high sensitivity with limiting analytes is required.

An ideal microarray scanning method will capture data at the high dynamic range typical of biological samples and do this at single molecule resolution. If single-probe molecules are imaged then the concerns about quantifying the label attached to each sample molecule we describe above become much less important. Hesse and colleagues have recently described a scanning system using a specialized CCD technique, time-delay integration, which is reported to generate extremely sensitive single molecule resolution with image acquisition times compatible with high-throughput microarray analysis (Hesse et al., 2006; Mir, 2006). They report scanning a 1 cm$^2$ microarray in approximately 1 h with almost five orders of magnitude dynamic range and a sensitivity that allowed expression data to be reliably obtained from less than 10$^4$ cells without amplification. Importantly, the data acquisition system permits visualization of labeled targets hybridized to single-probe molecules within an array element. Such a technique allows the number of molecules in the target population to be counted rather than inferred from the overall fluorescence signal across a probe spot. Although their imaging system requires a very thin array surface (they printed arrays on 150 $\mu$m thick aldehyde coverslips) and high quality microscopy, the potential for acquiring single molecule data is very attractive and it is hoped that the method will be developed further. Indeed, the developers of the single molecule system have recently reported an alternative fluorescence excitation method that promises improvements in sensitivity (Hesch et al., 2008). An alternative approach for single molecule imaging, dual-color total internal reflection fluorescence microscopy, has recently been described (Kang et al., 2007) and it is clear that this emerging area holds considerable promise if bench-top imaging systems for research labs can be produced.

We believe that the exciting developments in both imaging and new labeling techniques, coupled with a better understanding solid-phase hybridization, will provide much more quantitative microarray assays. We see no reason why microarray-based assays cannot be as sensitive and reliable as the next generation sequencing technologies currently being developed. Indeed the approach described by Hesse et al. (2006) is, in principal, much easier that sequence-based approaches since the path from sample extraction to detection involves far fewer molecular biology steps and thus less chance for bias. While new methods, particularly techniques for real-time monitoring of transcript levels in vivo, will undoubtedly be developed in the future, we feel microarray-based assays for gene expression and DNA analysis will continue to see action across a broad range of biological questions.

## 13.4 ANALYSIS

Of course, hand-in-hand with the development of improved labeling and imaging technologies, there will inevitably be improvements in the analytical and statistical methods underpinning microarray data processing. It should be obvious from Chapters 6 and 7 that we have seen considerable progress in all areas of data analysis and we are now a long way from the very simple filtering and twofold change in expression approaches used in the early days of microarray technology (Loring, 2006). Despite this progress, we recognize that current array data are still noisy and we require the development of more robust analytical methods. Improvements will undoubtedly come from more principled approaches to probe design and better understanding of hybridization kinetics, indeed progress in this area has already improved the analysis of Affymetrix GeneChip data by modeling the effects of probe nucleotide composition more effectively (e.g. Naef and Magnasco, 2003). Clearly, improvements in analysis methods will depend upon the direction taken by the technology. For example, if the individual probe molecule imaging techniques described above see widespread adoption, the analytical problems faced will be different from those generated by traditional spot segmentation approaches.

As datasets accumulate, especially those generated with modern array platforms that are more reliable, there is clearly a need for improved data-mining tools. When hundreds or even thousands of whole-transcriptome experiments are to be compared, the data abstraction and visualization tools we describe in Chapter 8 may be inadequate. Progress in this area will undoubtedly come from computer science and statistics, where there are already huge challenges in the analysis of complex financial and other high-dimensional data that mirror problems encountered with gene expression (Clarke et al., 2008; Wang et al., 2008) and high-throughput genotyping data (Liang and Keleman, 2008). The research in this area is at the cutting-edge and will be more widely applicable to so-called systems biology approaches, where high dimensionality data from a range of high-throughput experimental methods needs to be integrated and visualized. We believe there will be interesting and far-reaching developments in this area.

We contend that the future for microarray technology remains bright and imagine that relatively soon most research labs will use comparatively inexpensive versions of the technology to carry out routine gene expression analysis on even the smallest biological samples. While some have suggested that the costs of microarray-based assays are currently prohibitively expensive (D'Ambrosio et al., 2005) or argued that the technology is plagued by reliability and reproducibility issues (Draghici et al., 2006), we believe much of the technical issues are being addressed and, considering the amount of data generated, the technology is actually fairly cheap. A two million element array can be purchased and hybridized for around $500, much less than a cent for an individual hybridization assay. These costs have dramatically reduced over the past few years and will almost certainly continue to do so. Much in the same way as sequencing used to be a time consuming 'black art' practiced by individual labs but is now very much a commodity service, where commercial ventures sequence templates at fractions of a cent per base, so we see microarray technology advance. In addition, the use of the microarray assays coupled with microfluidics technology may soon see wide application in clinical and diagnostic settings. In 2005, the US Food and Drug Administration approved the first diagnostic tool based on microarray technology (the Roche AmpliChip CYP450), more recently the Agendia MammaPrint microarray assay for breast cancer diagnostics was also approved and it can only be a matter of time before many more such assays become available. Whether in the research lab or the clinic, microarrays will be with us for many years to come.

## REFERENCES

Bao P, Frutos A, Greef C, Lahiri J, Muller U, Peterson T, Warden L, Xie X. (2002) High sensitivity detection of DNA hybridisation on microarrays using resonance light scattering. *Anal Chem* **74:** 1792–1797.

Bhanot G, Louzoun Y, Zhu J, DeLisi C. (2003) The importance of thermodynamic equilibrium for high throughput gene expression arrays. *Biophys J* **84:** 124–135.

Binder H. (2006) Thermodynamics of competitive surface adsorbtion on DNA microarrays. *J Phys: Condens Matter* **18:** S491–S523.

Binder H, Preibisch S. (2006) GeneChip microarrays – signal intensities, RNA-concentrations and probe sequences. *J Phys: Condens Matter* **18:** S537–S566.

Bishop J, Chagovetz A, Blair S. (2008) Kinetics of multiplex hybridisation, mechanisms and implications. *Biophys J* **94:** 1726–1734.

Chen S, Su Y, Hsiu F, Tsou C, Chen Y. (2005) Surface plasmon resonance phase-shift interferometry: real time DNA microarray hybridisation analysis. *J Biomed Optics* **10:** 034005.

Clarke R, Ressom H, Wang A, Xuan J, Liu M, Gehan E, Wang Y. (2008) The properties of high-dimensional data spaces: implications for exploring gene and protein expression data *Nat Rev Cancer* **8:** 37–49.

Crut A, Geron-Landre B, Bonnet I, Bonneau S, Desbiolles P, Escude C. (2005) Detection of single DNA molecules by multicolor quantum-dot end-labeling. *Nucleic Acids Res* **33:** e98.

D'Ambrosio C, Gatta L, Bonini S. (2005) The future of microarray technology: networking the genome search. *Allergy* **60:** 1219–1226.

Draghici S, Khatri P, Eklund AC, Szallasi Z. (2006) Reliability and reproducibility issues in DNA microarray measurements. *Trends Genet* **22:** 101–109.

Fish D, Horne M, Brewood G, Goodarzi J, Alemayehu S, Bhandiwad A, Searles R, Benight A. (2007) DNA multiplex hybridisation on microarrays and thermodynamic stability in solution: a direct comparison. *Nucleic Acids Res* **35:** 7197–7298.

Francois P, Bento M, Vaudaux P, Schrenzel J. (2003) Comparison of fluorescence and resonance light scattering for highly sensitive microarray detection of bacterial pathogens. *J Microbiol Methods* **55:** 755–762.

Gerion D, Chen F, Kannan B, Fu A, Parak W, Chen D, Majumdar A, Alivisatos A. (2003) Room-temperature single-nucleotide polymorphism and multiallele detection using fluorescent nano-crystals and microarrays. *Anal Chem* **75:** 4766–4772.

Glasenapp C, Monch W, Krause H, Zappe H. (2007) Biochip reader with dynamic holographic excitation and hyperspectral fluorescence detection. *J Biomed Opt* **12:** 014038.

Gong P, Levicky R. (2008) DNA surface hybridisation regimes. *Proc Natl Acad Sci U S A* **105:** 5301–5306.

Halperin A, Buhot A, Zhulina E. (2006) On the hybridisation isotherms of DNA microarrays: the Langmuir model and its extensions. *J Phys: Condens Matter* **18:** S463–S490.

Heim T, Wolterink J, Carlon E, Barkema G. (2006) Effective affinities in microarray data. *J Phys: Condens Matter* **18:** S525–536.

Hesch C, Hesse J, Schutz GJ. (2008) Implementation of alternating excitation schemes in a biochip-reader for quasi-simultaneous multi-color single-molecule detection. *Biosens Bioelectron* **23:** 1891–1895.

Hesse J, Jacak J, Kasper M, Regl G, Eichberger T, Winklmayr M, Aberger F, Sonnleitner M, Schlapak R, Howorka S, Muresan L, Frischauf AM, Schutz GJ. (2006) RNA expression profiling at the single molecule level. *Genome Res* **16:** 1041–1045.

Holland M. (2002) Transcript abundance in yeast varies over sox order of magnitude. *J Biol Chem* **277:** 14363–14366.

Horne MT, Fish DJ, Benight AS. (2006) Statistical thermodynamics and kinetics of DNA multiplex hybridization reactions. *Biophys J* **91:** 4133–4153.

Kang S, Kim Y, Yeung E. (2007) Detection of single-molecule DNA hybridisation by using dual-color total internal reflection flourescence microscopy. *Anal Biochem* **387:** 2663–2671.

Karlin-Neumann G, Sedova M, Flakowski M, Wang Z. (2007) Application of quantum dots to multicolor microarray experiments. In: Bruchez M, Hotz C, editors. *Quantum Dots: Applications in Molecular Biology.* Humana Press, Totowa, NJ.

Levicky R, Horgan A. (2005) Physiochemical perspectives on DNA microarray and biosensor technologies. *Tends Biochem* **23**: 143–149.

Liang Y, Keleman A. (2008) Stastical advances and challenges for analyzing correlated high dimensional SNP data in genomic study for complex diseases. *Stats Surveys* **2**: 43–60.

Loring J. (2006) Evolution of microarray analysis. *Neurobiol Aging* **27**: 1084–1086.

McGall G, Labadie J, Brock P, Wallraff G, Nguyen T, Hinsberg W. (1996) Ligh-directed synthesis of high-density oligonucleotide arrays using semiconductor photoresists. *Proc Natl Acad Sci U S A* **93**: 13555–13560.

Mir K. (2006) Ultrasensitive RNA profiling: counting single molecules on microarrays. *Genome Res* **16**: 1195–1197.

Morin RD, O'Connor MD, Griffith M, Kuchenbauer F, Delaney A, Prabhu AL, Zhao Y, McDonald H, Zeng T, Hirst M, Eaves CJ, Marra MA. (2008) Application of massively parallel sequencing to microRNA profiling and discovery in human embryonic stem cells. *Genome Res* **18**: 610–621.

Naef F, Magnasco MO. (2003) Solving the riddle of the bright mismatches: labeling and effective binding in oligonucleotide arrays. *Phys Rev E Stat Nonlin Soft Matter Phys* **68**: 011906.

Ono N, Suzuki S, Furusawa C, Agata T, Kashiwagi A, Shimizu H, Yomo T. (2008) An improved physico-chemical model of hybridisation on high-density oligonucleotide microarrays. *Bioinformatics* **24**: 1278–1285.

Peterson A, Heaton R, Georgiadis R. (2000) Kinetic control of hybridisation in surface immobilised DNA monolayer films. *J Am Chem Soc* **122**: 7837–7838.

Peterson A, Heaton R, Georgiadis R. (2001) The effect of surface probe density on DNA hybridisation. *Nucleic Acids Res* **29**: 5163–5168.

Robelek R, Niu L, Schmid E, Knoli W. (2004) Multiplexed hybridization detection of quantum dot-conjugated DNA sequences using surface plasmon enhanced fluorescence microscopy and spectrometry. *Anal Chem* **76**: 6160–6165.

SantaLucia J. (1998) A unified view of polymer, dumbbell and oligonucleotide DNA nearest-neighbor thermodynamics. *Proc Natl Acad Sci U S A* **95**: 1460–1465.

Schaferling M, Nagl S. (2006) Optical technologies for the read out and quality control of DNA and protein microarrays. *Anal Bioanal Chem* **385**: 500–517.

Schena M, Shalon D, Davis RW, Brown PO. (1995) Quantitative monitoring of gene expression patterns with a complementary DNA microarray. *Science* **270**: 467–470.

Singh B, Hillier A. (2007) Multicolor surface plasmon resonance inaging of ink-jet printed protein microarrays. *Anal Chem* **79**: 5124–5132.

Smith E, Corn R. (2003) Surface plasmon resonance imaging as a tool to monitor biomolecular interactions in an array based format. *Appl Spectrosc* **57**: 320–332.

Timlin JA. (2006) Scanning microarrays: current methods and future directions. *Methods Enzymol* **411**: 79–98.

Timlin JA, Haaland DM, Sinclair MB, Aragon AD, Martinez MJ, Werner-Washburne M. (2005) Hyperspectral microarray scanning: impact on the accuracy and reliability of gene expression data. *BMC Genomics* **6**: 72.

Torres TT, Metta M, Ottenwalder B, Schlotterer C. (2008) Gene expression profiling by massively parallel sequencing. *Genome Res* **18**: 172–177.

Wang Y, Miller D, Clarke R. (2008) Approaches to working in high-dimensional data spaces: gene expression microarrays. *Br J Cancer* **98:** 1023–1028.

Wick LM, Rouillard JM, Whittam TS, Gulari E, Tiedje JM, Hashsham SA. (2006) On-chip non-equilibrium dissociation curves and dissociation rate constants as methods to assess specificity of oligonucleotide probes. *Nucleic Acids Res* **34:** e26.

Zhang C, Johnson L. (2005) Homogeneous rapid detection of nucleic acids using two-color quantum dots. *Analyst* **131:** 484–488.

Zhang C, Yeh H, Kuroki M, Wang T. (2005) Single-quantum-dot-based DNA nanosensor. *Nat Mater* **4:** 826–831.

# Index

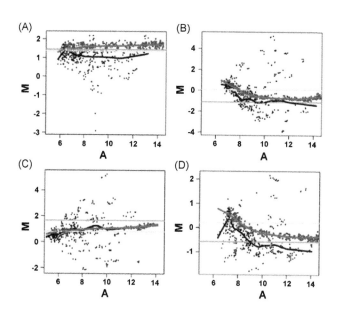

**Plate 1** (Figure 6.16 on page 166 of this volume)

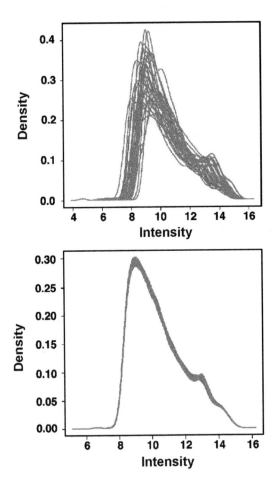

Plate 2 (Figure 6.18 on page 169 of this volume)

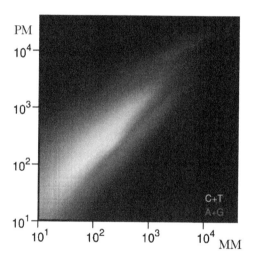

**Plate 3** (Figure 6.22 on page 181 of this volume)

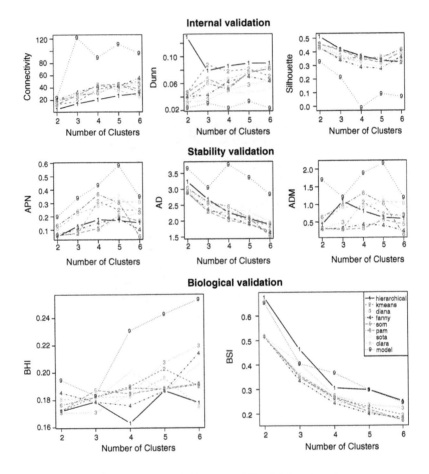

**Plate 4** (Figure 8.6 on page 289 of this volume)